Telecommunications and Networks

By the same Author:

Network Analysis, Prentice Hall, 1969
Development of Information Systems for Education, Prentice-Hall, 1973
Information Processing Systems for Management, Richard Irwin, 1981 and 1985
Information Resource Management, Richard Irwin, 1984 and 1988
The Computer Challenge: Technology, Applications and Social Implications, Macmillan, 1986
Information Systems for Business, Prentice Hall, 1991 and 1995
Management of Information, Prentice Hall, 1992
Artificial Intelligence and Business Mangement, Ablex, 1992
Knowledge-Based Information Systems, McGraw-Hill, 1995
Managing Information Technology, Butterworth-Heinemann, 1997

Telecommunications and Networks

K.M. Hussain
D.S. Hussain

BUTTERWORTH
HEINEMANN

Butterworth-Heinemann
Linacre House, Jordan Hill, Oxford OX2 8DP
A division of Reed Educational and Professional Publishing Ltd

 A member of the Reed Elsevier plc group

OXFORD BOSTON JOHANNESBURG
MELBOURNE NEW DELHI SINGAPORE

First published 1997

© K. M. and Donna S. Hussain 1997

British Library Cataloguing in Publication Data
A catalogue record for this book is available from the British Library.

ISBN 0 7506 2339X

Library of Congress Cataloguing in Publication Data
A catalogue record for this book is available from the Library of Congress.

Typeset by Laser Words, Madras, India
Printed in Great Britain

CONTENTS

Contents

Contents

ACKNOWLEDGEMENTS

The authors wish to thank colleagues for their helpful comments and corrections to the manuscript. These include Linda Johnson, Frank Leonard, Chetan Shankar and Derek Partridge. Thanks to Tahira Hussain for her help, especially in the preparation of the diagrams using the PowerPoint program. Any errors that still remain are the responsibility of the authors.

1

INTRODUCTION

The agricultural age was based on ploughs and the animals that pulled them; the industrial age, on engines and fuel that fed them. The information age we are now creating will be based on computers and networks that interconnect them.

Michael L. Dertouzos, 1991

Telecommunications is an old and stable technology if you think only of telephones and telegraph. But then in the 1960s came computers and the processing of data. Soon after, we needed data communications to transmit data to remote points; the connection of remote points by telecommunication is referred to as a **network**. Later, these points of communication increased in number, with the transmission no longer being limited to data but included text, voice, and even images and video. This extended use of telecommunications is the subject of this book. We shall examine the technology in the first part of the book, the management of the telecommunications in the second part, and in the third and final part the many applications that are now possible because of telecommunications.

Changes in technology

We start with an overview of the technology in Chapter 2. This will provide us with a framework in which we can then place the many components of the technology of telecommunications and networks. The first of these technologies to be examined is transmission. The earliest transmissions were by telephone for voice and telegraph for the written word. Telephones and telegraph were complemented by post and organized as a utility better known as the PT&T (Post Telephone and Telegraph). In the USA and UK, these services have been privatized and other countries may follow the path away from monopoly towards privatization and free competition. But this is a controversial question of politics and government policy-making, a 'soft' subject that

we chose to avoid here. Instead, we will confine ourselves to the more 'hard' and stable topics of technology: the management and applications of the technology.

Back to transmission. Early transmission was by wire, copper wires to be more precise. But copper is both expensive (and sometimes scarce) and bulky. It has been replaced by fibre optics, which uses thin glass fibres that are both cheaper and less scarce than copper. Fibre is also less bulky than copper and much lighter. One strand of fibre thinner than a human hair can carry more messages than a thick copper cable. A typical fibre optic cable can carry up to 32 000 long distance telephone calls at once, the equivalent of 2.5 billion bits of data per second. Recently, Bell Labs developed the rainbow technology that sent three billion bits of information down one fibre optic thread in one second, the equivalent of 18.75 million pages of double spaced text.

Fibre optics is less expensive than stringing wire across telephone poles and even less expensive in capital cost than cable. Its advantages are, however, restricted to the distance transmitted. For long distances, radio broadcasts and satellite are superior. But from the broadcasting and satellite station, the connection to the home or the office must still be made by wire or by fibre. With the increased volume and complexity of messages now being sent, the need for fibre is great and no longer in dispute. In the USA, the use of fibre for data communications has risen 500% during the period 1985–90. Many countries are turning to fibre, with Germany and Japan in the lead and the UK and USA not far behind. Fibre optics will be used for short distance transmission and will complement radio broadcasting and satellite.

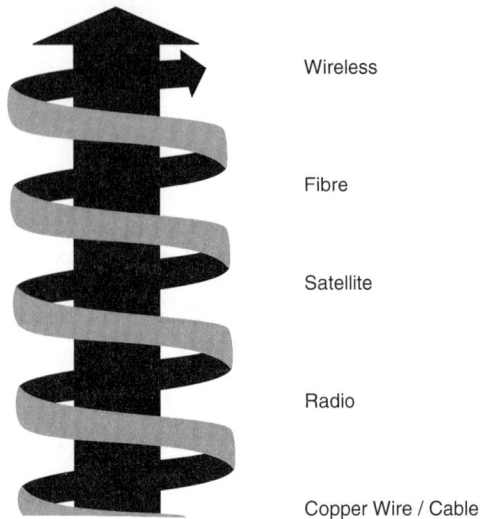

Wireless

Fibre

Satellite

Radio

Copper Wire / Cable

Figure 1.1   *Evolution of transmission media*

However, emerging strongly is the demand for wireless or cellular phones. They make transmission so much more portable that one can now transmit while driving a car, waiting at the airport, or even while walking the dog. This evolution of transmission is the subject of Chapter 3 and is summarized in the spiral of change shown in Figure 1.1.

Transmission is just one of the technologies enabling telecommunications. Other technologies include the many devices that make telecommunications possible by contributing to the transport of messages over networks. One set of such devices include the **bridge** that connects homogeneous (similar) networks and the **gateway** that connects non-homogeneous (dissimilar) networks.

One device that determines the route (path) that a **message** takes across **switches**, **bridges** (and/or) **gateways** is the **router**. This device contributes to the effectiveness of the transportation of the message, but the efficiency of the transmission depends largely on the message being transported. Many messages tend to have redundancies and even blanks (as in the sentences of this book). Eliminating these redundancies and compressing the message into a smaller sized message without losing any content is called **compression**. This makes the transmission efficient by taking less space (and time) to transmit the message. These devices are the subject of Chapter 4.

The technologies described above are fairly stable and have well established standards that are universally accepted. One technology that does not have universal acceptance and is very controversial is the international standard for an architecture and protocol for networks that is open to varying designs of hardware and software. This is the **OSI** (Open Systems Interconnection) model that was designed as a framework for the structure of telecommunications and networking. A description of the OSI and its competitors in the US (the **SNA** and the **TCP/IP**) are examined in Chapter 7. An international standard and one that is accepted globally is the **ISDN** (Integrated Systems Digital Network). ISDN will enable the transmission of analogue signals, which are now carried on telephone lines, as digital signals, like those used by the common desktop computer. This enables us to have just one signal, digital, instead of the two signals (**analogue** and **digital**) that we now carry, requiring equipment for interfacing and resulting in both inefficiencies and high cost. Integrated digital transmission is faster, and is easier and cheaper to maintain and operate, but transmission needs a **modem**, a device that translates from an analogue signal of, say, a telephone to a digital signal of a computer and vice versa.

Conversion to ISDN is expensive and slow but steady in the US, as reflected in the expenditures on ISDN which have doubled in the last three years since 1994. This conversion of the analogue world to the digital world has already resulted in the infrastructure becoming overloaded and overwhelmed. The demand for services to be transmitted has resulted in plans to extend ISDN to **B-ISDN** (Broad-band ISDN), which is now in the stages of getting international standards. Both ISDN and B-ISDN are the subject of Chapter 8. The evolution of these enabling technologies is shown in Figure 1.2.

We have mentioned networks as being interconnected points of communication. The earliest network was implemented by the US Department of Defense to facilitate the communication between their researchers and academics working on defence projects. These individuals were technical and inquisitive and became interested in developing a more reliable and efficient way of communicating not just their research projects but everything else including their daily mail. They unknowingly sowed the seeds of **e-mail** (electronic mail) and many other applications of telecommunications. These researchers (and later others in private

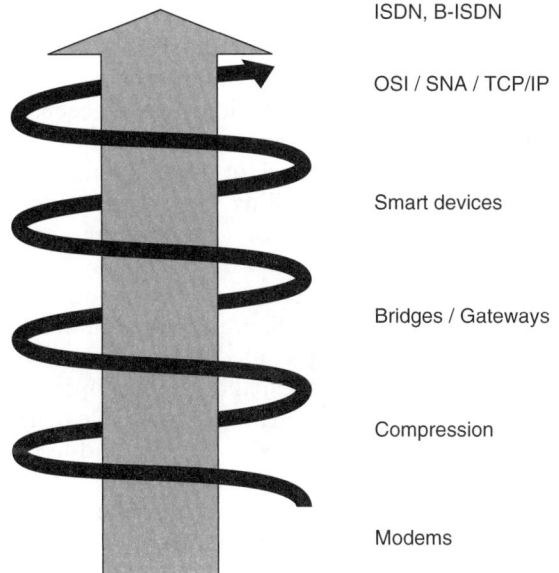

ISDN, B-ISDN

OSI / SNA / TCP/IP

Smart devices

Bridges / Gateways

Compression

Modems

Figure 1.2   *Spiral of enabling technologies*

1074 interconnected computers. In ten years, the number grew to 3.8 million and is still growing. The Internet is now being used not just by researchers but by individuals and also by businesses. The Internet is discussed as an application of telecommunications in Part 3, more specifically in Chapter 20.

Along the way to the Internet, the ARPANET contributed to the evolution of formal networking first in small local areas better known as the **LAN** (Local Area Network). This is the subject of Chapter 5. Networking in a broader geographic area is referred to as **MAN** (Metropolitan Area Network), and in a yet wider network the **WAN** (Wide Area Network). With large volumes of data, systems like the **SMDS** (Switched Multimegabit Data Service) will become more common in the future. The MAN and the WAN are the subject of Chapter 6. Their evolution is shown in Figure 1.3. These networks use architectures and protocols discussed in Chapter 8 and may or may not use the ISDN examined in Chapter 7.

industry) were also interested in developing a worldwide network of communications, and so ARPANET and the many technologies that it developed eventually led to the **Internet**. The Internet is a network of other networks. Despite no formal initiation or structure, it has become a very effective and popular means of communication. In 1984, the Internet had

Management of telecommunications

The technologies mentioned above (and defined operationally) are all examined in detail in

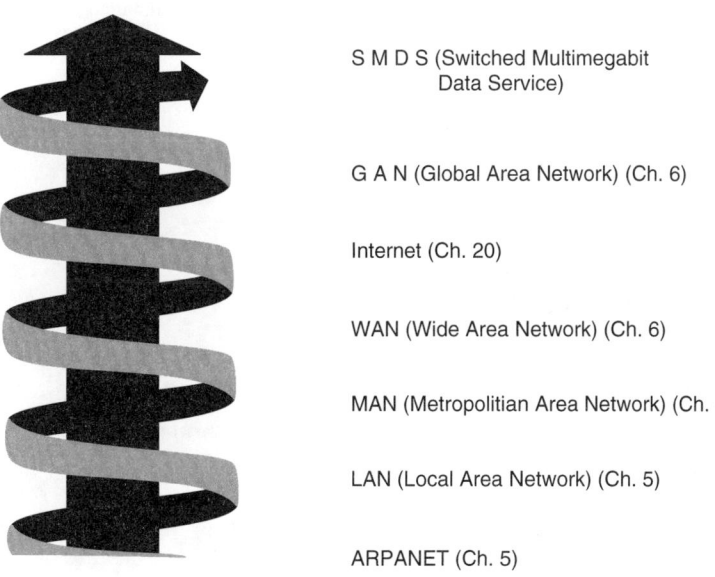

S M D S (Switched Multimegabit
            Data Service)

G A N (Global Area Network) (Ch. 6)

Internet (Ch. 20)

WAN (Wide Area Network) (Ch. 6)

MAN (Metropolitian Area Network) (Ch. 6)

LAN (Local Area Network) (Ch. 5)

ARPANET (Ch. 5)

Figure 1.3   *Spiral of networks*

Part 1 of this book. Part 2 is concerned with the management of these technologies. We start in Chapter 9 with the location and organization of telecommunications as part of **IT** (Information Technology) and as part of a corporate organization structure. The earliest organization structure was to centralize the large computer processors and mainframes that served all the local and remote users. This facilitated the economic use of the scarce resource of computer personnel as well as the expensive equipment. But then came the **PC**, the Personal Computer, and a parallel increase in the ability and desire for the centralized power to be decentralized to the remote nodes where the computing needs resided. This led to **DDP**, Distributed Data Processing. (At that time, most of computer processing was for data and only later did it extend to text, voice, video and images). PCs made distribution economically feasible and telecommunications which interconnected these nodes made it a feasible proposition. These organizational approaches are examined in Chapter 9.

Parallel to the growth of PCs was the dissatisfaction of the end-user of the centralized approach which was slow and unresponsive.

The **end-user** (ultimate user of computer output) was becoming computer literate, and no longer cowed by the computer specialist at the centralized and remote location. The end-users now had the desire (and sometimes with a passion) for the control of local operations. The end-users were willing to accept many of the responsibilities of maintaining and even selecting resources and developing systems needed at the remote nodes. They wanted the centre to do the planning of commonly needed resources (equipment, databases and even technical human resources) and the development of mission critical applications whilst leaving the computing at the nodes to the end-user. Thus evolved the **client—server system**, where the computer at a remote node is a **client** and the common computing resources (like data, knowledge and application programs) reside on computers called **servers**. Such a system requires solutions to special computing resources and the solution to many organizational and managerial issues. These are identified and discussed in Chapter 10.

The client—server approach is appropriate for a corporation or institution. But at a national and

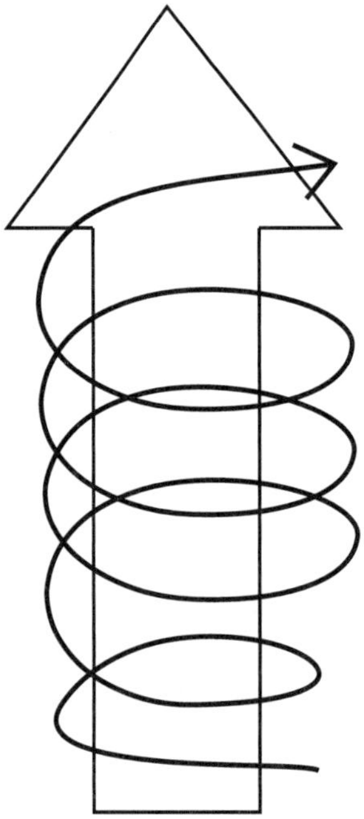

G I I (Global Info. Infrastructre) (Ch. 16)

N I I (National Info. Infrastructure) (Ch. 15)

Client–Server System (Ch. 10)

Distributed (Ch. 9)

Decentralized (Ch. 9)

Centralized (Ch. 9)

Figure 1.4 *Spiral of organization for telecommunications*

regional level an **infrastructure** for telecommunications is desirable that will not only meet the high demands of volume but the diverse demands of not just data but also voice and image. The carrying capacity has to increase. For example, one needs 64 000 bits per second capacity to transmit voice, 1.2 million bits per second to transmit high fidelity music, and 45 million bits per second to transmit video. Just as the infrastructure for road transportation changes from a city and local transportation to a motorway (freeway or autobahn) with all its interconnections, so also we need an entirely different set of transmission capacities and enabling technology for **interconnectivity** to connect and handle transmission. Such an infrastructure for national telecommunication (**NII**) is discussed in Chapter 15 and for a Global Information Infrastructure (**GII**) is discussed in Chapter 16. This evolution in the organization of telecommunications is shown in Figure 1.4.

The managerial issues of telecommunications is the subject of the following four chapters (Chapters 11–15). The management of standards is the subject of Chapter 11; and of security in Chapter 12. Chapter 13 is an overview of the management and administration of all of telecommunications and networking. The acquisition and organization of telecommunication resources is covered in Chapter 14.

Standards is one of the issues faced by management of telecommunications. **Standards** are agreed upon conventions and rules of behaviour are part of our daily life and certainly not new to IT where we have standards for hardware and software and even standards for analysis and design. We have all these types of standards for telecommunications plus a few more. It is important for telecommunications, if you consider the fact that telecommunications involves remotely located parties. In the case of global telecommunications this may be a continent away or across the oceans. If all of us were to pursue our own preferences in design and conventions for operations we would never be able to communicate with each other and there will be no compatibility and **interoperability** of devices and **protocols** (procedures). We do have agreement of many standards including international standards, but certainly not enough. One can experience that by going to another country and trying to plug in a computer. It is likely that your plug may not fit into the socket in the wall. We do not have all the international standards we need. They take a lot of effort and time. The international standards on network architecture mentioned earlier, the OSI, took ten years. But the timing was wrong. It came too late and had to face entrenched vested interests of manufacturers and suppliers of telecommunications equipment. But standards in high tech industries like telecommunications must not come too early before the technology is stabilized for that will 'freeze' and discourage newer approaches and innovations. Thus the task of telecommunications management is to correctly select the best technology and to assess the timing of adopting now and run the risk of being outdated or waiting and not benefit from existing advances in technology. The need for standards and the process of agreeing on standards by balancing the often conflicting interests is the subject of Chapter 11.

Another concern of telecommunications management is security. Again, as with standards, this is not new to IT management. But in telecommunications there are additional dimensions. The potential population of those who can penetrate the system is larger since there are now more people who have computers and know how to use them. Also, the temptation is larger. There is more data (and computer programs) that can be accessed and there is also more money that can be transacted across the lines of telecommunication. We thus need to control the access to networks by building **fire-walls** to protect our assets; **encryption** and other approaches are also needed to protect selected messages that are transmitted. The question for management (corporate and telecommunications) is not whether we need security but how much and where. Management must assess the cost of security and compare it with the risk of exposure. These subjects are discussed in Chapter 12.

Acquisition of telecommunications resources is the subject of Chapter 14. The process of acquisition is not new to either IT or to any corporation. What is new is the nature of resources that have to be acquired. At the corporate level the decision is that of selecting a LAN or MAN or WAN and not of selecting the devices for connectivity or the media of transmission which is part of the infrastructure. But there is the need to select the processors needed for accessing the network. In a client–server environment, the client may be

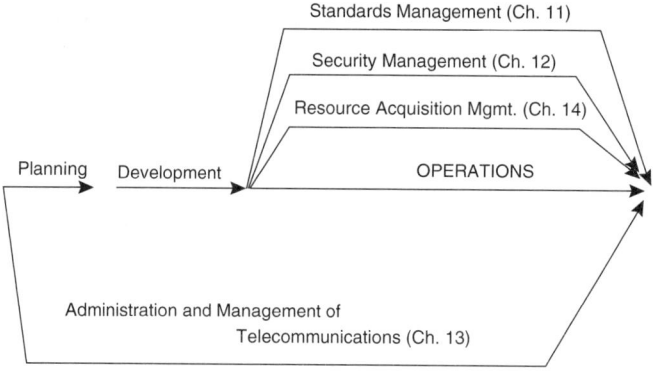

Figure 1.5 *Management of telecommunications*

a PC or a workstation. The server is a computer that could vary from a powerful PC to a mini or mainframe. But the servers for tomorrow have to be capable of handling not just data but multimedia. And so we need to consider not just **file servers** and **application servers** but also **video servers**. We have evolved from the stand-alone computer system to a system with a variety of computers that serve as clients or servers or both and are interconnected by telecommunications.

The chapter on management of telecommunications (Chapter 13) is more of a summary of all the related chapters. It is concerned with the planning, acquisition and maintenance of all the resources needed for telecommunications. A summary of these activities is shown in Figure 1.5. This includes personnel resources discussed in Chapter 9 on the organization at the corporate level.

The importance of telecommunications management can be gauged by the statistic that corporate spending on telecommunications in the US has more than doubled in the three years since 1994.

The last two chapters on management of telecommunications (Chapters 15 and 16) go beyond the corporate level. Chapter 15 is concerned with integration at the national level by providing an infrastructure for telecommunications much like we have an infrastructure for communications by road or plane. This infrastructure, often called the **information highway**, provides the interconnections for exchange of information and enables the integration of all the sources and destinations of information whether this be the home, office, business, school, library, medical facility or government agency. We need more

than standards for such integration and many of the issues that arise are not just technological but economic and political. These are examined in Chapter 15. We compound all these problems when we consider global communications and have additional issues of **transborder flow**, **global outsourcing** and the **protection of intellectual property**. These issues are examined in Chapter 16.

Applications of telecommunications

The next and final part of the book is concerned with applications of telecommunications and networks. Message handling is Chapter 17, multimedia is Chapter 18, teleworking is Chapter 19, the Internet in Chapter 20, integrated applications is Chapter 21, and a look into the future is Chapter 22. A graphic summary of this flow of topics is shown in Figure 1.6.

Our first discussion of applications will be on message handling applications in Chapter 17. Some of these applications have been around for a long time. An example of such late maturing applications is **e-mail** (electronic mail) that has suddenly 'taken-off' with high rates of growth and become a 'killer' application. It is so important an application that it will be discussed at great length later in Chapter 19.

Other applications of message handling are not so conspicuous but just as important. One is **EDI**, Electronic Data Interchange, which is used extensively for transfer of documents and files by businesses. Another application, also in business but restricted to financial institutions

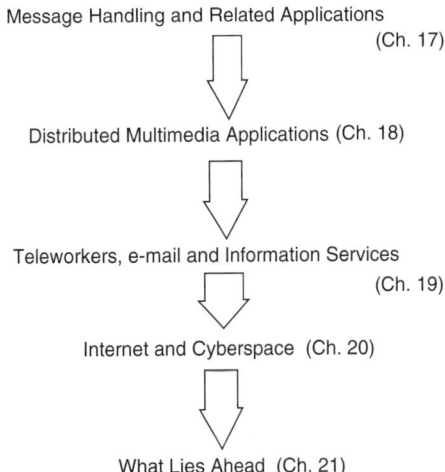

Message Handling and Related Applications (Ch. 17)

Distributed Multimedia Applications (Ch. 18)

Teleworkers, e-mail and Information Services (Ch. 19)

Internet and Cyberspace (Ch. 20)

What Lies Ahead (Ch. 21)

Figure 1.6 *Applications of telecommunications*

like banks, is the transfer of money by **EFT**, Electronic Funds Transfer. As James Martin once put it, 'Money is merely information, and as such can reside in computer storage, with payments consisting of data transfers between one machine and another.' Such money transfers are for billions of dollars a day all across the world. We take such electronic transactions for granted little realizing that if it were not for telecommunications our bank deposits and withdrawals would not be as easy or as fast as they now are. Of course, transactions may not be as safe either. These and related problems as well as their solutions are the subject of Chapter 17.

Another message handling application is that of **teleconferencing** but with the coming of **multimedia**, this may well evolve into **video-conferencing**. Other applications, like **home shopping**, **distance learning** and **electronic publishing**, are also becoming multimedia thereby greatly improving the quality of what is transmitted.

Applications still in the development stages include the delivery of **video-on-demand**, films, etc., delivered to the home at any time of the day or night. These applications along with interactive games will change the way we entertain ourselves, though we may want the ability for more self-control over the content.

Other exciting applications include the **digital library** that will enable you to read any article or book without having to go to the library, or browse through the contents of the Tate Gallery in London without having to physically visit the

place. This may well affect our learning as well as our patterns of how we spend our leisure time.

One final application of multimedia to be discussed here briefly is the use of telecommunications in medicine. It allows our entire medical record (in archives or observations taken in real time), including X-ray or CAT-scan pictures, to be transferred to an expert anywhere in the world for a second opinion. **Telemedicine** could also be valuable as a first opinion for those who may be located remotely (permanently or temporarily as when travelling). Again, as in many teleprocessing applications, there are problems of security, privacy and economics. These issues are examined in Chapter 18.

In Chapter 17 we mention e-mail. It is currently used extensively for correspondence (private and business) as well as for copying (**downloading**) computer programs residing at other computer server sites. It is much faster and more reliable than traditional mail, even air-mail. E-mail including foreign mail through the Internet is often available through local **information service** providers accessed by the telephone. These providers also offer many services that include entertainment, news, weather forecasts and education. Some of the services are interactive such as **chat sessions** where one can exchange views and information from someone that you may not know and someone who may be across the oceans. Information services could be customized so that you select what you want from the diverse options and do not have to take what is edited and passed down as is the case with the 12 000 newspapers and magazines and the many TV stations.

One service provider is CompuServe. It started by renting computer time from an insurance company that had purchased a computer and had unexpected excess capacity. In 1995, CompuServe was one of the three largest on-line service providers with around two million subscribers.

Information services may well be at its take-off stage approaching a killer application. In 1995 there were over eight million subscribers, with over two million subscribers joining just one information provider (AOL), in just one country (US). It may well become as ubiquitous as the telephone or TV. Its usage will increase as the usage of computers in the home increases. In 1995, 30% of all homes in the US owned computers and computer sales surpassed TV in annual sales for the first time ever.

Will computers and information services become as ubiquitous as the telephone and the TV? Will they be as end-user friendly and accessible as are telephones and TVs today? Will it take two to three decades to be accepted in the mainstream as it did for the telephone and TV? Must information services be regulated? Will all this information around threaten our privacy? Some insights into the answers to such questions will be found in Chapter 19, or in Chapter 20 which is on the Internet and cyberspace.

Before we get to Chapter 20, we discuss one other application that depends on computers and telecommunications. This is **telecommuting**, which is working at home using a computer and being connected to the corporate database through telecommunications. Telecommuters are also big users of information services, especially of e-mail.

With the boundaries of the workplace getting 'fuzzy', teleworking is a viable and attractive alternative to the crowded downtown office that must often be reached after fighting traffic jams and traffic lights. Telecommuting will require special resources and raises many issues especially of productivity and evaluation. These issues are the subject of Chapter 19.

The Internet is the subject of Chapter 20 and has been mentioned earlier as an outgrowth of ARPANET and LANs as well as in the context of information providers. If you cannot afford the monthly subscription of an information provider and do not have access to a LAN (through your employer or university) then you can always go to a café like the Café Cyberia in London where for an hourly payment you can surf the Internet.

Discussing the Internet will allow us to enter some of the space of **cyberspace**, where it is used not only by individuals but increasingly by businesses. Currently businesses do a lot of their communications and some of their advertising on the Internet but not much business in the sense of sales. This is because there is not yet any safe way to transact money on the Internet. There is much talk about **cybercash**, **cybermoney**, **digicash** and **digimoney**, but you are advised not to trust your credit card to cyberspace, at least not yet. The problems of security and privacy of information are among the issues to be examined in Chapter 20.

Chapter 21 is on integration towards a global systems through telecommunications. Without telecommunications we have problems (and solutions) of logical integration of files. With telecommunications we have problems of interconnectivity plus integration that can eventually lead to computer applications across

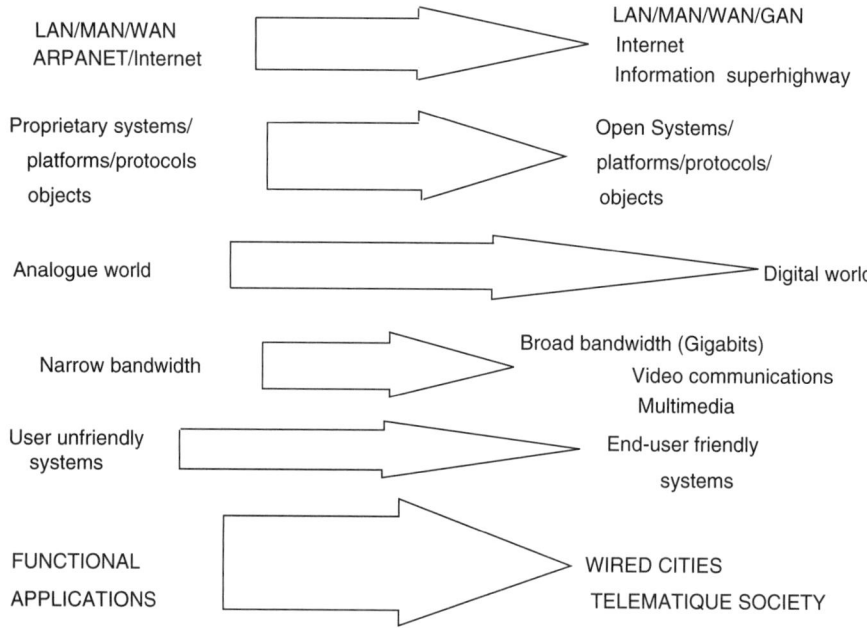

Figure 1.7 *Trends from past—present to present—future*

space and distance. We can then integrate applications not only in a corporation but in a city, and not just a city but a region and eventually anywhere in the world that we so desire. Of course this requires an international infrastructure and international standards and a few other prerequisites that will enable us to reach for a wired world or, as the French Norma and Mink called it, the **télématique society**.

Our final chapter also looks at the future from a historical perspective. One view is that there may not be any dramatic breakthrough in the technology of telecommunications in the immediate future but that we will continue to evolve on a steady growth curve consolidating much of what we have. Thus we will convert the analogue world into the digital world, enlarge the narrow band and single media to the broadband and multimedia, enhance LANs, WAN and the Internet, transform proprietary systems to open systems through standardization, and foster the growth from functional applications to the wired and télématique society through integration and the interconnectivity of telecommunications. These trends and evolution are summarized in Figure 1.7. We shall revisit this figure at the end of the book in Chapter 21 and then identify the technologies that led to each of the transformations.

Much of the future of telecommunications will depend of the response of the end-user and consumer as well as on the computing and telecommunications industry. This is difficult to predict because it requires predicting not only the technologies related to telecommunication and networking but also the environment where telecommunications and networks will be used. There will also be changes in related industries such as those of hardware, software, cable, telephone, and even the publishing and entertainment industries. Each of these industries are a multibillion dollar industry just in the US alone. Some companies have the cash to buy; some have the technology and experience to integrate; others have the connections into homes and offices. Possible combinations of firms in these industries are many and sometimes referred to as the **metamedia** industry. One entry into this new industry was announced in the US in early 1994 between the cable company TCI and the telephone company Pacific Atlantic. They were to have over 30 billion in assets and promised a 500 channel two-way interactive

video-on-demand entertainment. But then the regulated rates for TV were reduced and the telephone partner wanted a reduced price to buy. The cable company did not budge, and so before the year was out the proposed merger was dissolved. There may well be more failures, but meanwhile there is much experimentation going on in the US and in Europe with different media and a varied mix of services to test the public on what they want and what they are willing to pay.

All around the world new industrial structures are rising out of the deregulation of traditional PT&Ts (Post Telephone and Telegraph). In about 100 countries, various value-added information related services are opening up to competition. There is a growth in the demand of information related services leading to an information-intensive global society that is propelled by consumer markets. Competition and ease of entry will accelerate the creation of new services and new markets.

> ... The new technology creates new demand. ... While the telecommunications industry will provide the network for the information intensive society and the computer industry will provide the advance processing capability, the information service industry will continue to evolve with the passage of time. Although many of the information service providers are likely to emerge from the telecommunications and computer industries, new providers will create their own niches in the market in the future ... We are entering as era when global information networks and services will become reality. (Nazem, 1993: p. 19)

The fierce competition ahead may well result in a better integration of delivery and services offered to the consumer but the shakedown may take long and will most likely be painful and costly. Eventually there will be a winner and many winners before the new industry stabilizes. Whatever the final product offered and whatever the delivery mode, the process will be exciting.

In the next few years we will see improved technologies and infrastructures, new and enhanced servers with marked differentiation and specialization for different applications, and increasing competition in the telecommunications and network industries. This will enable us to communicate and cooperate with each other in ways that were not hitherto possible and could well result in the redefinition of the old paradigms of communication and work.

The problems still facing us are those of standards which is thwarting competition The high level of continuous technical communication in the telecommunications and computing industry has resulted in the industry refusing to settle down. This provides the end-users with more choice but for the network manager it is a greater challenge.

The next chapter is another summary and overview but only of the technologies to be examined in Chapters 3—8.

Case 1.1: Network disaster at Kobe, Japan

In 1995, an earthquake struck Japan at Kobe killing more than 5000 people and causing damage of over $100 billion. This damage included $300 million to the physical plant and infrastructure damage disrupting service to about 285 000 of the 1.44 million circuits in the region and knocking out over 50% of the overall services offered by the national telecommunications utility NTT, the Nippon Telephone and Telegraph Company. Many businesses were severely disrupted and even the recovery operations were greatly hampered because of the lack of telecommunications. Some businesses, however, survived because they had good network management and a plan for disaster recovery. One of them was an information provider that had planned for an earthquake in Kobe even though the odds of an earthquake there were very small. This firm had leased lines from NTT serving its main offices in four cities in Japan in addition to leasing domestic satellite services from a VSAT (Very Small Aperture Terminal) satellite installation to bypass the domestic network. It also had a back-up generator to ensure that the system could be up and running even if all the local power lines had snapped. In addition, it had a back-up centre, not in Japan, but in Singapore. When the earthquake hit, the firm started up its back-up generator and was in full operation within the day.

Source: *Data Communications*, July, 1995, pp. 47—8.

Case 1.2: Networking at the space centre

The space centre at Houston, Texas, has controlled the flights of all the early spacecraft. In 1995, a new command and control centre was partly operational and entered its beta phase of testing to replace the old centre and to prepare for the space shuttle into the twenty-first century. The new centre has over 19 kilometres of fibre optic cables connecting its hundreds of PCs and workstations with the larger computer systems owned by NASA. These computers and workstations are all interconnected in addition to being connected to the tracking stations all around the world.

One design specification of this complex and important networking systems was that almost all the equipment must be 'off the shelf'. This was specified in order to keep maintenance easy and not as costly as in the previous centre.

The design specification is a commentary on the state-of-the-art of telecommunications and networking. Even a large and in some ways very important real-time system can be constructed from products that are commonly available and are no longer 'high tech'.

Supplement 1.1: Milestones for network development

1969   The US Department of Defense commissions ARPANET for networking among its research and academic advisers.

1972   SNA by IBM offers the first systems network architecture for a commercial network.

1974   Robert Metcalfe's Harvard Ph.D. thesis outlines the Ethernet.

1974   Vinton Cerf and Bob Kahn detail the TCP for packet network intercommunications.

1976   X.25 is the first public networking service.

1978   Xerox Company, Intel and DEC give the first Ethernet specification.

1980   The FCC in the US deregulates telecom equipment at customer premises and allows AT&T to offer tariffed data services and computer companies to offer non-tariffed communications services.

1981   IBM introduces the personal computer, PC.

1982   Equatorial Communications Services buys two transponders and the Weststar IV satellites, giving birth to the first very small aperture service (VSAT) industry.

1984   AT&T divests ownership in local telecoms.

1984   The UK's Telecommunications Act authorizes the privatization of British Telecommunications Ltd. It is licensed and a regulatory authority is established.

1985   The Japanese government enacts the Telecommunications Business Law, which abolishes the monopolies of the country's domestic and international carriers.

1987   The Commission of European Community publishes the Green Paper which calls for open competition in the supply of equipment and the provision of data and value-added service.

1990   The ARPANET is officially phased out and the Internet is born. (For milestones in the life of the Internet, see Chapter 20.)

1993   British Telecom buys 20% of MCI and this marks the beginning of a truly global market.

Bibliography

Budway, J. and Salameh, A. (1992). From LANs to GANs. *Telecommunications*, **26**(7), 23–27.

Campbell-Smith, D. (1991). The newboys: a survey of telecommunications. *Economist*, 5 Oct. 1–52.

Doll, D.R. (1992). The spirit of networking: past, present, and future. *Data Communications*, **21**(9), 25–28.

Financial Times, 19 July, 1989, pp. 1ff. Special issue on Survey of International Telecommunications.

Malone, T.W. and Rockart, J.F. (1991). Computers, networks and the corporation. *Scientific American*, **265**(3), 128–136.

Nazem, S. (1993). Telecommunications and the information society: a futuristic view. *Information Management Bulletin*, **6**(1 & 2), 3–19.

Sankar, C.S., Carr, H. and Dent, W.D. (1994). ISDN may be here to stay ... But it's not plug-and-play. *Telecommunications*, **28**(10), 27–33.

Sproul, L. and Kiaster, S. (1991). Computers, networks and work. *Scientific American*, **265**(3), 116–123.

Tillman, M.A. and Yen, D. (Chi-Chung) (1990). SNA and OSI: three stages of interconnection. *Communications of the ACM*, **33**(2), 214–224.

Weinstein, S.B. (1987). Telecommunications in the coming decades. *IEEE Spectrum*, **23**(11), 62–67.

Part 1

TECHNOLOGY

2

TELEPROCESSING AND NETWORKS

In 1899, the director of the US Patent Office urged President William McKinley to abolish his department. According to the director, everything that could be invented had been invented.

Although the processing speed of a CPU is measured in micro-, nano- or picoseconds, users will not get the full benefit of this speed if tapes and disks on which input is recorded are physically transported to the computer for data entry. Likewise, the delivery of reams of paper output to the user can be time consuming, particularly when users are not located in the same building as the CPU, for example in a distant sales office, branch office or warehouse.

With **teleprocessing** (the processing of data received from or sent to remote locations by way of a telecommunications line, such as coaxial cable or telephone wires), input and output is instantaneous. This is the mode of processing for multiuser systems where people located in dispersed locations share a computer but need to input data and access up-to-date information at all times. You can see why the term 'teleprocessing' is often used as a synonym for **telecommunications, data communications** and **information communications**.

The technology of telecommunications, which links input/output terminals to distant CPUs, advanced in the 1970s to allow the linkage of workstations, peripherals and computers into networks. Networks are valued by organizations because they promote the exchange of information among computer users (many business activities require the skills of many people), the collection of data from many sources and the sharing of expensive computer resources. Networks may be:

1. Local area networks (LANs) which permit users in a single building (or complex of buildings) to communicate between terminals (often microcomputers), interact with a computer host (normally a mini or mainframe) or share peripherals.

2. Linked LANs within a small geographic area.
3. National networks such as ARPANET to link computer users in locations across the country. Database services also fit into this category.
4. International (wide area) networks, the most expensive networks because of long distances between nodes; the most difficult to implement because standards and regulations governing telecommunications vary from country to country.
5. A combination of the above.

This chapter surveys the technology or telecommunications, discusses the importance of telecommunications to business and looks at the problems of connectivity that corporate managers must resolve.

The rise of distributed data processing

When computers were first introduced, most organizations established small data processing centres in divisions needing information. These centres were physically dispersed and had no centralized authority coordinating their activities. Data processed in this manner were often slow to reach middle and senior management and frequently failed to provide the information needed for decision-making. Because of the scarcity of qualified computer specialists, the centres were often poorly run. In addition, they were unnecessarily expensive. By failing to consolidate computer resources, organizations did not take advantage of Grosch's law (applicable to early computers) which states that the increase in computational power of a computer is the square of the increase in costs; that is, doubling computer costs quadruples computational power.

The need for centralized computing facilities was soon recognized. Firms hoped that centralization would result in lower costs, faster delivery of output, elimination of redundancy in processing and files, tighter control over data processing, increased security of resources and greater responsiveness to the information needs of users. While consolidation of computing was taking place, computer technology was advancing. By the time third-generation computers were installed in computer centres, users no longer had physically to enter the centre to access the computer but could do so from a distant terminal connected by telecommunications. Time-sharing had also been developed whereby several users could share simultaneously the resources of a single, large, centralized computer. With centralized processing, teleprocessing (also called **remote processing**) became the norm.

However, not all of the expectations for improved service were realized when centralization took place. Complaints about slow information delivery and the unresponsiveness of the centres to user information needs were received by corporate management. Users resented the red tape that computer centres required to justify and document requests for information services. In turn, computer specialists at the centres chaffed at criticism, feeling overworked, underpaid and unfairly reproached by those with no understanding of the problems of systems development and the management of computing resources. This general dissatisfaction with operations led to a reorganization of processing once again – to **distributed data processing** (DDP) the removal of computing power from one large centralized computer to dispersed sites where processing demand was generated.

Although DDP sounds like a return to the decentralization of the past, it was not. By the time DDP was initiated, minicomputers with capabilities exceeding many former large computers were on the market at low cost. Computers were much easier to operate and maintain. Chip technology had increased CPU and memory capacity while reducing computer size. Desktop microcomputers were for sale. Strides in telecommunications meant that no processing centre had to be isolated, but could be linked to headquarters or to other processing centres (**nodes**) in a network. Furthermore, experience with data processing had given users confidence that they could manage and operate

their own processing systems without the aid (or intervention) of computer specialists.

Distributed data processing includes both the installation of stand-alone minis or mainframes under divisional or departmental jurisdiction and the placement of stand-alone microcomputers for personal use on the desktops of end-users. But it is generally associated with the linkage of two or more processing nodes within a single organization, each centre with facilities for program execution and data storage. (These nodes may be computers of all sizes, from microcomputers to mainframes.) Figure 2.1 shows sample DDP configurations. A host computer may provide centralized control over processing as in the star network or the nodes may be coequals. (Failure of the central computer impairs processing for the entire system if the host computer breaks down in the star configuration. The ring structure overcomes this problem because rerouting can take place should one processing centre or its link fail.) The hardware at each node is sometimes purchased from the same vendor, which facilitates linkage. But generally networks contain a mix of equipment from different manufacturers which complicates information exchange. This is discussed further later in this chapter.

We now look at equipment configurations and technology to support teleprocessing and networks.

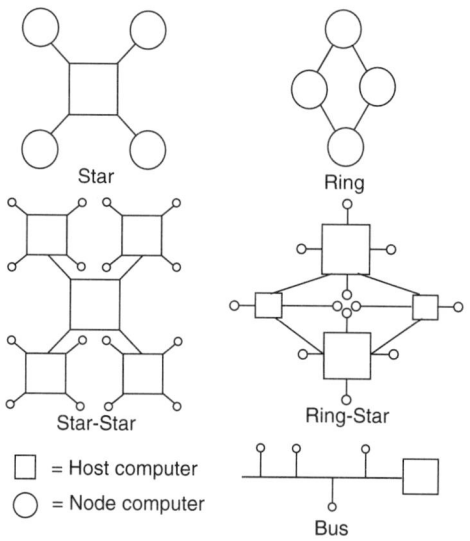

Figure 2.1 *Examples of DDp configurations*

Transmission channels

Data or information may be transmitted a few feet within a single office building or over thousands of miles. When planning for telecommunications, corporate management must consider what type of transmission channel is most appropriate for organizational needs and whether to use private or public carriers.

Types of channel

A **simplex communications, line** or **channel** enables communication of information in one direction only. No interchange is possible. There is neither any indication of readiness to accept transmission nor any acknowledgement of transmission received. A **half-duplex system** allows sequential transmission of data in both directions, but this involves a delay when the direction is reversed. The ability to transmit simultaneously in both directions requires a **duplex** or **full-duplex channel**, a more costly system. An advantage in computer processing is that output can be displayed on a terminal while input is still being sent. Figure 2.2. illustrates channel differences and lists applications for their use. Some channels carry voice transmissions, some data. Current technology allows voice and data messages to be carried long distances over the same line at the same time and at an affordable cost.

Transmission speed or signalling speed is measured in bits per second. In most communications lines, 1 baud is 1 bit (binary digit 0 or 1) per second. The capacity of the channel is measured in **bandwidths** or **bands**. These give a measure of the amount of data that can be transmitted in a unit of time.

A range of transmission options exists by combining different types of channel, transmission speeds and bandwidths — the cheapest and most limited being a simplex telegraphic-grade channel, the most expensive and versatile a full-duplex broadband system. Wire, cable, radio, satellite, telephone, television, telegraph, facsimile and telephoto are sample communications channels which vary in the types of data or information they transmit and transmission features.

Public or private carrier?

In the USA, telecommunications lines that serve the public are licensed by the Federal Communications Commission (FCC). There are over 2000 telecommunications carriers available to the public (called **common carriers**) such as AT&T for telephone and Western Union for wire

Type	Transmission direction	Graphic representation	Example
Simplex	One direction only	A ⟶ B	Radio
			Television
Half-duplex	One direction only at any one time. Can be in both directions in sequence	 A ⟶ ⟵ B	Walkie-talkie Intercom
Duplex or full duplex	In both directions simultaneously	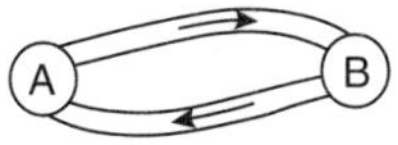 A ⇄ B	Picture-telephone Dedicated separate transmission lines (such as a presidential 'hot-line')

Figure 2.2 *Types of channel in telecommunications*

and microwave radio communications. Some provide point-to-point service on a dedicated line; others, switched services, routing data through exchanges and switching facilities, sometimes in a roundabout route to reach a final destination. A variation of the latter is a **packet-switching service** which breaks a data transmission into **packets**, each containing a portion of the original data stream, and transmits the packets over available open lines. Upon arrival, the data in the packets are reassembled in their original continuous format. Packet-switching networks can support simultaneous data transmission from thousands of users and computers.

A shared rather than dedicated point-to-point communications line reduces the outlay of a company for long-distance communications circuits. Most packet-switching services have another advantage as well: they support a standard protocol (rules governing how two pieces of equipment communicate with one another) like X.25 which is vendor independent. That is, they will transmit data to and from equipment sold by many different manufacturers. As a result, a user at a single terminal can access non-homogeneous hardware connected in the network.

An alternative to a common carrier transmission facility is a **private data network**. Such networks are economically feasible over short distances, which explains why they are called **local area networks** (LANs). Some LANs are vendor specific: that is, they support connectivity only between hardware manufactured by one manufacturer or manufacturers of compatible equipment. Examples of such networks include IBM's token ring network, Wang's Wangnet, and Xerox's Ethernet. (See Figure 2.3 for an illustration of Ethernet use.) Some LANs are all-purpose networks. Connectivity and protocol support is provided for the equipment of many vendors. **Private branch exchanges** (PBXs), primarily telephone systems that connect hardware, are a third option.

Choosing between these options is a difficult task that involves many technical issues, including speed, capacity, cabling and multivendor support. The network must fit into the existing environment and meet the organization's functional needs. Furthermore, no company wants to invest in a system that will require the replacement of existing hardware or the addition of costly interfacing equipment; nor does any corporate manager want a system that will quickly

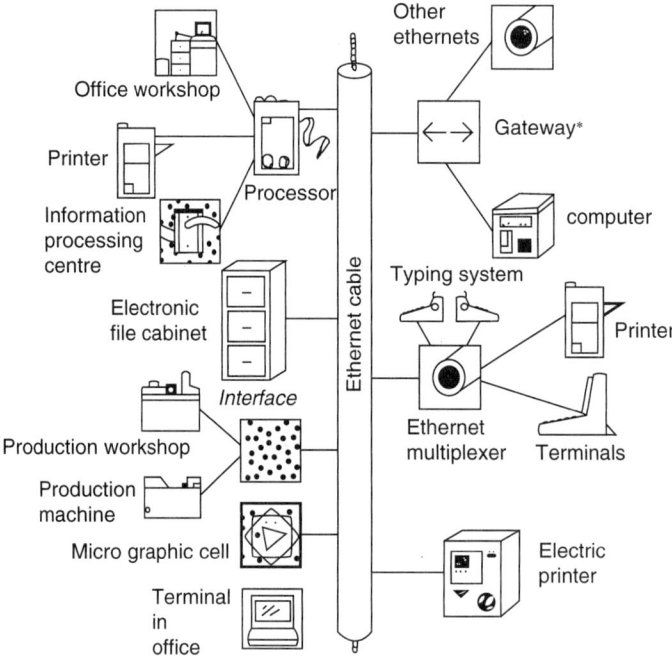

Figure 2.3 *Ethernet*

become obsolete or outgrown. The variety of LAN products adds to the dilemma and the intense competition among vendors to sell LAN systems puts pressure on corporate managers at the time a network decision is being made.

Interface equipment

To transfer information by telecommunications, many computer systems must add **interface** components: that is, hardware and software to coordinate the receipt and delivery of messages.

To illustrate, most terminals produce **digital signals** (pulses representing binary digits), whereas many telecommunications lines transmit only **analogue signals** (transmission in a continuous waveform). As a result, equipment is required to convert digital data to analogue signals (a process called **modulation**) when a message is sent and to reconvert the waveform

back to pulses **(demodulation)** at the receiving end (see Figure 2.4). A peripheral called a **modem** performs this conversion, a name derived from modulation and demodulation.

In addition, a **multiplexer** may be added to combine lines from terminals that have slow transmission speeds into one high-capacity line (see Figure 2.5). Sometimes a number of terminals share a channel (or channels). A **concentrator** is equipment that regulates channel usage, engaging terminals ready to transmit or receive data when channels are free or sending a busy signal. For long-distance networks, a **repeater** acts like an amplifier and retransmits signals down the line. A **bridge** has a similar interface function but retransmits between two different LANs of homogeneous equipment. A **router** not only retransmits but determines where messages should be forwarded. A **gateway** connects networks that use different equipment and protocols. (Although managers should be familiar with these terms, they rely on the

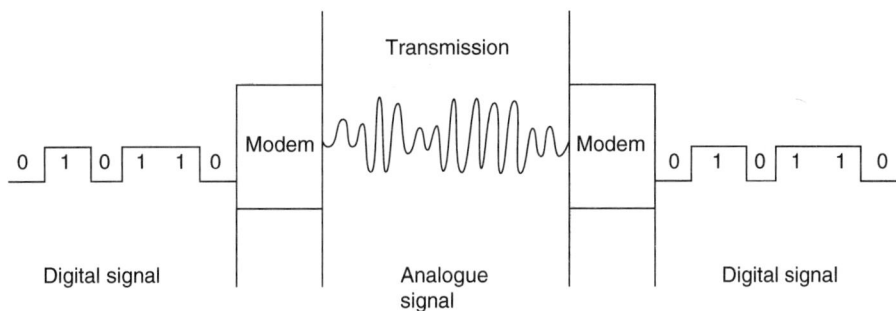

Figure 2.4 *Digital and analogue signals*

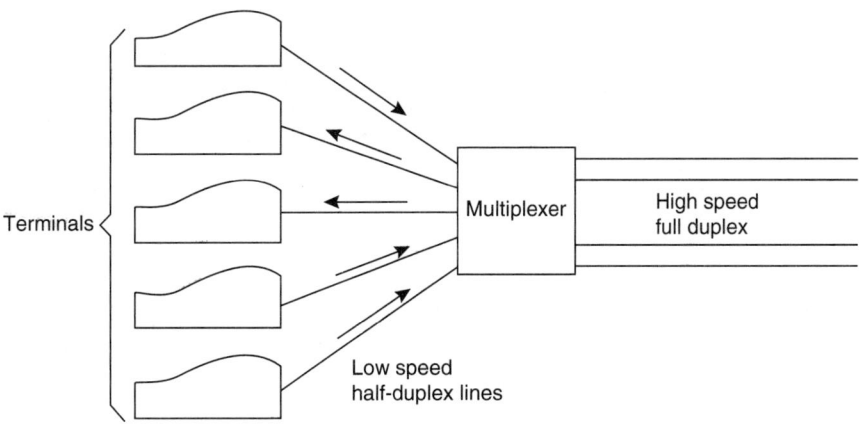

Figure 2.5 *Multiplexer*

expertise of telecommunications specialists for network design.)

A LAN of microcomputers, peripherals and interconnections with other networks may have a component that caters to all the requests of the networked computers. For example, a **disk server** is a component that acts like an extra disk drive: it is usually partitioned so that each computer can access a particular private storage area. A **file server** is more sophisticated, allowing access to stored data by file name.

Large mainframe computer systems generally include a **front-end processor** programmed to relieve the CPU of communications tasks. For example, a front-end processor may receive messages, store transmitted information and route input to the CPU according to pre-established priorities. It may validate data and preprocess the data as well. Another major function of front-end processors is to compensate for the relatively slow speed of transmission compared with the processing speed of the CPU. Front-end processors may also:

1. Perform message switching between terminals.
2. Process data when teleprocessing load is low or absent.
3. Act as multiplexers and concentrators.

4. Provide access to external storage and other peripherals.
5. Check security authorizations.
6. Keep teleprocessing statistics.
7. Accept messages from local lines with mixed modes of communication.
8. Facilitate the use of the CPU by several users in a time-sharing system.

Figure 2.6 illustrates a sample teleprocessing system that incorporates some of the equipment described in this section.

Interconnectivity

Each computer system may have a unique configuration of computing resources such as computer speed, file capacity and peripherals that include fast printers and optical scanners. As the management of each computer system may not be able to afford all the resources that they need, it is desirable to be able to share resources when they are not being fully utilized. This can be achieved through interconnectivity, the linking of computer systems by telecommunications and networks.

It is telecommunications that provides the link and connectivity between computers that enables the sharing of resources and communication between users of different systems. When

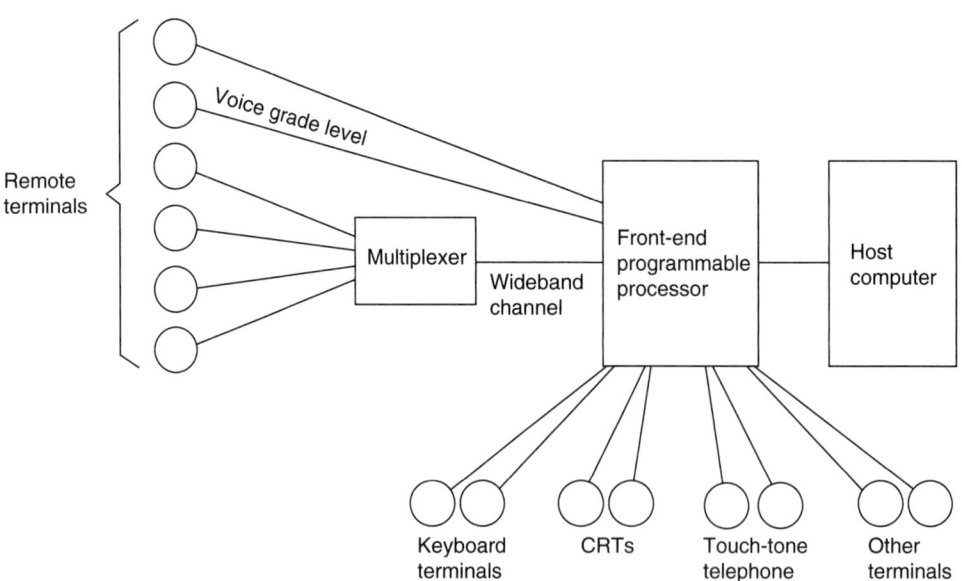

Figure 2.6 *Example of a teleprocessing system*

interconnectivity creates a network at the local level we call it a LAN, local area network. This is discussed in great detail in Chapter 5. When the interconnectivity is within a metropolitan area, we have a MAN, metropolitan area network; when extended to a wide area it is known as a WAN, wide area network. And when the interconnectivity is global, we have a GAN, global area network. The MAN, WAN and GAN are compared in Chapter 6.

A GAN providing international connectivity is also known as the Internet and is discussed in Chapter 20.

Networks in the 1990s

The 1980s was a decade in which a large number of LANs were installed. In the 1990s, many of these LANs will be joined into national and international networks. Already the rewiring of Europe and the USA is under way to create a coast-to-coast network to carry voice, images and data messages simultaneously over the same line at low cost.

How will these integrated networks affect business communications? Information transmission will be faster. For example, phone companies are installing digital computer switches and supplementing low capacity copper transmission lines with microwave and high capacity fibre-optic cables that will transmit information more than seven times faster than current rates.

A single network will often suffice. Many companies are currently part of several communications networks, each serving a different purpose (eg. LANs, telephone traffic, facsimile machine).

New services will become available such as cellular and mobile phones. An engineer at a construction site will be able to look at electronic blueprints simultaneously with the architect at the office who drew up the plans. A reporter covering the earthquake in Japan may send photos to London headquarters for distribution in an electronic newspaper delivered to the computer screen of subscribers. Salespeople may be able to sell and deliver their products without ever having to make personal calls to customers.

Telecommunications will become more reliable. When equipment breaks down at a location, transmission will be routed to avoid the bottleneck.

The price of telecommunications services will drop. Although the development and installation costs of integrated networks are staggering, revenues generated by telecommunications are high. The market is growing and competition in the telecommunications industry is fierce, factors that traditionally favour the customers by leading to lower costs.

The 'bottom-line' measure of the worth of telecommunications and networks in the 1990s will be in its applications. Some applications like EFT (Electronic Fund Transfer), EDI (Electronic Data Interchange) and e-mail (electronic mail) will not change conceptually but they will be used more creatively and for a wider range of uses. For example, in 1996 for the first time the US troops abroad (in Bosnia) were communicating daily with their families at home by e-mail. The US government provided equipment and training facilities to do this. These communication services are part of message handling, a subject discussed in detail in Chapter 17.

E-mail and many applications in telecommunications in the early 1990s were textual. Soon the stream of text will be integrated with other media such as voice and pictures giving us multimedia applications that could be very useful in education, medical services, entertainment and many a business where communications will no longer be by letter or even e-mail but by teleconferencing and video-conferencing. These applications are discussed in Chapter 18. Multimedia will also be part of the discussion on information services and telecommuting, subjects discussed in Chapter 19; and in the discussion of the Internet, a subject examined in Chapter 20.

The Internet is perhaps an area where the growth was much larger and faster than any one had predicted. Controlling the content and privacy on the Internet and improving global communications, not just for transfer of data but for selling and buying products with secure international transfer of funds on the Internet, will be a high priority in the late 1990s.

The most popular design for a global network is called **Integrated Services Digital Network** (ISDN) initiated in 1984 by the International Telecommunications Union, an organization of the United Nations comprising telephone companies around the world. However, this group has not yet agreed on standards for hardware and software – standards that are required if computing power, information and telecommunications are to be integrated in a single transportation system. It may take years (possibly decades, according to some detractors) before differences can be resolved.

Nevertheless, technical and market tests for integrated digital networks have been made by the state-owned telephone companies in Germany and Japan. Numerous ISDN trials conducted by American telephone companies are also under way. The success of ISDN will affect all companies with a vested interest in telecommunications. For instance, a private data network run by IBM permits the exchange of information between companies with compatible IBM machines. If this network were meshed with ISDN, the network would be able to expand its services. For computer manufacturers, the success of ISDN may accelerate sales. Companies already in telecommunications and computer companies wanting a share of the telecommunications market are likewise following ISDN projects with interest, looking for ways to attract telephone defectors to network services of their own.

Issues facing corporate management

The quality of decision-making by managers should improve with integrated data networks because more information and more timely information will become available on which to make decisions. However, telecommunications add to the responsibilities of corporate management as explained next.

Organization of information resources

The duty of corporate management is to plan for data access, cost-effective usage of computing resources, and the sharing and distribution of information within and between departments. A computer network does facilitate data collection, processing and information exchange at low cost, but is not the only option. In fact, the array of options in the organization of computing resources is what makes the manager's role so difficult.

To illustrate, a multiuser system that uses time-sharing to link terminals, called a **shared-logic system** may be preferable to the installation of a local area network. (A shared-logic system utilizes terminals connected to a centralized computer in which all processing occurs. Local area networks tie otherwise independent computers, usually microcomputers, together.) A manager must be familiar with the strengths and weaknesses of both shared-logic systems and LANs in order to evaluate their relative benefits and trade-offs. In choosing an appropriate system, the following questions should be asked:.

1. How are computing resources used in the company? If the system is primarily for high-volume transaction applications, shared-logic technology should be favoured. If for general use, such as word processing and spreadsheets, then a LAN is appropriate. (If files do not require constant updating by many different people, perhaps a multiuser system is not necessary after all.)

2. Are concurrent requests for information from databases likely? Shared-logic systems are better able to respond to such requests. In addition, most provide file and record locking and offer transaction logging and recovery facilities.

3. Is peripheral sharing a primary requirement? Sharing is convenient and cheap with a LAN. In addition, the incremental cost of adding resources to a LAN is low whereas expanding a shared-logic system may require a complete change in the CPU.

4. Is ease of applications development important? Many users who write their own programs find development tools for personal computers easier to use than shared-logic applications development tools.

5. Is growth expected? LANs can be upgraded with ease since each added workstation brings its own CPU resources.

6. How dispersed are users? LANs are not designed for wide area access. Most serve a single building.

7. Is data security an issue? A LAN allows decentralized data under user jurisdiction. With a shared-logic system, a security officer can impose strict control over data use and storage.

8. Are users willing and able to take responsibility for systems operations? If not, a shared-logic system would be advisable.

9. Are gateways to other computer networks required? Although much work is currently being done on gateway technology, applications that must be integrated with

other large systems are better served at present by shared-logic systems.

10. Will the network contain products of different manufacturers? Many LANs enable such connectivity whereas connectivity between products of different makes is minimal with most shared-logic systems.

In general, dissatisfaction with the current operations is the driving force towards the establishment of LANs. In organizations with a proliferation of personal computers, a LAN is considered when users need to share data, management wants better processing control, the need to integrate new systems exists, and input/output inefficiencies are a concern. In a shared-logic minicomputer environment, a LAN might prove the answer to poor response time, excessive downtime, high costs, user pressure for personal computers, and the lack of application availability when needed.

There are other resource configurations to consider still. For example, a microcomputer can be hooked up to a mini or mainframe with excess capacity and used to create data, upload data to mainframe storage or download data for microcomputer processing. Or employees with compatible hardware might pass around disks holding files to be shared.

These organizational structures are not mutually exclusive. A single firm may have one or more LANs to supplement a shared-logic system. In addition, stand-alone microcomputers may be on the desktops of some workers. In-house computers may also be linked to external computer resources. Thus the organization of computing resources can be tailored to the unique operating environment of each firm. It is management's responsibility to decide how telecommunications can best serve the company's long-term interests.

Organization of telecommunications and networks is crucial to the orderly operations and growth of most computing. Its organization structure is examined in Chapter 9 with a popular configuration, the client–server approach, examined in Chapter 10. Whatever the organization structure, there are tasks and issues that face network managers. One of these tasks is the acquisition of the necessary resources needed for telecommunications and networking. In today's market, the network manager has a spectrum of choices in both hardware and software and must decide how best they are used for working

as individuals or in groups. Selecting most (if not all) of the resources from one vendor is tempting for it will eliminate problems of interfacing with vendors and the incompatibility of resources. But in the real world, computer products are put together as and when the budget allows. Many corporations have developed systems one at a time, so they often represent different generations of technology with the problems of connectivity and compatibility not having been addressed. The result is a mix of systems with dissimilar architectures and operating systems unable to exchange information without 'patch-work' and inefficient interfaces.

One solution for interconnectivity, at least on the hardware side, is to have industry standards for hardware manufacture such as the ISDN mentioned elsewhere. However, reaching agreement on standards is a slow and difficult process. The problem is further complicated in networks and telecommunications because, for meaningful communication in a worldwide market, telecommunications have to be global and standards have to be not just nationally agreed upon but agreed upon internationally. This subject is examined in Chapter 11.

Security is also an issue with computer systems especially when they use telecommunications and networks for now they are exposed to many sources of infiltration and systems violation. Data/knowledge must now be protected from unauthorized modification, capture, destruction or disclosure. This problem is addressed in Chapter 12.

Network management is the subject of Chapter 13. The resources managed are examined in Chapter 14.

Thus far the discussion concerns network management at the corporate level. However, corporations must communicate with other corporations and individuals within the country and need a national infrastructure. This is the subject of Chapter 15. Communicating across national borders is becoming increasingly important in our worldwide economy, and is the subject of Chapter 16.

Summary and conclusions

In the 1980s, many business organizations installed local area networks to supplement their computer systems by having interconnectivity and the capability of sharing resources. In the 1990s, more and more businesses (and non-profit organizations, including government agencies) will participate in regional and national networks of linked LANs. The future trend is towards integrated digital networks extending nationally and internationally. The result will be faster, more reliable telecommunication services for the business community far beyond the present day e-mail (electronic mail), EFT (Electronic Fund Transfer), teleconferencing and access to on-line remote database services used in offices today. These applications are the subject of Chapter 17–20.

Telecommunications provide managers with more information and more timely information than in the past which should improve decision-making. But telecommunications also adds to management's responsibilities in areas such as the organizations of information resources and making them operational and secure. Such management of computing resources are examined in Chapters 9–16.

To enable us to discuss applications of telecommunications and the management of telecommunications and networks needed for these applications, we need to know more about the technology of telecommunications. In this chapter we took an overview of some of the basic telecommunication technologies. The details will be the subject of the Chapters 3–8.

This chapter is in a sense an overview of this book with an introduction to some of the basic technology of telecommunications and networks. In this chapter we looked at the front-end and the back-end of a telecommunications system which are discussed in detail in Chapter 4. In between are the transmission channels such as the telephone, radio, cable, and the wireless and cordless channels. These are the subject of our next chapter.

Case 2.1: Delays at the Denver Airport

In 1995, the new airport at Denver in the US opened after long delays and a cost of $5 billion. It was designed to be the state-of-the-art structure designed for air transportation well into the next century. The airport had a sophisticated network that automated many subsystems at the airport

in addition to maintaining telecommunications not only in the airport but with pilots in the air, travel agents in town and airports around the world.

One subsystem was designed to deliver baggage from the plane to the airport building even before the passengers were ready to claim their baggage. The $300 million subsystem used ATM technology (to be discussed later in this book) along with 55 computers and was designed to handle 30 000 items of luggage daily. This subsystem delayed the opening of the airport and the contractor for the subsystem claimed that they were rushed and they needed more time to install the system to start with but were not given that time. How much longer had they wanted? Around 16 months which happens to be about the time for which the opening was delayed.

One lesson that has been drawn from this sad story is that tomorrow's technology should not be installed today without adequate preparation and good risk assessment. It has also been argued that in this situation the risk was worth taking. If there were not some managers who took calculated risks in computing and telecommunications, then we would not have many of the applications that we now have today.

Source: *Data Communications*, July 1994, p. 30, and *International Herald Tribune*, Feb. 28, 1996, p. 4B.

Supplement 2.1: Top telecommunications companies in 1994

Company	Revenue (Million US$)	Mainlines (Millions)
NTT (Japan)	68.9	59.8
AT&T (USA)	43.7	–
Deutsche Telekom (Germany)	37.7	39.2
France Telecom	23.3	31.6
BT (UK)	21.3	27.1
Telecom Italia (Italy)	18.0	24.5
GTE (USA)	17.4	17.4
Bell South (USA)	16.8	20.2
Bell Atlantic (USA)	13.8	19.2
MCI (USA)	13.3	–

Source: *International Herald Tribune*, Oct. 11, 1995, p. 12.

Bibliography

Cerf, V.G. (1991). Networks. *Scientific American*, **265** (3), 72–81.

Derfler, F.J. Jr. (1991) *PC Magazine Guide to Connectivity*. Ziff-Davis Press.

Derfler, F.J. Jr. and Freed, L. (1993). *How Networks Work*. Ziff-Davis Press.

Dertouzes, M.L. (1991). Communications, computers and networks. *Scientific American*, **265** (3), 62–71.

Doll, D.R. (1992). The spirit of networking: past, present, and future. *Data Communications*, **21** (9), 25–28.

Financial Times, 19 July, 1989, pp. 1ff. Special issue on 'Survey of International Telecommunications'.

Flanagan, P. (1995). The ten hottest technologies in telecom: a market research perspective. *Telecommunications*, **29** (5), 31–41.

Interfaces, **23** (2), 2–48. Special issue on 'Telecommunications'.

International Herald Tribune, 4–11 October, 1995. Special series on 'Telecommunications in Europe'.

Soon, D.M. (1994). Remote access: major developments in 1995. *Telecommunications*, **28** (1), 57–58.

3

TRANSMISSION TECHNOLOGIES

The new mobile workforce doesn't so much need computer devices that communicate as they need communication devices that compute.

Samuel E. Bleeker

Introduction

In this chapter we will look at the transmission media needed for communication. The most common (and oldest) is the copper wire and its variations. Such wiring is best for short distances and small capacities. However, for longer distances and for a variety of traffic such as voice and video, we need glass fibre optic cables. Even the glass fibre has limitations for distance and then we need microwave or satellite capability. Each of these media will be discussed in turn for their advantages, limitations and applications.

A more recent transmission media is the cordless and wire-less person-to-person communications. It is aptly described by Arnbak as a '(R)evolution'. We will examine the evolution of this technology and its revolutionary implications for the way we may communicate in the future.

Wiring

The oldest and still commonly used transmission media is the copper wire. It comes in one of many forms: solid or stranded; unshielded, shielded, or coaxial. The shielding is required to protect the conductor from outside electrical signals and reduces the radiation of interior signals. The conductor carrying the electrical pulse that represents a message itself can be **solid** or **stranded** or **twisted**; the stranding and/or twisting of a pair of wires provides a shielding reducing the absorption and radiation of electrical energy. Shielded twisted wires are relatively expensive and difficult to work with. They are also difficult to install.

Whether single or stranded, a shield could be of woven of copper braid or metallic foil which

has the same axis as the central conductor and hence is referred to as a **coaxial cable**.

It is easy to install connectors to a coaxial cable but the connectors must be good since a bad connection can adversely affect the entire transmission system. Such connectors are often made of tin or silver; the latter is more expensive but more reliable.

The main problems with copper wiring are threefold: it has a low capacity, it is slow and it is adequate for only a short distance. As distances increase and as larger demands are made on the capacity (by volume as well as the nature of traffic, such as video demanding greater capacity), and as the need for greater speed becomes relevant, then copper cables are inadequate. Another cable made of **glass fibre** is more appropriate. A glass fibre is thinner than a human hair, stronger than steel, and 80 times lighter than a copper wire of the same transmission capacity. The capacity of a fibre is one billion times the capacity (in bits/second) of a copper wire (for the same cross-section).

Glass fibre is made of silicon, a substance as common as sand. Fibre transmits pulses of information in the form of laser emitted light waves. It is not only fast in transporting data but it is effective for greater distances than is copper; and fibre is also more reliable. The distance for fibre links is more than 11 times the maximum distance for coaxial cable and 15 times the distance for some twisted wire systems. Even for short distances, fibre is used because it can carry a mix of multimedia traffic that includes data, text, images and voice. Thus even for the short distances as for internal wiring in an aircraft, fibre is used to carry voice and music.

Fibre is also reliable because it does not pick up extraneous electrical impulses and signals.

Table 3.1   *Comparison of wiring approaches*

	Unshielded twisted wire	*Shielded twisted wire*	*Coaxial cable*	*Fibre optics*
Speed and throughput:	Fast enough	Very fast	Very fast	Fastest
Average cost/node:	Least expensive	Expensive	Inexpensive	Most expensive
Media size:	Small	Large	Medium	Tiny
Maximum length:	Short	Short	Medium	Long
Difficulty in installation:	Difficult	Difficult	Moderate difficulty	Relatively easy
Protection from electrical interference:	No protection	Some protection	Good protection	Very good protection

These signals are picked up by copper which becomes an antenna and absorbs energy from radio transmitters, power lines and electrical devices. Also copper develops voltage potentials to the electrical ground resulting in interference. In contrast, glass fibre cables are immune to electrical fields and so they do carry clean signals that never spark or arc and add to the reliability of fibre as a conductor.

The light waves in glass fibre can be precisely controlled and is less vulnerable to unauthorized access compared to the electrical pulses in a copper cable. This adds to the security of the systems and is very important when confidential messages are being transported.

Glass fibre is much lighter and smaller than a copper wire but it is much more expensive. The average cost in the US of wiring a home with fibre is roughly $1500 compared to $1000 with copper wire. Note that the increment is only $500 per home but the total cost of having copper and then replacing it with fibre costs $2500. There are two observations worth making. One is that replacing copper wire represents a loss of investment to the carrier owning the copper wire who should then be expected to oppose fibre in order to support his investment; and two, the laying of fibre in new homes represents a savings of $1000 per home over replacing the copper wire (and including the sunk cost). This explains why it is cheaper (in total cost terms) to install an advanced technology starting from scratch (without an infrastructure) than replacing an old infrastructure. This explains the advantage that developing countries without any infrastructure have. But this advantage can also apply to developed countries like France that had an outdated telephone system in Paris. Fibre and advanced switches were installed and a free computer terminal was given to every household

with a telephone, instead of a telephone directory (which at the time of the initial planning cost just as much as a terminal). As a consequence, Paris today has one of the most advanced telephone systems and a infrastructure basic to a wired city. More on infrastructure and more on a wired city later. We must get back to transmission technologies.

A comparison of wired technologies for transmission is summarized in Table 3.1.

From the above discussion, one can conclude that the different media of transmission do have their distinct advantages and limitations. Their use would depend on the carriers responsible for the transmission and will vary with countries depending on their applications, be they telephone, cable TV or PCs (personal computers). The density of these applications vary with countries. Statistics for a sample of geographic areas is shown in Table 3.2.

A recent application of telecommunications is the transmission of video. The characteristics of video compared to those of telephone, cable and PCs are shown in Table 3.3.

The different modes of wiring having different applications result in a coexistence of all or most of these forms of transmission in many a telecommunications environment. One is shown in Figure 3.1, where the transmissions for long distances requiring large capacities and carrying

Table 3.2   *Services in selected parts of the world*

	US	*Japan*	*Europe*
Telephone lines/1000 people	48.9	42.2	42.2
Cellular phones /100 people	2.6	1.2	1.2
TV households with cable (%)	55.4	13.3	14.4
Personal computers/100 people	28.1	7.8	9.6

Source: Stix (1993: p. 104)

Table 3.3   *Comparison of services in the US*

	Telephone	*Cable*	*PC*	*Video*
Cost as % income per capita: (Drop over the years)	14−2	4−1	11−2	3.5−1
Usage:	Communication	Entertainment	Computing	Entertainment
Transmission media:	Copper wire	Broadcasting	Fibre	Co-axial cable
Period for adoption in home:	1876-1950	1950-1991	1975-1995	1971-1988

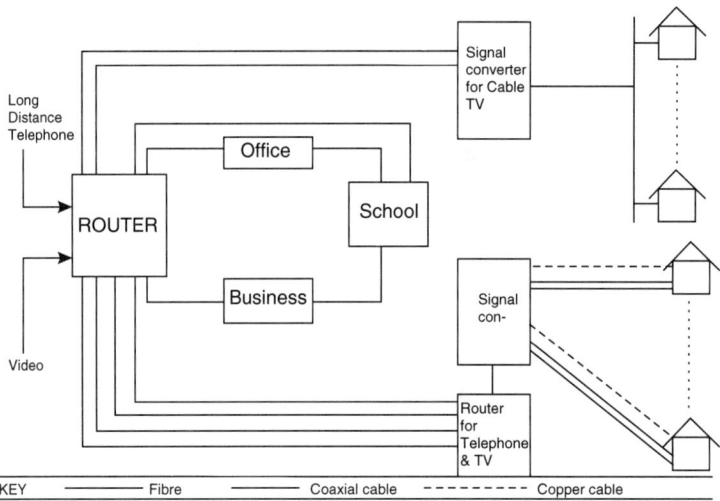

Figure 3.1   *Use of different transmission media*

a mix of traffic will use fibre, say for the connections to offices, schools and businesses. For local connections for local TV, coaxial cable is used; and for connections to PCs and telephones in homes, we use coaxial cable or copper cable.

Microwave

A problem with wiring is that it is limited in distances for which it can transmit effectively. Beyond that distance, microwave transmission is necessary. Earthbound or terrestrial **microwave** systems use high-frequency electromagnetic waves to transmit through air or space to microwave receiving stations. These stations, however, must be within 'line of sight' of each other. They are placed close enough to each other, roughly 30 miles apart (on land) and the

transmission towers are tall enough so that the curve of the Earth, hills or other tall obstructions do not interrupt this line of sight. Weather conditions like humidity and dust can also affect transmission. Thus the microwave alternative is not ideal, though it was a great innovation when it was first invented. It is now a reasonable economical option only for distances up to around 20 miles offering signalling speeds of 1.54 megabits/second.

Satellite

A satellite in space can overcome the 'line of sight' problem besides being less sensitive to the weather. When a satellite is placed in geosynchronous orbit in space, its speed allows it to match exactly the speed of Earth's rotation, so that it is stationary in relation to any point on

Earth. Microwaves are then sent to the satellite some 22 500 miles away and then sent back to Earth stations anywhere on Earth. Each satellite can transmit and receive signals to slightly less than half the Earth's surface; therefore at least three satellites are required to effectively cover all the Earth.

The return journey by satellite could take up to half a second, which may not be desirable for real time applications but good enough for many applications, especially for day-to-day private and business data communications and even voice communications by telephone. One configuration of satellite transmission using Earth stations between an office and a remote factory is shown in Figure 3.2.

Wire-less/cordless systems

Whether by microwave or by satellite, or by other means of transmission, there always is a telephone wire at some end. Eliminating this wire gives us the cordless system. It is considered a new technology because it is recent in time but

it is an old technology since it really is still basically microwave broadcasting. Even the name 'wireless' is as old as the Marconi's Wireless Telegraph system of 1900. To avoid the confusion, the new system is sometimes referred to as **wire-less** or **cordless**.

The great advantage of the wire-less system is that it no longer is fixed and restricted to a physical space. You can now 'roam' around (**roamers** are subscribers in locations remote from their home-service area) anywhere you wish and still communicate with a phone that has a transmitter. Of course, the longer the distance, the different the technology employed. Thus, within a building or home, you may use a wired network with laptops and palm top computers. These are connected through a wall-mounted **transceiver** (which connects thin and thick cables like the fibre and coaxial) to a wired LAN and establishes radio contact with other portable networked devices within the broadcast range. For wider ranges, small antennas on the back of a PC or within PCs communicate with radio towers in the surrounding area and are referred to as **cellular** phones. For a wider range of distances,

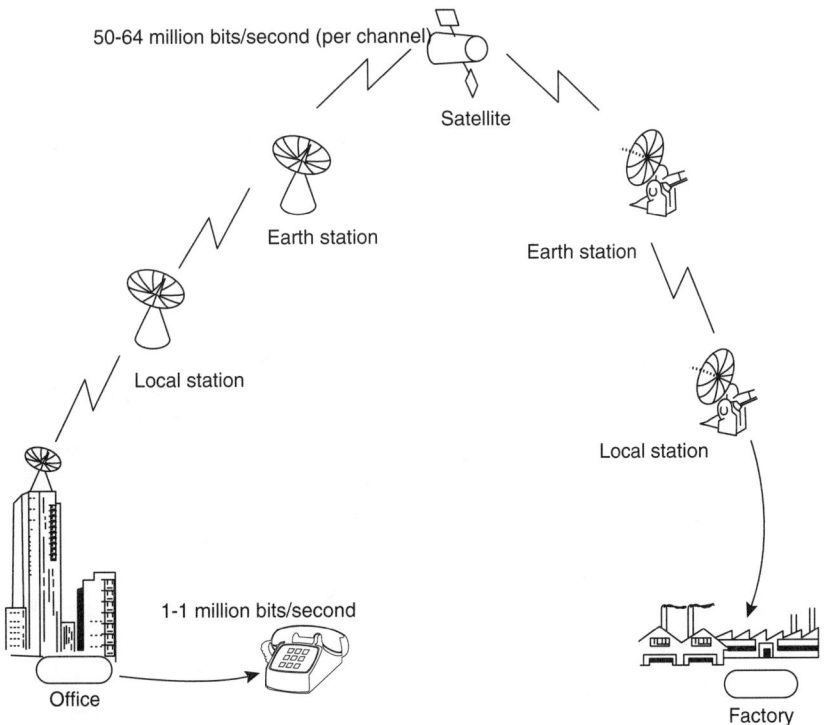

Figure 3.2 *One configuration of a satellite communication*

even around the world, satellites are used to pick up low powered signals from mobile networked devices.

Besides distance, the cordless system has a wide spectrum of other combinations like local or long distance; portable and personal (PCS); digital and analogue; and two-way or one-way. Some have choices of their own. Thus the digital computer could be either a palm-top, a lap-top, or a **PDA** (Personal Digital Assistant). The one-way (unidirectional) may be either radio or message service like paging or downloading of data from either a local or a wide area network. The two-way (bidirectional) allows for a dialogue and its manifestations are either the cordless telephone, the portable computer, the wireless LAN, or the cellular radio PCS (Personal Computer System), and the CDPD (Cellular Digital Packet Data), which transmits 'packets' of data through sparsely used radio channels or during gaps in conversation. All these types are shown in a taxonomy of cordless systems in Figure 3.3. This classification is not mutually exclusive. We shall not discuss each in detail, but instead we shall examine three popular systems: the cellular radio, the packet radio and the PCS.

Cellular radio

A cellular radio transmits radio signals to a base-station to a cellular switch through broadcasting that connects the wire-less network with the fixed network by a cellular switch with the PTSN (Public Switched Telephone Network). The PSTN is connected by fibre or coaxial cable to devices like telephones or other cellular phones, or to modems in host computers which may be minis or mainframes or other PCs. The base-station can serve tens of wire-less terminals, whilst the cellular switch may serve up to 100 base-stations. One configuration of the cellular phone is shown in Figure 3.4.

Cellular technology had its birth in the 1970s in the labs of Bell in the US. In 1993, a European standard GSM (Groupe Special Mobile) was adopted by 32 countries. There are other standards in the US and Japan, as well as the British standard, the CT-2, the Cordless Telephone – Version 2.

With developments leading to the second generation cordless technology, we see a change from stand-alone consumer items to elements of a geographically dispersed network. Network control remains distributed, with wireless terminals competing with little or no central coordination for available channels ... The second generation networks, all using digital speech transmission, will employ a variety of access technologies including narrowband time division with eight channels per carrier (GSM), narrowband time division with three channels per carrier (IS-54), frequency division (CT-2), and time division with twelve channels per carrier (DECT). (Goodman, 1991: pp. 33–4).

Packet radio

A packet radio is an approach that uses a switching and **PAD** (Packet Assembler and Disassembler) where the message being assembled

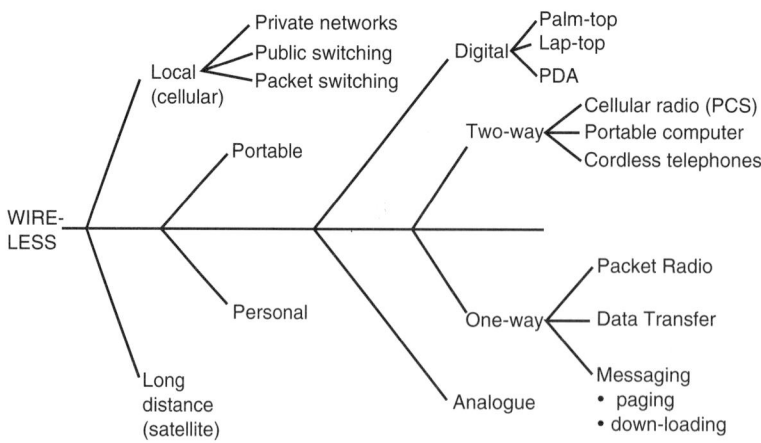

Figure 3.3   *Taxonomy of wire-less/cordless transmission*

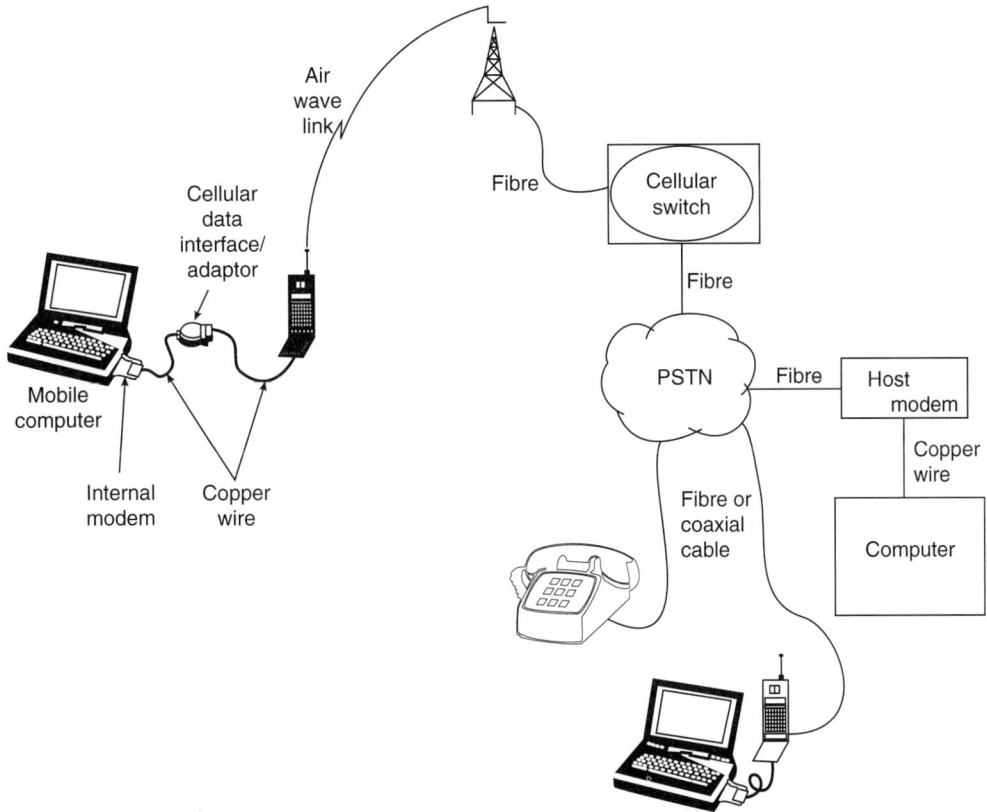

Figure 3.4   *Cellular phone connection to corporate computer*

(or disassembled) into **packets**, which are self-contained sets of information. The packets are then relayed by the PAD through a satellite to the possible destination by broadcasting to different base-stations. The base-stations not being addressed ignore the broadcast while the base-station that is being addressed accepts the message and passes it on to the ultimate destination, which may be a device or host computer. Such a configuration is shown in Figure 3.5.

PCS (Personal Computer Systems)

A **PCS** is a **LEOS**, a Low Orbit Satellite System. It is based on a digital architecture. A defining technical characteristic of PCS is its

> high capacity and spectral efficiency. Assigned spectrum is divided into discrete channels, which are utilized by grids of low-power base-stations with relatively small cell contours.

Because PCS cell contours are relatively small, PCS handsets can operate at low power and still be small, light and inexpensive. PCS also will be useful for private in-building or campus-based wireless PBX systems because of the potential to assign frequencies to relatively discrete areas...Data ports can be inexpensively built into PCS handsets to allow direct transmission by bypassing the handset's voice coder, thus permitting PCS handsets to act as wireless modems for portable computer and facsimile machines. (Wimmer and Jones, 1993: p. 22).

In the mid-1990s, the PCS was still considered experimental in the US though it was well advanced in Europe, especially the UK. In the US, there was much debate and argument about the assignment of frequency spectrum which hampered the development of the PCS, but the infrastructure for a PCS was well planned. It was projected that

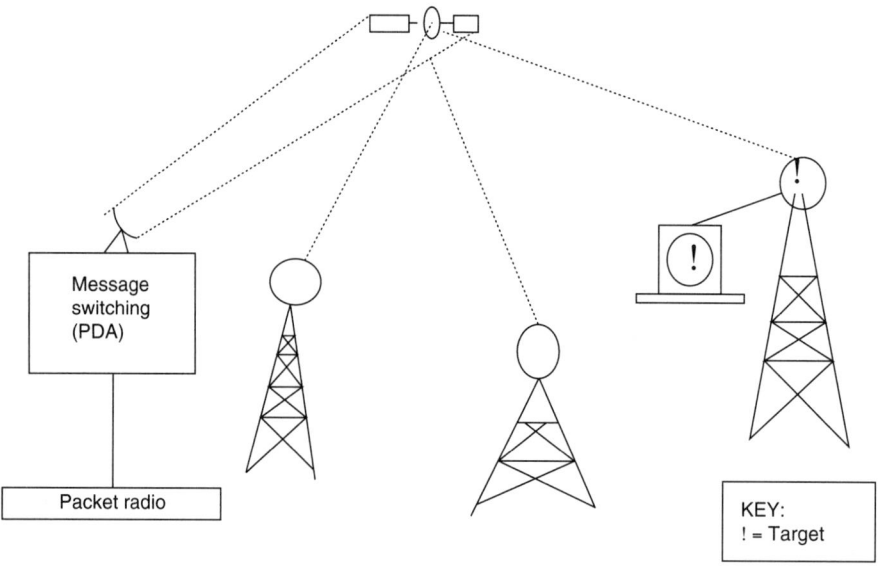

Figure 3.5   *Packet radio satellite broadcast*

the cellular industry will spend $20 billion on PCS infrastructure alone. Because of the high frequencies, 2 gigahertz versus 800 megahertz, PCS requires five times more towers to provide the same service as existing cellular phones ... Once the new spectrum becomes operational, PCS will deliver a technological advance that will draw multiple converts. The lure: lower mobile phone bills. PCS is not a separate or competing network with cellular, but a way for large cellular operators to expand their networks. New spectrum to play with and insurance that

they will have bandwidth to expand.   (Flanagan, 1995: p. 32).

Comparison of transmission systems

A comparison of the two cordless systems (the packet radio and the cellular phone) with two other popular cord systems (the microwave and the mobile satellite system) is shown in Table 3.4.

Table 3.4   *Comparison of transmission approaches*

	Applications	*Advantages*	*Disadvantages*
Microwave	"Long" distance transmission	Easy installation Wide band Good compatibility	Poor penetration (e.g. of walls and floors) Poor mobility
Mobile satellite	Mobile and long distance transmission	Saturation coverage Reliable	Expensive Poor data rates Limited capability
Cellular	Telephone Circuit Switching Roaming modem	Voice/data supplier Mobility Matured technology Existing infra-structure	Expensive Poor penetration Not universally available
Packet radio	Packet switching Roaming location	Mobility User pays only for packets transmitted	No voice support Coverage limited to large metropolitan areas

These four systems may survive the shakedown of the industry as it matures. Meanwhile, and even thereafter, these technologies may well coexist for they all have distinct functions to perform.

There are a number of applications of wire-less computing that include POS (Point-Of-Sale), dispatching of repairmen or taxis, delivery of services varying from pizzas to medical help, etc. However, to some extent, the cordless and wire-less transmission is in fashion and even a status symbol. What is needed to bring it all together into a useful and productive technology are two things. As identified by analyst Jay Batson, they are:

> Number one is a critical mass in a specific area so that the entire metropolitan area is covered. Number two, there have to be PC applications that know how to better use wireless data capabilities. There hasn't been a kind of consciousness raised all the way up to the software level that people use on PCs, or pagers. The secret is for a compelling market use to evolve. (Flanagan, 1995: p. 38).

Wire-less technology does increase productivity by using productively the time we idle away whether this be at a waiting room, in a car (as a passenger of course), or even while walking. It enables us to perform when we think of it or when we need to make a transaction and not have to wait till we get to our office. It could be a tool for professions like the a salesperson or a medical professional. And it can be 'life-saving' for a rural resident or in an emergency like a car breakdown, or a medical emergency while 'roaming'.

Despite the many applications of the cordless and wire-less devices, they may never replace the transmission done from a fixed site like a work-site whether it be at home or the office. The workplace is ingrained in many cultures and their work ethic, partly because they have all our worktools together and partly because the workplace creates an atmosphere conducive to work. Transmission and telecommunications capability is fast becoming a tool for the office and part of our office culture. Soon it may be difficult to distinguish between computers with telecommunications capability and telecommunications devices with computing capability. To complement office computing, there may well be handsets that deliver not just data, but also voice and even video. Mobility and portability may create a new class of applications that combine computing and personal electronics. The symbiosis of computing and communications has applications and implications that we have yet to discover and appreciate.

Summary and conclusions

The main media of transmission are the wires that string our telephone poles; the cables that connect long distances including below the oceans, the microwaves that connect longer distances; the satellite that enables much longer distances to be traversed quickly and efficiently, and the cordless phone. These transmission technologies are shown in Figure 3.6. Each of the main transmission technologies has variations.

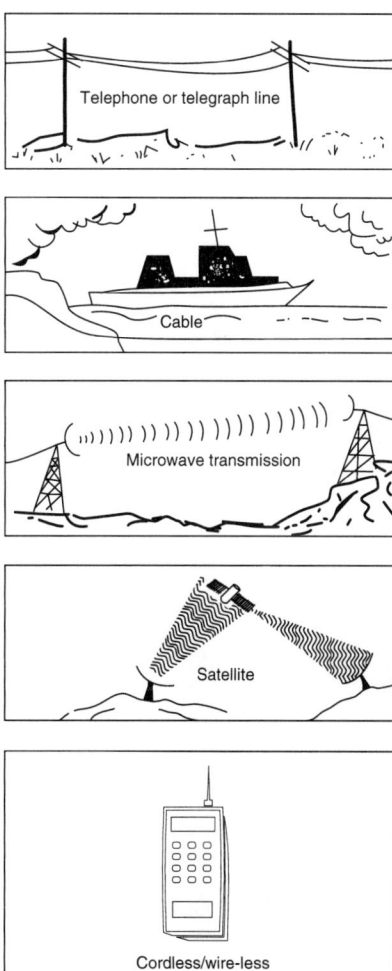

Figure 3.6 *Micromotive transmission waves*

A taxonomy of the main important variants are shown in Figure 3.7, a different view of Figure 3.3. It is the permutation and combination of these variants that is most suitable for any one environment of telecommunications. Selecting the right permutation and combination is the function and responsibility of the manager, sometimes the technical person but increasingly the corporate manager. To make such decisions it is necessary to understand some of the technical aspects of these technologies, which is the purpose of this chapter.

Looking at the future, it is quite possible that we may see personal cordless communicators which will enable us to communicate with each other in an office building (no matter how tall or large it may be) and with people outside the building (wherever they may be in the world). One such scenario is shown in Figure 3.8.

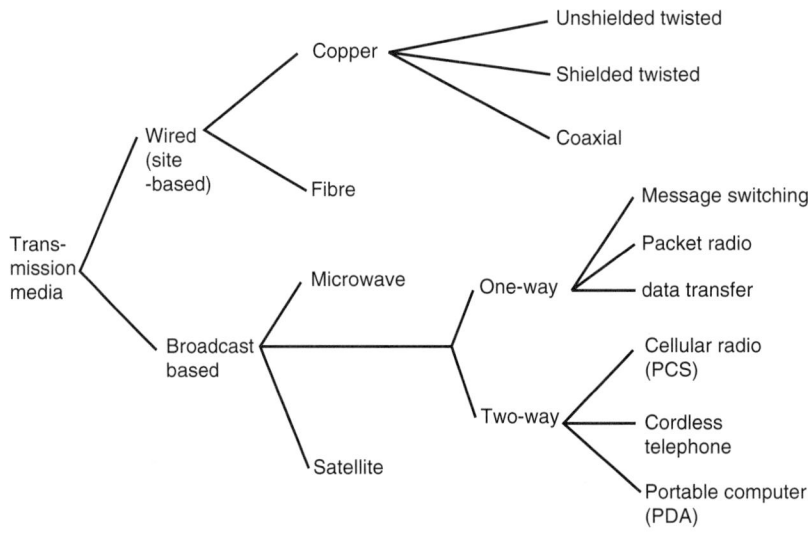

Figure 3.7 *A Taxonomy of transmission media*

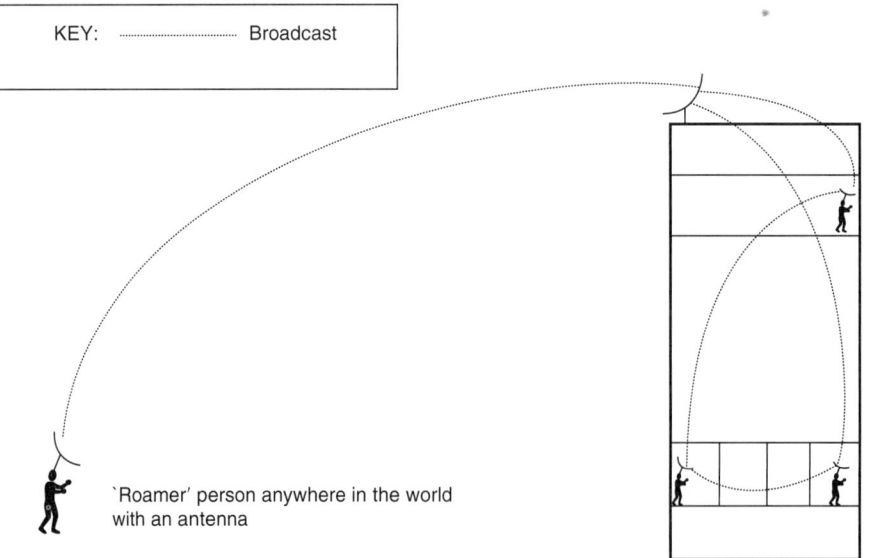

Figure 3.8 *A view of transmission in the future*

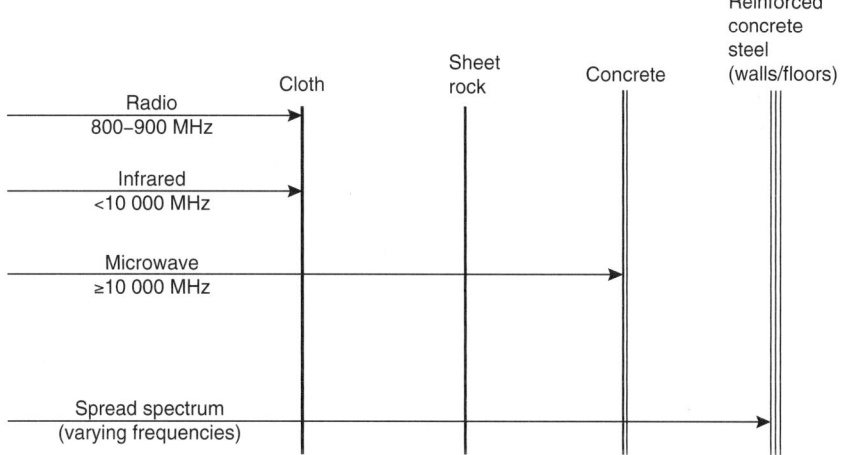

Figure 3.9 *Varying penetrations by transmission media*

The constraint to current cordless communication seems to be the limits of penetration through opaque objects like walls and floors in a building. Cloth limits the penetration of radio and infrared; concrete but not sheet rock limits microwaves; and reinforced concrete steel (walls or floors in a building) limits **spread spectrum**, which is the use of multiple frequencies to achieve reliable communication with relatively weak signals. These penetration limits are shown in Figure 3.9. The current restrictions are not a problem for most of us and the current technology seems quite adequate. What is needed is greater access to such equipment and its integration with other applications.

In this chapter we have examined transmission technologies. But there are other related and enabling technologies for telecommunications that are needed. One of these we have mentioned frequently: switches. In the next chapter we will examine switching and enabling technologies of telecommunications.

Case 3.1: The Iridium project

The Iridium project is a global satellite communications systems designed by Motorola (US) that allows customers to call and be called at any time and from any place using hand-held telephones directly through a constellation of 66 LEO (Low Earth Orbit) satellites in six orbital planes and in an orbit of 780 kilometers. It has two bands,

the L-Band that is from the satellite to earth and the Ka-Band that is an intersatellite link with gateways and feeder link connections.

The Iridium satellites are designed for Low-Earth Orbit (LEO) using the TDMA (Time-Division Multiple Access) technology. The Iridium project is expected to be operational in 1998. It has been financed in 1993 by a first round of equity offering of $800 million of which Motorola purchased a minority interest. The total cost is expected to be around $3.4 billion. Other owners in this private enterprise adventure are: Bell Canada Enterprises; Krunichev Enterprise, the Russian manufacturer of the Proton rocket; China Great Wall, the Chinese manufacturer of the Long March rocket; Nippon Iridium, a consortium of 18 Japanese's companies; STET, the Italian cellular and radio carrier; Mawarid Group, a Saudi investment group; Muiridi Investments of Venezuela, United Communications Industry Co., a Thai cellular and paging operator; Lockheed, a US launch provider; Raytheon, a US telecom equipment manufacturer; and Pacific Electric and Cable, Inc., a Taiwan equipment manufacturer.

Case 3.2: CT-2 and PCs in the UK

The UK has been a world leader in the adoption of the CT-2 (Cordless Telephone—Version 2) and the allocation of spectrums to the PCS (Personal Communication Systems).

The initial decision of assigning market share to individual CT-2 licensees was too small

> to provide wide coverage with high quality and low cost ... United Kingdom licensees implemented a telepoint service, in which subscribers had to look for signs to determine where they could place a call. Subscribers typically find this inconvenient and demand more ubiquitous coverage ... The advent of a new telecommunications service in the United Kingdom encouraged greater price competition and the creation of valued new services by existing market participants. (Wimmer and Jones, 1992: p. 22)

The British Department of Trade and Industry made a study of the spectrum from 470 MHz to 3.4 GHz, and found that

> personal and mobile communications will increasingly displace fixed radio relay services in the 1−3 GHz region as the latter move to less-congested, higher frequencies, or are replaced by alternate means of communication. (Wimmer and Jones, 1992: p. 24).

Source: Wimmer and Jones (1992: pp. 22−7).

Case 3.3: Transmission at a trillion bits per second

All through the first four decades of computing, systems were input/output bound because the speed of the CPU computing was much faster than the speed at which data could be fed as input and results could come out as output. With telecommunication, the bottleneck remained because transmission speeds were also much slower than CPU speeds. The bottle neck was likely to get worse since CPU speeds were still on the increase. And so the industry set an informal goal of increasing transmission speeds to a rate of a trillion bits per second through an optical fibre. The date of the breakthrough seemed feasible for around the end of this century.

In early 1996, almost five years before the deadline, at a conference on Optical Fibre Communications, three research teams announced that they has independently achieved the Holy Grail of high speed transmission: the terabit threshold of transmission at a trillion bits per second. This was the equivalent of transmission 'the contents of 30 years' worth of daily newspapers in a single second, or conveying 12 million telephone conversations simultaneously'. In relative terms, consider the speeds achieved as being some 400 times faster than the commercial systems in use today which carry 2.4 billion bps, or 2.5 gigabits a second. Trillion bps capacities are necessary for carrying video and multimedia signals for applications like films and video-on-demand. The time for commercializing these high speeds is estimated at five years.

The developments of high speed transmission were done by Fujitsu Ltd., the Nippon Telegraph and Telephone Company and a team from the AT&T Research Labs. The three teams used different approaches but all had one thing in common: 'instead of sending one stream of light through the fibre, as is done today, they sent multiple streams of light, each at a slightly different wavelength, thereby multiplying the amount of information that can be transmitted.'

One potential drawback in all the three approaches is that the transmission will most likely be limited to relatively short distances of around 50 miles. This is because light signals degrade with distance and have to be amplified and reconstituted en route. This is very expensive for multiple light streams.

Source: *International Herald Tribune*, 2/3 March, 1996, pp. 1 and 4)

Bibliography

Amedesi, P. (1995). Satellite communications in the year 2000: the view from Europe. *Telecommunications*, **29**(4), 47−48.

Arnbak, J.C. (1993). The European (r)evolution of wireless digital networks. *IEEE Communications*, **31**(9), 74−82.

Cox, D.C. (1995). Wireless communications: what is it? *IEEE Personal Communications*, **2**(2), 20−35.

Da Silva, J.S. and Fernandes, B.E. (1995). The European research program for advanced mobile systems. *IEEE Personal Communications*, **2**(1), 12−17.

Derfler, F.J. (1997). Sky links. *PC Magazine*, **16**(1), NE1−NE12.

Dunphy, P. (1992). Myths and facts about optical fibre technology. *Telecommunications*, **26**(3), 33−40.

Flanagan, P. (1996). Personal communications services: the long road ahead. *Telecommunications*, **30**(2), 23−28.

Goodman, D.J. (1991). Trends in cellular and cordless communications. *IEEE Communications Magazine*, **29**(6), 31–40.

Gunn, A. (1993). Connecting over the airways. *PC Magazine*, **12**(14), 359–382.

Imielinski, T. and Badrinath, B.R. (1994). Wireless computing. *Communications of the ACM*, **37**(10), 18–28.

Johnson, J.T. (1994). Wireless data: welcome to the enterprise. *Data Communications*, **23**(3), 42–58.

Jutlia, J.M. (1996). Wireless laser networking. *Telecommunications*, **30**(2), 37–39.

Laurent, B. (1992). An intersatellite laser communications system. *Telecommunications*, **26**(7), 90–92.

Lauriston, R. (1992). Wireless LANs. *PC World*, **10**(9), 225–244.

Saunders, S. (1993). The cabling cost curve turns toward fibre. *Data Communications*, **22**(11), 55–59.

Steinke, S. (1997). Rehab for Copper Wire. *LAN*, **12**(2), 57–62.

Stix, G. (1993). Domesticating cyberspace. *Scientific American*, **269**(2), 101–110.

Wimmer, K.A. and Jones, B. (1992). Global development of PCS. *IEEE Communications Magazine*, **29**(6), 22–27.

4

SWITCHING AND RELATED TECHNOLOGIES

If every instrument could accomplish its own work, obeying or anticipating the will of others ... if the shuttle could weave, and the pick touch the lyre, without a hand to guide them, chief workmen would not need servants, nor masters slaves.

Aristotle

Introduction

In the previous chapter we examined various media for transmission. But having a transmission media (and a computer) is not sufficient any more than having a car and a road will get you to where you want to go. There is much information and enabling technology that is needed. For transporting information, you need an address of the destination and detailed instructions on how to get from the origin to the destination. If you are travelling a distance that involves many changes and switches that you must make on, say, a motorway (or autobahn or freeway) you must know about all the ramps to get on and off and know how to interpret the many signs and instructions that you will meet. And, of course, you must follow the rules of the road. Can you imagine what would happen if an American driving on the right and an Englishman driving on the left were to share the same lane of a road and go in opposite directions? There would certainly be a loss of life. (Many Swedes and Finns lost their lives before they agreed to drive on the same side of the road.) In networks and telecommunications there will be no loss of life but probably a loss or at least distortion of messages being transmitted. The messages must of course have an address. This address cannot be interpreted by humans as in road transportation but by a computer and so it must be in machine readable form and be complete as well as specific without any danger of misunderstanding or ambiguity. The address is interpreted by **a router** which includes

software that determines the best path (or only feasible path) between origin and destination and may involve many **switches** that direct traffic between networks and LANs. Also, there are rules of telecommunication called **protocols** that must be followed meticulously. There is one more important concept involved. Let's go back to the car transport analogy. Suppose that you had to pay toll according to volume and you had a compact car worth of passengers and luggage. Would you take a truck or even a large car and pay for the empty space? Of course not. You would pack your car tightly and get rid of all the redundant space in a truck or large car. The same choice occurs with telecommunications. You must pay by volume (kilobytes) of message transmitted and so it behoves you to pack your messages tightly. This is known as **compression** and is important not only for reducing costs but also for reducing the time of transmission. Compression is often done by a router which along with a **bridge** and a **gateway** are facilities that connect networks.

It is these components and elements of a telecommunications systems which we will examine in this chapter. The router, the bridge, and gateway are the first topics to be examined followed by addressing and protocols like the TCP/IP protocol commonly used even by the ubiquitous Internet (over 25 million users worldwide). ATM which is used not only for data and text but also images and video will be mentioned but deferred to a later chapter. We conclude with a short discussion of these enabling technologies being smart and intelligent. This chapter is somewhat of an overview introducing many basic concepts that will be examined later in the context of their interrelationships.

Router, bridge, repeater and gateway

A **router** is primarily an interconnection to a network. It is a portal device, where a **portal** is the meeting point between local and long distance services. (The equivalent in business is a mail-room or a shipping dock.)

If there is a message that is initiated at the router, and there are many possible paths for that message, then the router will find a 'best' path for the message. What is best will be a function of variables like cost, transit delay, undetected errors, mishaps in transmission, size and priority assignment.

Consider the simple case shown in Figure 4.1. The path from origin A to destination B has two intermediary nodes D and E in route 1, has a direct by route 2; and has one intermediary node C in route 3. Is route 2 the best because it is the most direct? Maybe not, because it may be slow, costly, error prone, etc. Besides, it may be busy and congested and hence not a feasible path anyway. So all possible routes have to be looked at and the 'best' for the assigned objective function chosen, given the constraints such as cost and availability. There are many routing algorithms. Tanenbaum discusses eight, and that was long before the ubiquitous Internet (Tanenbaum, 1989: pp. 197–203).

The message may be composed of **frames** which are logical entities; or, if the message is too large, it may be broken up into **packets**. In that event each packet can be sent on a different route and reassembled at the other end. This would mean that each packet must be labelled to identify its entity and destination. Sometimes the packet has to be fragmented so that the packet for one network is not too large for another. The router should also be able to assign priority to one message over another (maybe a message to a hospital for a dying patient!).

One important function of the router is to read (and follow) the instructions in the header (at the front of the packet or frame) that concern the router, to provide addressing for the next destination, to strip the data that is no longer necessary, and then to repackage and retransmit the message.

For performing its many functions a router has a high overhead. For a lower overhead, one may consider the simplest portal device: the **repeater**, which is like a small box which you can hold in your hand and connect the two segments of a network cable. The repeater retimes and regenerates digital signals before forwarding them. However, even a string of repeaters cannot extend a LAN beyond a few thousand yards. You then need another network portal connection called the **bridge**.

A bridge has a higher throughput than a repeater and even higher thruput (no, that is not misspelled, it is merely a short-hand used in the trade) than a router but has a less efficient routing and flow control. The bridge exercises great discrimination over passing traffic by refusing passage to the other side of the bridge if the destination is not on the other side of the bridge. This reduces non-essential traffic. However, a bridge may also not guarantee the delivery of the frames

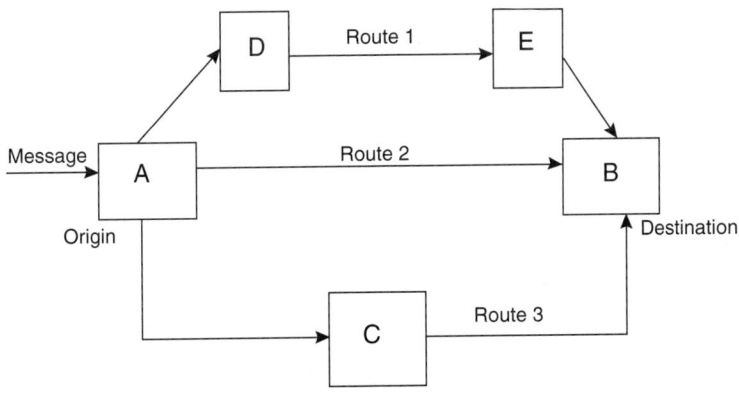

Figure 4.1 *Routing*

nor its mishaps. For example, if the frame arrives at the bridge faster than its processing rate, then a congestion of links and the bridge may occur with frames being lost unintentionally. Also, if a frame exists beyond the maximum transit time allowed, the frame may have to be discarded. However, like the router, the bridge does check for the frame size allowed by the network to be used and allows for priority. Unlike a router, a bridge is protocol independent.

But there are multiple protocol routers, called MPR, which is 'simply a router that supports more than one protocol. Without the appearance of advanced hardware technology like high speed microprocessors, it is now possible to provide software support for routing a number of high-level protocols simultaneously and within a single system.' The MPR can be combined with a bridge to give a bridge—router system, which 'is a device that simultaneously combines the functions of bridges and routers to give the "best of both".' Bridge/routers work by routing the routable protocols and bridging the non-routable ones.' (Grimshaw, 1991: p. 32).

Like the router (and the repeater), the bridge is concerned only with networks that share the same protocols. If the protocols are different, then we need a gateway.

A gateway is the most sophisticated and complex of the facilities that connect networks and allows for different protocols at one or all of the layers of the systems architecture (Chapter 8 is devoted to systems architecture).

Because of the different protocols, a gateway must perform three distinct functions:

1. It must translate addresses for protocols using different address structures (addressing discussed later in this chapter).
2. It must convert the format of the message because it may be different in character codes, structure of format and maximum message size.
3. The protocol must be converted to replace the control information from one network with control information required to perform the equivalent functions in another network. This can be a complex problem if the protocols at each level of the systems architecture have to be converted.

Whatever network linking facility is used, it is desirable that the message be as short as possible and may need compression. But what is compression?

Compression

We discussed compression implicitly when we discussed stripping redundant data by the router. Compression by stripping is one approach to increasing throughput through a transmission media. The other approach would be to increase the capacity of the transmission channel which would involve increasing the bandwidth. This is important in wide-area networks and will be discussed later in Chapter 6. But there are limitations to increasing bandwidth and so instead we reduce the flow traffic. We do this in road transportation in a city where we cannot broaden the road and so we couple two parallel roads and make each go one way only. In telecommunications we do not change the direction of flow but instead reduce the redundancies in the message. This is what compression is about.

Compression is not new to data processing. In transactional processing we do not always process an entire file with all its fixed information, like the name of a corporation, the date of its foundation, etc. Instead, we process only the variable data in addition to an identification which appears as a header to the set of records to be processed.

Compression is also used in DSSs (Decision Support Systems) where we have a matrix to process. Sometimes this matrix has a ordered set of cells that are empty. In the case of Figure 4.2, which shows the cells of an $m \times n$ matrix, the cells with values are those in the left hand triangle while the right hand triangle is empty. The compression solution to this matrix is to eliminate the empty portion and thereby greatly reduce the message to be transmitted.

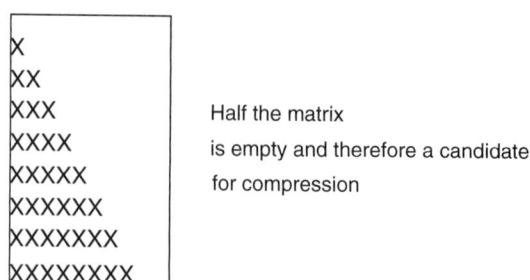

Figure 4.2 *Matrix for compression*

For a matrix in, say, a linear programming or input—output analysis problem with $m = 100$ and $n = 150$ (which is not too large for such problems) and a value of four significant digits, the savings would be around 60 000 bytes of data.

Compression is also important in the transmission of images and voice where whole blocks of the image and voice being recorded can be empty. This compression problem is of great interest to the movie industry where standards have been developed. The Motion Picture Experts Group (MPEG) has agreed on two standards, MPEG1 and MPEG2, which have reduced the full video signal from 250 million bits per second to 1.5 million for VCR quality. The problem with the MPEGs is that the encoding is very expensive and time consuming. There are, however, approaches being researched which include algorithms based on fractals and wavelets. Another solution may lie in combining compression techniques with display technologies. Researchers suspect that a video display with a web of thousands of tiny computers — one for each picture element of a video screen — might be able to generate video images with far less data than conventional videos currently required.

Whatever is transmitted (video, voice, data or text), the process of compression is the same as illustrated in Figure 4.3. The message is compressed before the message is sent and decompressed at the other end. Sometimes, an error checking character is attached to ensure that no error creeps into the compression process. This error checking could be part of the header in the address, a subject that we will now examine.

Addressing

For a message to be transmitted, it is necessary that it has all the information required to find the destination along with control information. If the message is a long one, it is divided into packets, and each packet has its own header so that the router interprets the header and routes the packets on one of the different paths available, with the message being assembled back into its original form at the other end.

An example of a header for a frame (a unit of information transmitted as a whole) which has fixed sized fields is shown in Figure 4.4. The headers will vary in format (fixed or variable fields), number of fields and content of the fields depending on the media of transmission and the technique of communications used. It could be more complex than in Figure 4.4, but lest a potential e-mail user is intimidated by the header format it should be stated that an e-mail address format is much simpler. For

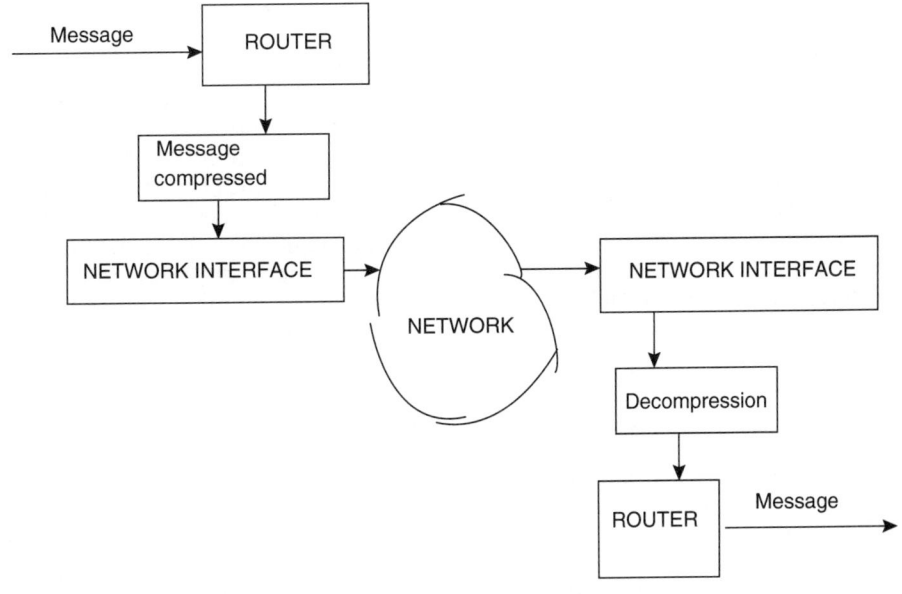

Figure 4.3 *Routers and networks*

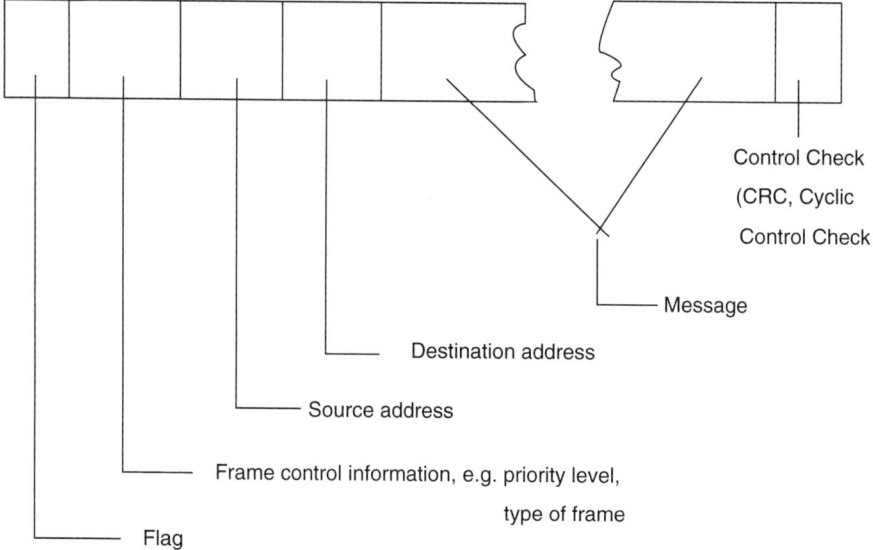

Control Check
(CRC, Cyclic
Control Check

Message

Destination address

Source address

Frame control information, e.g. priority level,
type of frame

Flag

Figure 4.4 *Field in a packet*

example, the e-mail address of the author once was KHussain@ix.netcom.com. Note that the fields are not fixed in size but the end of a field is identified by a separator such as @ or a dot. Note also that the name is not complete, with an initial missing. This is because there is a limit to the size of the name field. Note also that lower case is used for the address. Sometimes the whole name is also required to be in lower case. We shall discuss the anatomy of an e-mail address later in the chapter on e-mail, but suffice it to say that it is fairly end-user friendly. It is translated into a machine readable equivalent for interpretation by the router.

There are standards for addressing; one standard for global addressing is X.500 by CCITT, an early international organization for standardization.

One crucial device for transmission and telecommunications is the modem, which we now revisit.

Modems

We discussed modems briefly in Chapter 2 and saw how they enable us to traverse freely between the digital universe of a computer to the analogue world of telephones. This traverse between the AC (alternating current) of a digital computer and the DC (direct current) world of telephones

is achieved by the RS-232C, which is a serial connection consisting of several independent circuits sharing the same cable and connector. It is shown in Figure 4.5 as a box on top of a computer and can be 'added on' to a computer when telecommunication capability is needed. The newer PCs come with a built-in modem so that it is transparent (invisible) to the user.

So what else is new with the modem? Why all the interest in an old device that has been around since the 1950s and 1960s? The answer is that the modem has greatly evolved not just in speed but in other capabilities. In speed, the modem with a microprocessor first introduced in 1981 now has speeds of 57.6 kbps (kilobits per second), 192 times the speed of the original Bell 103 modem. The faster speed of the modem may cut the phone bills by nearly half and with the price of modems dropping, the payoff period may be just a few months. The computing power in a modern modem with computers replacing the microprocessor is awesome, some are more than early mainframe computer systems. In fact, they are so fast that the other equipment in the system cannot keep up with them including many of the computers that they connect. Also, the infrastructure may not support the fast speeds of new modems like the 28.8 kbps. Many public phone circuits (even in advanced countries) have limits in their bandwidth and lose signal quality in long loops (even local loops). Furthermore,

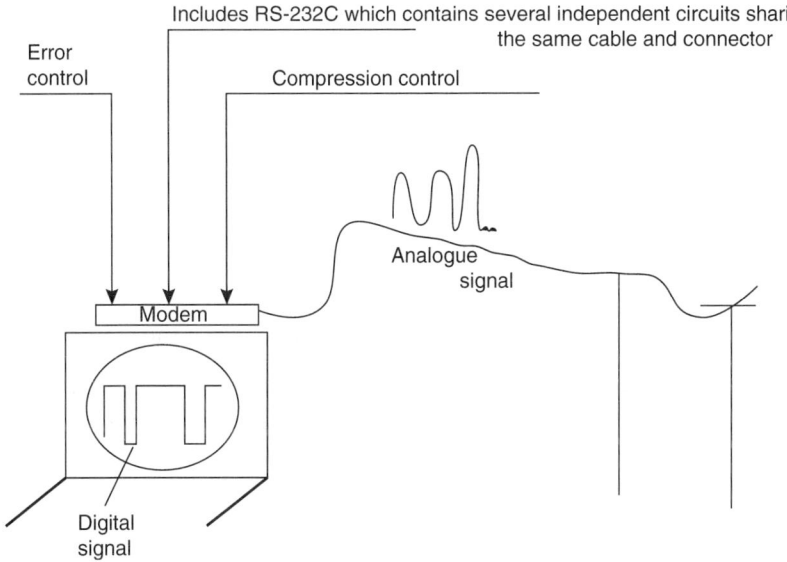

Figure 4.5   *Modem converts signals*

the switching is often poor, with companies overloading their circuits.

The advances that came in 1984 did not come from industry but from the nudging of the ITU (International Telecommunications Union), the successor of CCITT. The new standard was designated the V.22bis and is still in wide use today though it is being steadily replaced by the successor V.32 (at 9.6 kbps) which is currently in wide use. The V.34 standard probes the line to diagnose its qualities and weaknesses. The V.42 standard is concerned with error control and the V.42bis with data compression. (The V symbol stands for modem standards while the numbers is part of a block code for modems.)

Modern modems have improved file-transfer performance, better diagnostics and better documentation. They are more reliable and have good technical support for detecting and solving operational problems. There is also greater interoperability between the different models of modems from different manufacturers.

Another advance in the modem world is the fax modem which came into international prominence when the government of the USSR had unexpected trouble with its rebels because of the effective communications amongst the rebels through the fax. The faxes are strings of 0 and 1 bits that represent a graphical image of the transmitted page which are processed at both the sending and receiving ends by computers. The

sending end may also have a mechanism that compresses the data before transmission. The processing capability at both the sending and the receiving ends must be very computer intensive in addition to consuming storage in vast quantities if you want a reasonable resolution of the image.

Fax images by computer (even PCs) are sharper than the regular fax machine, but the downside is that the computer system must be turned 'on' and have the fax program in resident memory of the computer at all times. But the fax in a computer is so easy to operate. You select on the screen what you want to fax, enter the address where it is to be sent, and off it goes. Of course, if what you want to send what is not in computer memory, then you must have an optical scanning device for input. Some computers coming out in early 1996 came with in-built scanners which will greatly increase the use of fax by computer.

Smart and intelligent

A smart device is one that has a microprocessor or microcomputer to do its computations. An example would be a router that needs a computer to do all its many computations for determining the optimal or near optimal (or even a better) route for messages. This is not a trivial problem as any practitioner in Operations Research and

Management Science will tell you, for they have long been concerned with the Travelling Salesman problem, which is to determine an optimal route for a salesman that must make a specified set of stops on his sales trip. To calculate the routing problem, computing capability is embedded in the device which then makes it smart.

In contrast, an intelligent device also needs a computer but makes computations that involve making inferences. Consider the router and the problem posed in Figure 4.1. (You perhaps thought that it was a trivial diagram but there is 'a method in this madness'.) Further, let us assume that comparisons showed that route 2 was better (by a given set of objectives) than route 1 and that route 1 was better than route 3. What can we say about route 2 compared to route 3? If we applied a decision rule of transitivity (if APB (A preferred to B), and BPC, then APC), then, since route 2 is better than route 1 and route 1 is better than route 3, one can safely infer that route 2 is better than route 3. This ability to make inferences given data (of preferences in our problem) and decision rules or heuristics (rules of thumb) enables us to make inferences that reflect intelligence and hence the device can be called **intelligent**.

You may consider this a trivial problem, so let us consider a modern modem. One is shown in Figure 4.6 with a computer chip which makes the modem smart. If there were no computer capability, merely the RS-232C, then the modem could be called a dumb modem. Now suppose

that the modem is a fax modem with all the capability of discriminating voices and images. This capability requires pattern recognition which uses heuristics developed by practitioners of AI (Artificial Intelligence). It also needs a computer that performs the many necessary calculations. But it is the capability of making inferences about sounds and images that makes the device intelligent. In our example, the fax modem is both smart and intelligent.

This exercise in smartness and intelligence is designed to make the reader appreciate the possible future of the world of telecommunications. It is going to be much smarter and more intelligent as we learn more about making intelligent choices even if the variables are 'fuzzy' and not discrete (like the words 'good', 'better', 'high', 'cheaper'). We are learning much about fuzzy systems and this is being incorporated in many daily household and industrial goods. As these intelligent and fuzzy systems are adapted to telecommunications we can expect more effective and efficient devices. We can also expect optimal models for routing for minimum cost as part of the work being done on optimization in telecommunications (Luss, 1989). There is even research being done on an AIN system, the Advanced Intelligent Network (Brim, 1994) which will allow the system to perform logical operations after a call and determine the credit status of a caller, all in a fraction of a second. This has important relevance to on-line real-time reservation applications, be it for the airlines, car rental or the theatre. BT in the UK, Deutsche Budespost

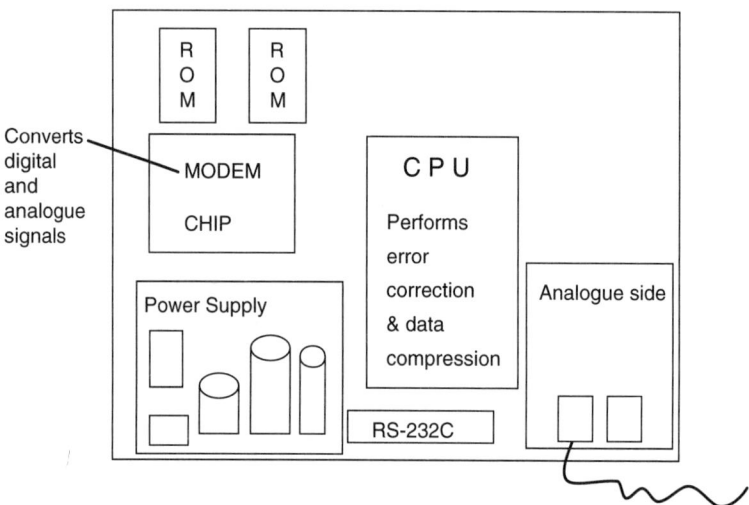

Figure 4.6 *Layout of a modem*

Telekom in Germany and American Express in the US are all working on AIN applications.

Protocols

Whether smart, intelligent or dumb, once you have the necessary devices to transmit and have the transmission media it is time to have protocols, which are agreed upon rules of behaviour. These are rules for sharing telecommunications, media and devices on it. For example, we cannot have two messages going in opposite directions on the same channel! Protocols are conventions for representing data in digital form and procedures for coordinating communication paths. We need protocols that are ground rules for interaction, sharing, routing and distribution in the relaying of information.

We have had protocols for telephony from the early days of Morse and Bell, but in telecommunications protocols for information exchange have only emerged since the 1970s. There were *de facto* protocols from IBM and Xerox Corporation and more formal ones from the DOD, Department of Defense in the US. They had the problem of having a large inventory of computers of different models and different manufacturers that did not interoperate and wanted software protocols that would enable transmission across the many configurations of computing. The DOD established the TCP/IP, the Transmission Control Protocol/Internet Protocol. Soon computer companies in the computer and telecommunications industry started following the TCP/IP or else they could not do business with perhaps the largest user of computing, the DOD.

Basically, the **TCP** is responsible for assembling packets for transmission and to properly order and reassemble them upon request. The **IP** part of the protocol addresses routes and delivers packets to the destination network and host computer. We shall have more to say about TCP/IP later on in Chapter 8.

The TCP/IP should not be confused with IP/TCP. Both are concerned with control of transmission but IP/TCP was designed for a specific type of network, the Ethernet. It has options for five protocols: routing control, error control, echo control (largely for diagnostics), packet exchange control and sequence packet control.

Besides TCP/IP and IP/TCP, there are other network protocols such as IPX, DECnet, NetBIOS, AppleTalk, and XNS. It is such proliferation of protocols that led manufacturers and designers of protocols to a recognition in the 1980s that there was a need for open protocols, that is, protocols that do not favour any one single manufacturer. What was needed was the ability by users to mix and match networking hardware and software so as to create customized networks. This would enable one to plug a cable into a wall socket and achieve interoperability of networking; this is also known as **plug-and-play** capability.

TCP/IP is an attempt towards an open protocol and a telecommunications lingua franca. It is also the favourite choice of an enterprise-wide backbone protocol where a backbone is the highest hierarchical level. Once connected to a backbone there is a guarantee of interconnections. We will discuss backbones in later chapters. In its early version, TCP/IP was most appropriate for data transmission and primarily for local connections. For media other than data, such as voice or audio, and for long distance transmission, other approaches are desirable including the ATM, a subject that is best discussed along with long distance and wide area networks. But before we get to the wide area communications, we need to discuss local communications and local area networks, the subject of our next chapter. Before we do that, we need to discuss one other important facility: the hub.

Hubs

An important facility relating to switching is the hub. We have hubs in daily life like the hub at some airports. Some airlines have a set of points that they want to serve but, instead of serving each point to point directly, they connect each point to at least one hub and then switch the traffic to the desired point through one or more hubs. The airlines have their main maintenance and administrative facilities at the hubs with minimum facilities at the other points of service, thereby greatly decreasing the overhead costs and increasing efficiency. This arrangement may inconvenience the passenger somewhat but it is tolerable if you are not pressed for time and the money savings are worth the delay. If you are pressed for time and have direct traffic, then you have a special line dedicated to your service.

Much the same happens in telecommunications where a hub is where messages are switched to their desired destination. It has heavy duty LAN switches that can handle heavy traffic with high bandwidth demands and must handle sharing of bandwidth with varying traffic like data, e-mail, multimedia and real-time like video-conferencing with varying demands of high bandwidth and low latency.

All traffic must be ideally routed seamlessly and efficiently. The hub must also be able to identify and respect priority traffic such as a CAT scan for a dying patient. All traffic must also be analysed and monitored for all segments of the network to identify potential bottlenecks, congestions and delays.

There are many ways of handling traffic. One may be the store-and-forward but then the hub must decide whether to store all the traffic on a segment to stack up all the traffic or only part of the traffic, and if so what part. Some facilities are dedicated to certain traffic while other facilities are shared. Some hubs are modular and others are not. Some hubs have a standard backbone; others do not. Some hubs allow FDDI links, some allow ATM links, while others do not care which link you use. Some hubs are fixed-port; others handle multiple nodes for each port. Some hubs handle all topologies, while other hubs restrict the type of topology. Some hubs use optimal algorithms such as for routing, while others use satisfying algorithms and simulation to find a route. Some hubs are intelligent in that they have the ability to make intelligent decisions and handle 'fuzzy' variables. Most hubs are hybrid and have a varying combination of the many features available. Such a hub can have multiple switching and network connections in one box, and so can interconnect conventional network modules on one floor, connect stackable and modular hubs on another floor, and switch traffic directly at high speed. All this can be done from one central management point to be monitored by a single console, sometimes even remotely.

Summary and conclusions

In this chapter we have examined the protocols and devices necessary for telecommunications. An elegant way to summarize the discussion on bridges, repeaters, routers and gateways is to compare the different layers of network architecture, as is done in Figure 4.7. In doing so, the author has violated a cardinal rule of not using concepts or terms (layers of network architecture) without having defined the terms. There may be no solution to this problem when discussing telecommunication and networks, but we shall handle the problem by minimizing the terms not yet defined and scrupulously define and explain them later. Hopefully, by the end of the journey in this book, all these terms will come together and make sense.

There is a similar problem with the term 'network'. We have come across the concept many times but dare not use it too often because

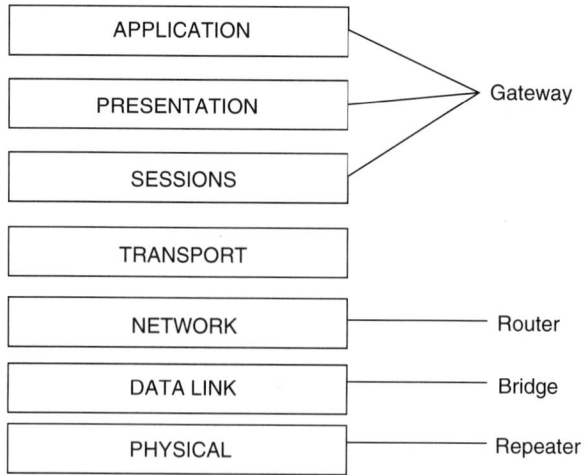

Figure 4.7   *Repeater, bridge, router, and gateways*

it has not yet been well defined or explained. We now liberate ourselves from this constraint and discuss the concept in detail as the topic of our next chapter.

Case 4.1: Networking at the space centre

The space centre at Houston, Texas, has controlled the flights of all the early spacecraft. In 1995, a new command and control centre was partly operational and in its beta phase of testing to replace the old centre and to prepare for the space shuttle into the twenty-first century. The new centre has over 19 kilometres of fibre optic cables connecting its hundreds of PCs and workstations with the larger computer systems owned by NASA. These computers and workstations are all interconnected in addition to being connected to the tracking stations all around the world.

One design specification of this complex and important networking systems was that almost all the equipment must be 'off the shelf'. This was specified in order to keep maintenance easy and not as costly as in the previous centre.

The design specification is a commentary on the state-of-the-art of telecommunications and networking. Even a large and in some ways very important real-time system can be constructed from products that are commonly available and are no longer 'high tech'.

Bibliography

Brim, P. (1994). The advanced intelligent network: an overview of markets and applications. *Telecommunications*, **28** (12), 51−54.

Bryan, J. (1994). LANs make the switch. *Byte*, **19** (9), 113−120.

Govert, J. (1992). New techniques for testing switched 56 services. *Telecommunications*, **26** (1), 19−20.

Grimshaw, M. (1991). LAN interconnections technology. *Telecommunications*, **25** (2), 25−32.

Halfhill, T.R. (1994). How safe is data compression. *Byte*, **19** (2), 57−74.

Johnson, J.T. (1992). Coping with public frame relay: a delicate balance. *Data Communications*, **21** (2), 33−35.

Luss, H. (1989). Optimizing in methodology, algorithms, and applications. *AT & T Technical Journal*, **68** (3), 3−6.

Robinson, E. (1995). V.34: off the starting blocks. *PC Magazine*, **13** (5), 241−247.

Stalling, W. (1991). Streamlining high speed internetworking protocols. *Telecommunications*, **25** (2), 43−47.

Saunders, S. (1995). Next generation routing: making sense of the marketectures. *Data Communications*, **24** (12), 52−65.

Stone, D. (1994). When is a fax not a fax. *PC Magazine*, **13** (15), 229−241.

Tannenbaum, A.S. (1989). *Computer Networks*. Prentice-Hall.

Tolly, K. (1994). Testing TCP/IP software. *Data Communications*, **23** (2), 71−86.

5

LANS: LOCAL AREA NETWORKS

Just because FDDI is faster than Ethernet does not spell the end of Ethernet; nor does the even faster HIPPI spell the end of FDDI. The lesson learned from the Token Ring/Ethernet war was that we can have multiple types of data links working together to provide a coherent network.

Carl Malamud, Stacks

Networks connect people to people and people to data.

Thomas A. Stewart

Introduction

In our society we have many networks as in roads, railways and even grids for electrical power. In each case, a network is an interconnection of nodes. In telecommunications a network is more than interconnectivity of nodes: it is a system with software and protocols that enables the exchange of information and computing resources. In the early 1980s this was localized, as in a building, or a set of buildings, like on a university campus, a company headquarters or a manufacturing plant. This is referred to as a Local Area Network, or a LAN: an interconnection of computing resources within a limited geographic area. But there are many situations where the interconnectivity must extend beyond a local environment. We have a MAN, a Metropolitan Area Network, and a WAN, a Wide Area Network. They need special software, protocols and even hardware and switches for directing the transmission along the desired route. MANs and WANs deserve a separate chapter and will be the subject of our next chapter, Chapter 6.

In this chapter we will examine the nature and desirability of interconnectivity of computing, the characteristics and objectives of networking, networking as a paradigm for computing, and the protocols and switching methods necessary for a LAN. Also discussed are the many approaches to access in a LAN including the Ethernet, token ring, circuit switching, FDDI, frame relay, SONET and the wire-less network.

Interconnectivity

Computers these days are both powerful and cost-effective. They are everywhere, doing anything and everything that involves computing. There are supercomputers manipulating billions of commands per second, working on chemical reactions, forecasting weather, and analysing complex images in medicine and industry. Computers are handling billions of messages in offices and businesses and helping with design and production in manufacturing plants. There are over 50 million personal computers with thousands of software packages along with wireless devices based on satellite and cellular systems which enable us to work at home, in the car, or even while walking or jogging. Interconnecting this computing power will dissolve the temporal and geographical barriers and enable us to work and play with partners that are distant and even in other countries and continents. It will enable video-conferencing and conveying video images by businesses across national borders, remote medical diagnostics, the exchange of research results, and sharing of information and resources that are otherwise isolated and disconnected. It may even affect our social lives and bring the world closer together if the increase in e-mail (electronic mail) demand is any indication. The impact on business and society of such interconnections and the problems in global interconnectivity are the subjects for later chapters; suffice it to say that interconnectivty can have profound implications on the way we work and live in the years to come.

Whatever technology emerges for the future, there is great certainty that systems will not be merely stand-alone systems. In the increasing complex information systems environment, there will be a need for the interconnecting of processors, not only to gain access to a more powerful processor, but also to gain access to another data/knowledge-base or software package, or even another expensive peripheral. One such schema for interlinking of processors is shown in Figure 5.1. In the real world, the processors of the same type may also be interconnected to each other in addition to being connected with other types of processors within an organization as well as with processors connected to a LAN. All these possible connections are not shown in Figure 5.1 in order to keep the diagram simple.

Figure 5.1 shows the interrelationship between computers and telecommunications. It is telecommunications that integrates computer systems (often disparate in terms of capacity, capability and even design) into a network that enables the sharing of computing resources by all those connected to the network.

The configuration for networking used in Figure 5.1 is a bus and of course there are other configurations of networking. Networking and telecommunications are one of the fastest growing sectors of the computing industry. One factor that is holding back a greater acceptance and use of networking is a lack of standards, standards that range from architecture to the format of a message. Once standards, and preferably international standards are adopted, networking will offer the access necessary for the sharing of processors and other scarce computing resources.

In addition to sharing the resources of other computers, it is equally desirable that one would want to share peripherals. Many peripherals are very expensive and a department within an organization or even the organization itself cannot afford to have one of its own and must share. This is often true of image processors, optical scanners, voice processors, or even a fast printer. Another peripheral may well be a storage device like an electronic cabinet or as archival storage device. Many of these devices can be accessed directly, once they are attached to a network or through another multiplexer or another computer. If not available on one network, another network can be accessed through a bridge or a gateway. One such configuration is shown in Figure 5.2.

One approach to networking is to have services (including documents or files or a peripheral like a printer) resident on dedicated computers called 'servers', and then the end-user, called a 'client', accesses the server that has the required service. This configuration is called the 'client–server' approach and is discussed in Chapter 10 as an alternative organizational option for computing.

Physical linking is, of course, not sufficient for accessing other computing resources. There is the

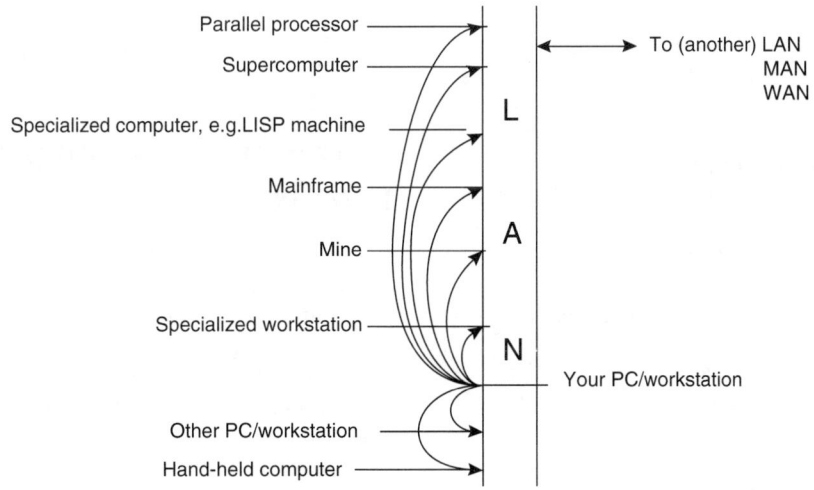

Figure 5.1 *Interconnectivity of a pc/workstation*

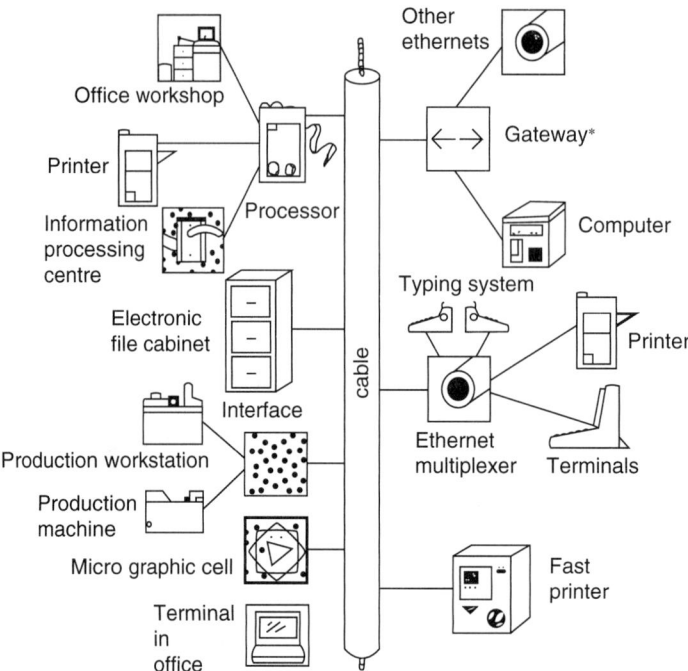

Figure 5.2 *Peripherals served on a LAN*

need for software and protocols that facilitate and enable access. Once this is available, a network is imbued with certain characteristics that are associated with networks and others that are desirable and depend on the application. We will now examine such characteristics.

Characteristics of networks

Characteristics and performance objectives of networks, more specifically LANs, are as follows:

- fast response time
- low delays which should be bounded
- notification of estimated delays when they occur
- high throughput (thruput)
- high channel capacity
- fairness of protocol in assigning access (within a priority scheme)
- ability to add or remove a station easily
- low and fast maintenance
- interoperability

The terms used above are common to many information systems and will not be discussed further,

except the last one. Interoperability requires agreements such as the one for operations between routers. There is not yet a good agreement between router-to-bridge unless your router can pretend to be a bridge. So many issues of interoperability in networking are still to be resolved but progress is beginning to be made. 'With full device interoperability, national and international standards, and a simplified way to order digital lines, plug-and-play networking may yet be possible.' (Fritz, 1994: p. 130).

Networking as a computing paradigm

Given the inherited characteristics and some desired performance objectives, one can safely say that networking is now a new and different computing paradigm. It is compared with the traditional paradigms of computing in Table 5.1.

In comparing networking with batch processing one finds that there is little duplication in characteristics or on application appropriateness. It seems logical then that these two modes of processing will coexist. However, this is not true

Table 5.1 *Comparison of computing paradigms*

	Batch processing	*Time-sharing*	*Desktop (stand-alone)*	*Networking*
Main target audience	End-user of output	Access for end-user	'Owner'	Anyone bonafide
Connection	None	Telephone	None	Telephone and network
User status	Subservient	Dependent	Independent	Unrestricted
Objective	Computation	Access	Computation	Communication
Operations	Processing in batches acccess	Varied processing	Varied	Remote
Applications	Customized reports	Shared computing resources	Varied, but often localized	Varied, from e-mail to downloading programs

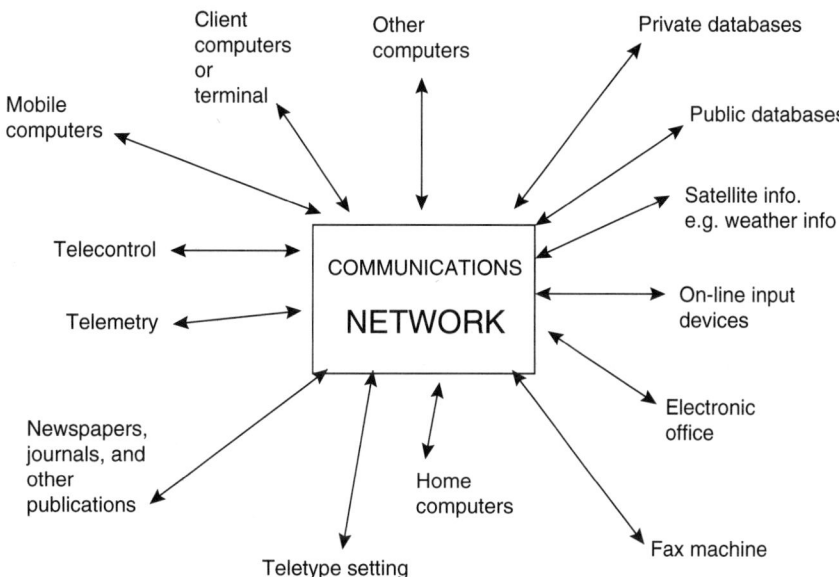

Figure 5.3 *Inputs to a communications network*

with time-sharing. Here networking has broader scope and extension for sharing computing resources and so time-sharing will be subsumed by networking. As for stand-alone computer systems, there is a commonality in equipment access: both use a desktop. But while the desktop provides stand-alone computing with perhaps a local database, a network offers much wider access, not just to other computers and peripherals, but also to other databases. Thus desktop computing may well adopt the networking connection. Networking therefore emerges as a new and important paradigm of computing. It has many applications that would otherwise not be possible. These are discussed in three later chapters. Networking also has many inputs made possible because of its remote connections. These inputs are shown in Figure 5.3.

Important to all networking, including a LAN, is the access approach for navigating the network. There are many access approaches and they depend on two things: first, the topology of the nodes in the network, and second, the switching mechanism chosen.

Topologies and switches

There are three basic topologies, configurations formed by connections between devices in a LAN. These are the bus, the star and the ring. (There are many combinations of the basic three topologies not shown in order to focus on the basic topologies.) It is generally accepted that the bus is simple to comprehend, is easy to change (add or subtract nodes), and has the ability to bypass a faulty node. The ring also has the bypassing capability but, unlike the bus, if one side breaks down, it has access from the other side. It is therefore more reliable, especially when compared to the star, where if the centre breaks down then the entire system collapses. However, the star is easiest to control, test, maintain and manage (since it is so centralized), but difficult to change.

A comparison of the three basic topologies is summarized in Table 5.2 and in Figure 5.4.

There are two basic types of network switching methods: circuit switching and packet switching. Circuit switching is what is used in our daily

Table 5.2 *Topologies compared*

	Bus	*Star*	*Ring*
Routing	Requires full duplex modem No routing	Centralized	Bidirecttional path possible
Control	By convention	Centralized	By convention
Nodes	Minimum distance between nodes	No restriction	Restricted
Robustness	Failed nodes bypassed	If centre fails, all fails	Has an alternate path if one fails
Modifications	Easily done	Easily done	Not easily done

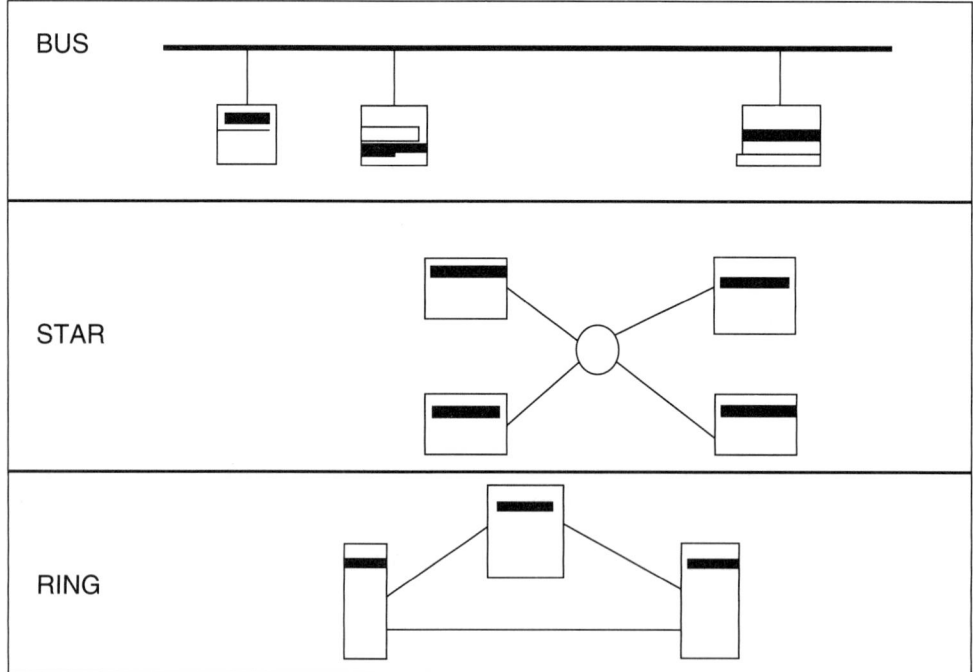

Figure 5.4 *Basic topologies in networking*

telephones: you dial, you converse, you hang-up. After you try for a connection (if you get it), you have the exclusive use of the circuit and can stay as long as you want. When you are finished, you hang-up and break the circuit connection, and that is the end of it. Because of the importance of the connection, this approach is called connection switching.

Circuit switching is shown in Figure 5.5. Note that a message from origin 1 to destination 6 does not go directly through A and E but with a diversion through A, D and E. Thus, the connection need not be direct but whatever is feasible at the time. This indirect path may be the only available path and is what often happens in our daily telephone conversations. Your author once had to make a phone call from Paris to Marseilles in the south of France. The connection he got was through New York.

Circuit switching is appropriate for voice messages and where human interaction is required, but it is not appropriate for the transmission of data as needed for most computer processing. Here, the traffic is bursty, with bursts or surges of data sent at high speeds for short periods of time. Bursty traffic has a high variation and unpredictability in transmission rates (it varies from

100 bps for a terminal to a million bps for many a processing job).

For such traffic, packet switching is most appropriate. Here, a message is broken up into packets which are units of information travelling as a whole between devices conceptually similar to a bus. The packets share the transmission channel and may not all go together. That does not matter much because each packet is uniquely identified and can be reassembled at the other end. Each packet also has a destination address which is read and used to route by a packet switch which could be a programmable computer.

Circuit switching is illustrated in Figure 5.5 using a very simple segment; a segment of another net for packet switching is shown in Figure 5.6.

The packet switching process is analogous to a motorway with cars representing packets. Cars for different destinations share the same road until such time as their exit comes and then they leave on another road towards their destination. Knowing where to get off, being in the correct lane for the desired exit, and getting off precisely when it is so necessary requires considerable human cognition and attention, especially if the traffic is heavy and the possible exits are many. Fortunately, in telecommunications the switching is all automatic, with all the

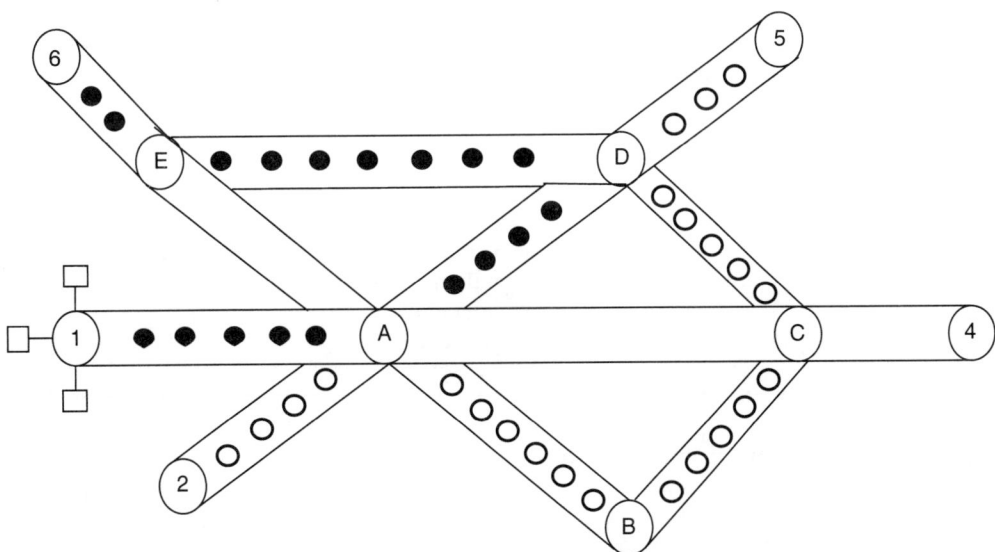

● From 1 to 6 via A, D, E

○ From 2 to 5 via A, B, C and D

Figure 5.5 *Circuit switching*

PACKET SWITCHING

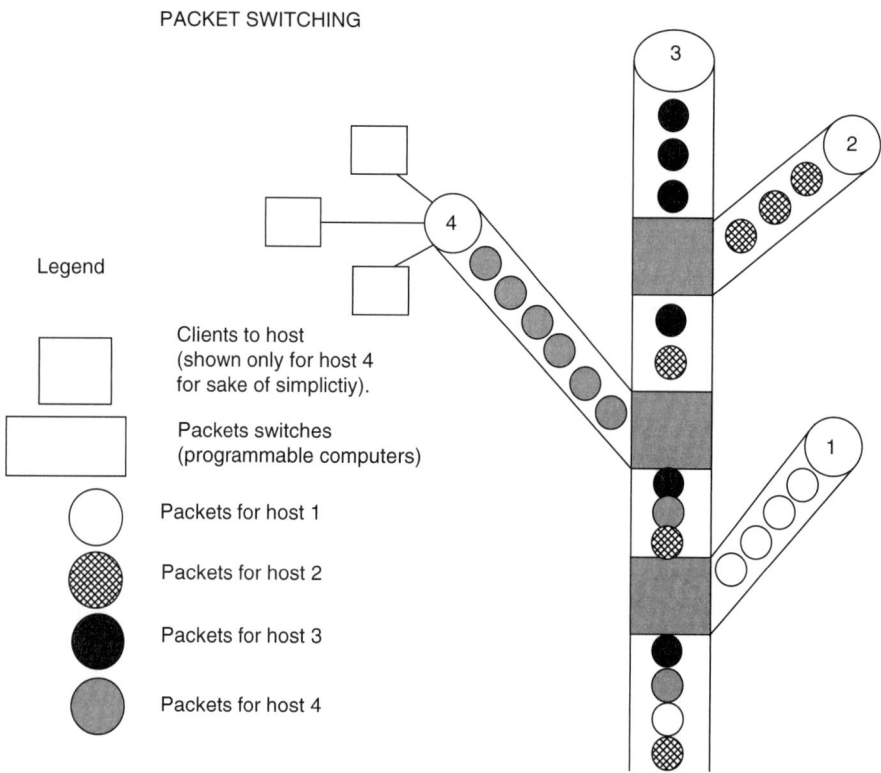

Legend

Clients to host
(shown only for host 4
for sake of simplictiy).

Packets switches
(programmable computers)

Packets for host 1

Packets for host 2

Packets for host 3

Packets for host 4

Figure 5.6   *Packet switching*

Table 5.3   *Summary of circuit switching and packet switching*

Circuit switching	*Packet switching*
Requires a connection	Is connectionless
Each circuit call is one or more messages	Each message is one or more packets
Logic required at switching centres	Logic for control at each node
Circuit is for exclusive use	Channel is shared and more than one message is moved simultaneously
Interactive	Not interactive
Data can be a continuous stream	Data is bursty
No 'store-and-carry'	Can be 'store-and-carry'

decision-making and choices being made by the network software. If the traffic is very busy, the message is delayed and the system will store and forward the message when passage is possible. This does not happen in circuit switching where a message can be lost when facing heavy traffic and congestion.

The protocols needed for packet switching are contained in X.25 which is approved by CCITT. The X.25 protocol was so important and dominant that packet switching was often referred to as the X.25 network. Being connectionless, packet switching may take different paths from origin to destination, depending on what links are available at the time.

Packet switching and circuit switching is summarized in Table 5.3. Error control is summarized in Figure 5.7.

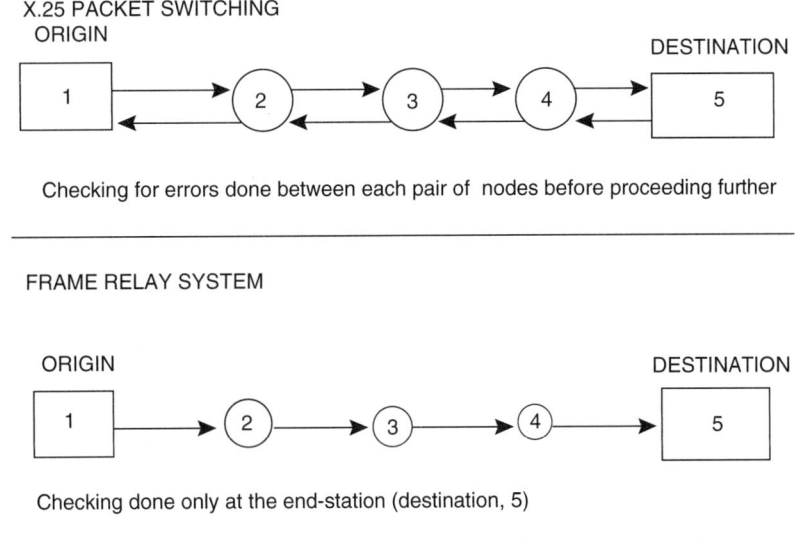

Figure 5.7   Error control in frame relay and packet switching

Access methods

There are many approaches to accessing a destination in a network. It is a combination of topology and a switching method. Each is selected for the appropriateness of a topology and a switching method for each set of applications. We shall discuss these approaches starting with the oldest: the Ethernet.

Ethernet

Ethernet is a combination of packet switching on a bus topology and was first used by ARPANET, the network implemented by the Department of Defense in the US to facilitate communications between its many researchers in government business and academia. In parallel with ARPANET, similar projects were started in Europe, especially by the National Physical Laboratory in the UK and the Institut d'Informatique et d'Automatique in France. In 1970, the Research Center for Xerox announced the Ethernet standard which has since been adopted by many private and public networks all around the world. Thus, it is not only the oldest LAN system, but perhaps the most used network system.

In an Ethernet system, the network broadcasts a signal over a coaxial cable over a distance of more than a mile. If another transmission is detected then transmission ceases and tries again after a random interval of time. This approach is called the SCMA, Carriers-Sense-Access Method.

The token ring

Another approach is the token ring which combines the ring topology with packet switching method. Before transmitting a message, a token (short series of bits) is passed. If accepted, then the computer accepting the token is free to transmit one or more packets and other computers wait until they get the token. This avoids clashes and establishes a mechanism for sharing a transmission media of cable.

There are many variations of the token ring approach including the Cambridge Ring developed by the University of Cambridge in the UK.

Circuit switching

Circuit switching approach creates an end-to-end path before invoking the flow of data between two nodes. It is a simple technique and used extensively by the PT&Ts and other telephone carriers. However, it is not always the most efficient approach since setting up a connection may take longer than the message itself to which many of us making a phone call can attest.

FDDI

FDDI, Fibre Distributed Data Optical Interface, combines the token ring approach to the high capacity of fibre optics achieving high rates of up to 400 million bps. The system is organized with two parallel rings, so that if one fails, recovery is made on the other ring. This approach is also called the Dual Ring Approach.

Frame relay

A frame relay uses packets in a circuit switching environment. Furthermore, it is similar to a virtual circuit which delivers packets in order but with variable delays, except that virtual circuits are determined at the time subscribers are connected to the system.

Frames are 64−1500 bytes in length. The start and end of a message is identified by a flag. Frames can also handle video and voice traffic. They are popular partly because they are attractively priced.

Both the frame relay and the X.25 for packet switching, when of variable length, had to constantly adjust the flow and timing of messages. However, with reliable digital circuits in the late 1980s, designers stripped off many of the X.25 functions, reduced the overhead, and subsumed the X.25 into the frame relay. Firms that are new (relatively speaking) to telecommunications, like MCI and Data Communications, do not even offer the X.25; instead they rely on the frame relay.

Frame relay provides faster service than X.25 and at the same time provides much of the same communications facilities including flag recognition, address translation, recognition of individual frames, and filling in the interframe times. Some of the benefits of the frame relay are:

- higher network productivity;
- reduced network delay;
- savings in bandwidth;
- better and more economical hardware implementation.

The frame relay can be used as a high capacity backbone for the X.25 access networks and for LAN interconnectivity. Its advantages are:

- high speed interconnectivity are lower costs;
- minimum delays since the network does not terminate protocols;
- no significant new software by bridges, routers and gateways;
- and evolutionary path for high-speed LAN interconnection services (Bushman, 1994: pp. 42−3).

Frame relays differ from packet switching and is best described in the comparison of Figure 5.7.

SONET

SONET is the acronym for Synchronous Optical Network. It supports a multiplexed hierarchy of transmission on speeds ranging from 51 to 2400 bps. The SONET system 'allows data streams of varying transmission speeds to be combined and extracted without first having to breakdown each stream into its individual components.' (Cerf, 1991: p. 78).

The SONET is used by virtually all important carriers except perhaps the largest, AT&T, partly because SONET offers its own optical algorithm. It offers superior performance and monitoring and can route around network failure points in 40 to 60 milliseconds.

SONET may well be the gigabit backbone for networks, especially long distance wide area networks, the subject of our next chapter.

Wire-less networks

Wire-less LANs are capable of transmitting 10 Mbps within a room or building. It is similar to cellular technology for data transmission. Cellular phones rely heavily on analogue broadcast techniques, so using them to move digital data will be inherently difficult. One approach is the CDPD, Cellular Digital Packet Data. The CDPD is similar and yet different from the circuit switched network. A comparison of the two is shown in Table 5.4.

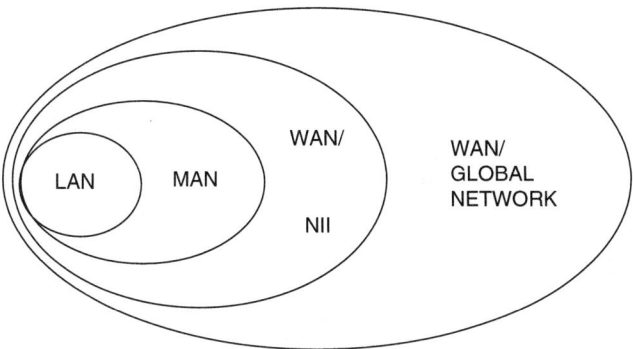

Figure 5.8   *LAN as a subset of other networks*

Table 5.4   *Circuit switched vs. cellular digital packed radio*

Characteristics	Circuit switching	CDPD
Store & forward	No	Yes
Large data files and fax	Yes	No
Host log on	Each session	Once a day
Cellular channel	Dedicated	Shared
Application profile	Large transactions	Small transactions
	File transfer e-mail	Dispatching e-mail messaging

Summary and conclusions

In this chapter we examined the need for connectivity and examined how connectivity leads to networking as a new and important paradigm in computing.

We also noted that there are three basic topologies: the bus, the star and the ring. Also, there are two basic types of switching: the circuit switching and packet switching. Each has its own advantages and limitations. An application environment would make it desirable to combine one topology with one switching method. This gives us a variety of methods of navigating the LAN: the Ethernet, the token ring and its variation the Cambridge ring, FDDI, frame relay, wire-less, and the SONET. Each is briefly described and a comparison is made.

An important approach is the ISDN which is discussed in a later chapter, Chapter 7.

With the proliferation of computers and the increase of computer literacy and experience

among the end-users, we have seen a great demand for LANs. However, this demand will soon require that the boundary limitations of the LAN be relaxed and extended to a MAN and a WAN to provide remote services. The LAN then becomes a subset of a MAN/NII (National Information Infrastructure)/WAN. This relationship is shown in Figure 5.8. We shall explore these possibilities in later chapters. First, however, we need to discuss the MAN and the WAN, the subject of our next chapter.

Case 5.1: A network in the UK

Racal Electronics agreed to buy British Rail Telecommunications for £132.8 million, 'giving the company a data link passing through most major British cities along British Rail tracks'. British Rail accounts for about 80% of the optical fibre network's customers. But Racal said, 'usage at present was only 20% of the network's capacity'.

The acquisition gives Racal strength in the field of voice and data communications, its core activity that accounts for more than half of the company's revenue. Racal now is, in the words of its Chief Executive, 'a carrier's carrier.'

Source: *International Herald Tribune*, Dec. 6, 1995, p. 17.

Case 5.2: The Stentor network in Canada

Stentor is an alliance of Canada's major telephone companies with integrated local and long distance services. It also offers seamless services with North America through its alliance with MCI.

Stentor maintains the world's longest, high density and fully digital fibre optic network stretching approximately 4800 miles across Canada. It services include digital switching, intelligent network services, high reliability (second to Japan) and a 58% telephone penetration.

'In 1994, Stentor announced the Beacon Initiative, a project that will bring the information highway to 80 to 90 percent of Canadians by 2005.' The activities of the Bacon Initiative will include a $8 billion upgrade of the telephone networks over the next 10 years; a 500 million enhancement programme over six years to provide national interconnectivity; the creation of a new multimedia company that will provide and distribute multimedia services; and venture capital for the development of multimedia applications and products for the information highway.

Source: *Stentor*, Dec. 1995, p. 3.

Case 5.3: A network planned for China

The People's Republic of China is planning a network infrastructure that will have mail-hub servers in twelve of its largest cities by the end of 1996. The country-wide network will connect on-line many corporate and governmental groups with one another and the rest of the world with e-mail.

The servers are to be provided by Control Data Systems Inc. whilst the network service provider is Rayes Technology Co.

Source: *Computerworld*, Nov. 6, 1995, p. 76.

Bibliography

Abeysundara, B.W. and Kamal, A.E. (1991). High-speed local area networks and their performance: a survey. *ACM Computing Surveys*, **23**(2), 221−261.

Boyl, P. (1996). Wireless LANs: free to roam. *PC Magazine*, **15**(4), 175−202.

Bushman, B. (1994). A user's guide to frame relay'. *Telecommunications*, **28**(7), 42−46.

Brueggen, D.C. and Yen, D. (Chi-Chung) (1990). Local area network connectivity. *Computer Standards & Interfaces*, **11**, 103−114.

Cerf, V.G. (1991). Networks. *Scientific American*, **265**(3), 72−81.

Derfer, F.J. Jr. (1992). LAN Fundamentals, Part 2. *PC Magazine*, **11**(7), 229−250.

Francis, B. (1991). Linking LANs with laptops. *Datamation*, **37**(10), 61−63.

Gareiss, R. (1993). Tommorow's networks today. *Data Communications*, **24**(13), 55−65.

Gifford, J. (1995). Wireless local loop applications in the global environment. *Telecommunications*, **29**(9), 35−37.

Lane, J.L. and Upp, D. (1991). SONET: the next premises interface. *Telecommunications*, **25**(2), 49−52.

Miller, A. (1994). From here to ATM. *IEEE Spectrum*, **31**(6), 20−24.

6

MAN/WAN

No technological imperative determines how to link LANs or even whether to do so.
Ben Smith and Jon Udell, 1993

If you want to connect Ethernets to ATM, you may need a Ph.D. to figure out the long term implications.
Paul Strauss, 1994

Introduction

LANs in the 1980s were very successful, but its inherent constraint of being confined to a local area is its greatest limitation for the 1990s. While in the 1980s, industry matched resources to applications, in the future, applications may well have to be matched to networks. Future applications are moving away from the legacy applications of processing data to the processing of real time data, text, voice and images. Processing needs are shifting from the desktop networking to enterprise-wide strategic computing; from store-and-forward computing to strategic computing; from off-line computing to real-time computing; and from local processing to remote processing.

Networking is being transformed from the early environment of military systems tied to private networks that were switched mechanically to analogue and digital devices of today with broadband fibre optically transported messages that are electronically switched by sophisticated software and intelligent network systems.

Businesses are becoming less centralized and more distributed. Businesses are no longer operating exclusively within their national boundaries but going global where physical boundaries are no longer an issue. Traffic in telecommunications is shifting from a LAN to a metropolitan MAN and to a wide area WAN.

We must avoid a communications gridlock if we are to benefit from open and worldwide markets. To fully benefit from the information age we must have networks that interconnect the millions of computers and thousands of computer installations world-wide to enable useful and meaningful information exchange. A step towards this goal is a step beyond the LAN and towards the MAN and WAN, the subject of this chapter.

In this chapter we will examine the nature of processing in a MAN/WAN and compare it with the LAN. We will also examine the planning and performance management of a WAN which includes bandwidth management and switching management. The discussion of the management of switches will lead to a detailed discussion of the ATM.

MAN and WAN

MAN stands for Metropolitan Area Network and is a computer network that typically covers a part or all of a metropolitan city. It usually encompasses a compact area with an area that ranges from one to a few dozen miles in radius. In contrast, a WAN (Wide Area Network) is for a much larger area, much larger than either a LAN or a MAN. Typically, the distance may be 100 or a few thousand miles in radius. In practice, a WAN may cover a nation, a continent, and be international and world-wide.

A MAN is mostly copper and fibre cables like the digital circuits based on the local phone company and connected to LANs with cable or with microwave connections to a nearly microwave system.

The MAN is not heard of much these days perhaps because it mostly uses a stable and proven technology developed by the telephone companies. It is, however, an important basis of the so-called intelligent building, intelligent not in the AI (Artificial Intelligence) sense, but in the sense that it is (potentially at least) equipped by computers. Each room is wired for plug-and-play equipment. You acquire your hardware and software, plug the computer into the wall socket, and are connected to the world through networking. This may not be a common scenario today, but many professionals in computing see that is a viable scenario.

Planning for a WAN

In planning any network, one must consider some of the facts of the changing environment:

- an ever increasing number of computers and workstations with regular upgrading of technology and all of them quite powerful;
- an increasing number of end-users and organizations using networks;
- an increasing awareness and increase in computer literacy of the end-user, which includes the manager, worker and the home-owner;
- the end-user certainly able and desirous of end-to-end interconnections and integration of applications;
- increasing complexity of businesses that demand remote processing for their input or output or both;
- increasing need of multimedia processing including data, voice, images and video with possibilities of films-on-demand in the near future;
- the extent of remoteness of processing is continuously enlarging to include the entire world.

One can safely say that in our world the need for computing and telecommunications is demand driven. The demand drives the computer industry which in return drives the telecommunications industry. The cycle is complete in that the telecommunication industry drives the computing industry which in turn influences the demand. This cycle is shown in Figure 6.1.

We can see the demand driven technology in the area of multimedia where the end-user is no longer satisfied with printed reports in batch as in the 1960s and 1970s but demands graphic and image processing especially in the factory with CADD (Computer Aided Drafting and Design). But soon that was inadequate and there was a

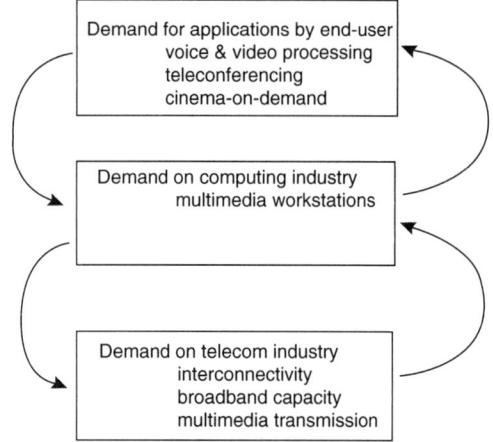

Figure 6.1 *Interaction between demand and technology*

demand for voice processing, both analysis and synthesis. And then that too was not sufficient and there is now a demand for audio processing. To add to all this, there is a demand that this be done in real time with animation and in colour. So, the level of aspiration is steadily nudging up with the traffic mix being very multimedia and in real time as shown in Figure 6.2. All this must be done quickly and seamlessly (smoothly) across remote distances. The end-user cannot tolerate a waiting time of even a fraction of a second. Thus, the demand on the computing industry and on the telecommunications industry to transport all this information is steadily increasing. With each advance there is demand for more advances. With each reduction of response time there is demand for more reductions and faster processing. With each inclusion of media there is a demand for more and better service and for these applications to be integrated. Output in printed form is no longer sufficient; output should be as voice with a possible input response also as voice and all in real time.

Voice is shown as only one of four media in Figure 6.2 but it is more important in relative terms. It is important as a stand-alone technology and has many applications for itself; in addition, it is part of video processing (and voice processing) and sometimes part of real-time processing. So it is important. It is of special interest to the computing professional because it is an analogue signal whilst computers are digital. That is why an integrated digital system is of such interest and is the subject of a separate chapter, Chapter 7.

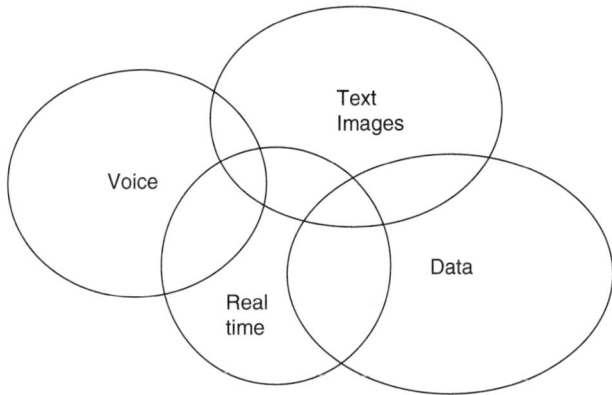

Figure 6.2 *Traffic mix in telecommunications*

Performance of a MAN/WAN

Whether we discuss voice processing or not, and whether we talk about a MAN or a WAN, we must design and implement a system that has good if not high performance. But what is performance for telecommunications and networks? It depends on the perspective, but both end-users and technicians will agree that delays are important. One approach may be to have a secure system with low losses and high reliability. This is a subject that we will discuss in Chapter 12, on the Security of Networks. One important performance measure is to reduce delays. This can be done by increasing bandwidth, by adding additional bandwidth, reducing the load on bandwidth without reducing traffic (by compression perhaps), or by increasing the speed of the links. These strategies are all of what is called bandwidth management, which we shall discuss below. We shall also discuss another reason for delay and even loss of reliability, and that is switching management. This includes the strategy of switching such as the hub and spoke method and of course the reliability and robustness of switches that can operate reliably in heavy load as well as in a variety of traffic conditions.

Bandwidth management

There is considerable empirical evidence which supports conventional wisdom that delays increase as the system gets loaded and its utilization increases. This is shown in Figure 6.3

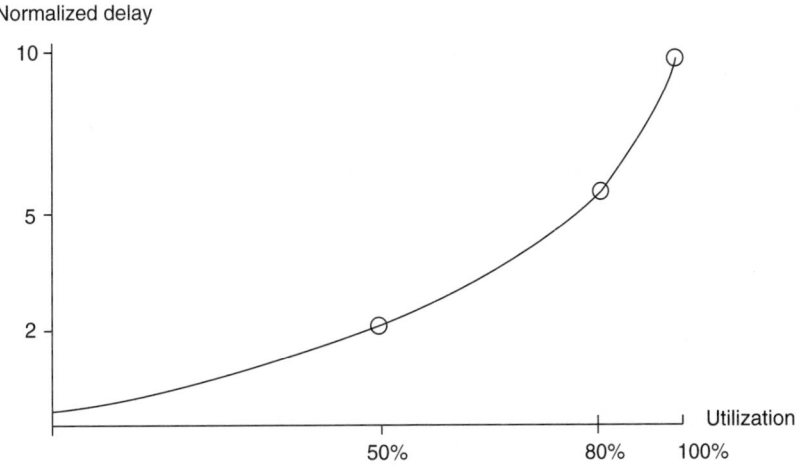

Figure 6.3 *Link utilization and delays (adapted from Weiss, 1990: p. 58). Not to scale*

Table 6.1 *Link speed and time required for a 640 kbyte load application, assuming 100% efficiency and no contention*

Link speed (kbps)	Transmission time (s)
9.6	5460
56	960
112	47
1500	3.5

Source: Adapted from Weiss (1990: p. 58)

where we can see that beyond a 50% utilization delays increase rapidly and increase much more rapidly after a 80% utilization. A quick answer to this problem would be to add another channel. Multilink or bonding (sometimes also called **bandwidth-on-demand**) are ways to add channels to a network. Multilink is software-based while bonding is hardware-based. Multilink does pose a problem: it may deliver packets out of sequence. TCP/IP accepts packets out of order and rearranges them in proper order. The designers of Multilink did not anticipate this problem arising and have no good solution for it.

Another set of approaches is to reduce the load on the system by compression and the elimination and control of unnecessary data transmitted. Such filtering also eliminates unwanted network traffic. For an enterprise network operating with a wide variety of protocols filtering can add up to 25−30% more effective bandwidth.

Yet another approach would be to increase the speed of the transmission link. This can produce a dramatic effect as shown in Table 6.1 where the transmission time can drop from 5460 seconds to 3.5 seconds by an increase in the link speed from 9.6 kbps (which is the speed of many modems on a LAN) to 1.5 Mbps which is the speed on some WANs. The moral is clear: keep the local traffic on slow and cheaper channels but transfer the long-haul loads on the faster channels even though they may be more expensive because they are more cost-effective.

Switching management

One problem is large switching problems like those faced by a WAN and whether or not to have many point-to-point connections with few switching in between, or to have all the local traffic funnelled into a hub and then transmit the load over long distances on a high speed WAN. We have a similar problem in air traffic. Some airlines, as in the US, funnel traffic on small planes to large hubs and then transfer them on non-stop long hops across the ocean sometimes taking up to 12 hours. Besides the unhappiness of some passengers, the technical problem is that such switching at hubs gets quite complex with disastrous results when timings are not just right. This complexity increases not just with the increase in traffic volume but also with the change in traffic mix.

In earlier chapters we have drawn the analogy in telecommunications to highway (motorway or autobahn) traffic with ramps and instructions for making changes at intersections. In packet switching we can see the parallel of packets (passenger) sharing the same channel (lane of road) and then being shunted off when they must make a connection. The problem is relatively easy when one considers just one type of traffic: data traffic. But now consider a mix of traffic: data, voice, images and video with some in real time. This is analogous to having not just auto and car traffic but also having train and waterway traffic all at the same place. Complexity increases very quickly. This actually happens at Slossen in Stockholm where roads, trains, a canal and pedestrians come together at the same time and yet all the traffic flows seamlessly. The problem in telecommunications must also be solved not just for a mix of traffic but for heavy traffic that can well be expected in the days to come. Part of the solution lies in a robust switching technique, and this is where we see the arrival of the ATM.

The ATM

ATM stands for Asynchronous Transfer Mode where asynchronous means that the signal is not derived from the same clock, and therefore does not have the same fixed timing relationship. ATM is supported internationally and in 1988 it was chosen as the switching and multiplexing technique for B-ISDN (Broadband ISDN, to be discussed in a later chapter). It offers a high data rate (Gbps) as well as a low latency rate (latency is the time between access time and transfer time).

It was partly in response to the need for large capacity and partly in response to the need to handle voice and video in telcommunications that the ATM was developed. And why is handling of voice and video different from data?

For one thing, the processing is different. The ATM technology provides a common format for bursts of high speed data and the ebb and flow of the typical voice phone. With voice, periods of silence can be edited out without any loss in the message. And with video, only changes in the image are sent; the rest is already available and can be used for reconstruction of the message. Also, voice and video must be done in real time in order to avoid any losses in synchronization.

ATM is a connection-oriented technology so that each cell is specified before the connection is made. ATM uses the cell structure which has a fixed message field of 48 bytes and a header of 5 bytes. The header in the cell contains all the information a network needs to relay the cell from one node to the next over an established route. The header also contains control bytes as shown in Figure 6.4. This ATM structure is compared with the frame relay structure in Figure 6.5 which

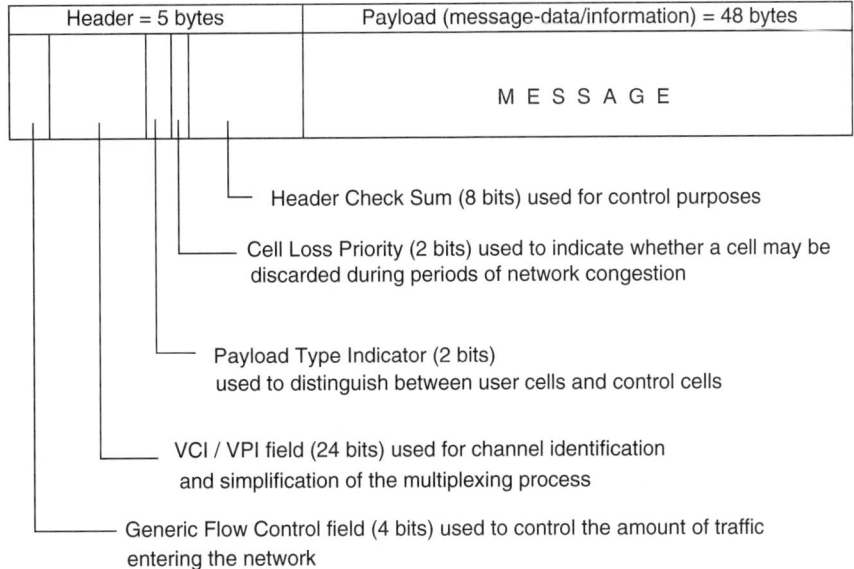

Figure 6.4   *Structure of an ATM cell*

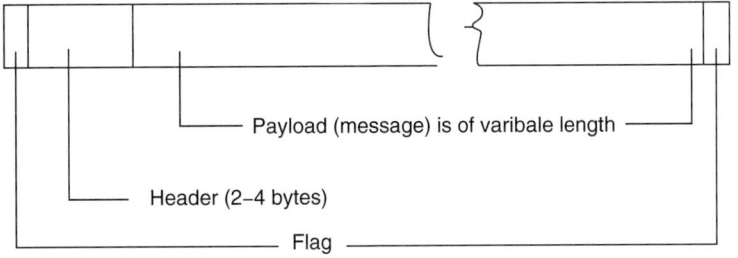

Figure 6.5   *Comparison of ATM and frame structure*

has a smaller header and a variable length message field. (The flag in a frame is a series of bits that indicates the start and the end of a payload message.) In Ethernet, the packet for switching is 64 bytes long. The header is larger because of the complexity of the variable length message that follows the header but also because it provides data that the software needs to recognize where the message starts and where it ends. Fixed length packets, in contrast, can be implemented completely into hardware which is faster.

ATM 'users send bursts of as many or as few cells necessary to transfer their data; they also pay only for the cells they send, not for the speed of a dedicated facility they may use only part of the time'. (Lane, 1994: p. 43). The bursts are sent through a centralized ATM switch that has dedicated connections to its end-users who may have a PC or a workstation, in addition to connection to a computer that may be a server. This is illustrated in Figure 6.6. The dedicated connection with the ATM switch which routes messages and controls access in the event of contention, conflict when more than one user wants to use the same resource simultaneously. This arrangement is analogous to the PBX (Private Branch Exchange) used for voice calls.

ATM establishes virtual connections between each pair of ATM switches needed to connect a source with a destination. Up to 65 536 virtual channels can be multiplexed into a virtual path. These connections are termed 'virtual' to distinguish them from the dedicated circuits

used in the STM, Synchronous Transfer Mode. 'It is the virtual nature of ATM services that will provide greater efficiencies in the future. Today, most communications capacity is idle. A voice circuit is only 30–40 percent efficient; most of its time is spent listening.' (Lane, 1994: p. 43).

Connections can also be virtual as in PVC (Permanent Virtual Connection) where parameters (but not necessarily routes) are established in advance in contrast to the SVC (Switched Virtual Circuit) which provides resources as required.

We have characterized ATM as a connection-oriented technology, but that does not preclude it from handling connection-less traffic as well. This traffic on an ATM has often been identified as being heavy because of the nature of the traffic, which includes images, voice and video. To give the reader an appreciation of the magnitudes involved, we shall examine some applications for their data intensity. We start with text, a recent book written by your author of the rough size as this book occupying over 12 million bits. In contrast, a full diagram like many in this book, would take around 1 million bits each. Images are data intensive. An example from the real world would be a medical imaging application such as the rendering and transmission of a diagnostic X-ray would involve 2–10 billion bits of information (Vetter, 1995: p. 31). More such applications can be found in scientific research with high definition three-dimensional images in real time. In industry, examples of such images would be in CADD, computer aided design and

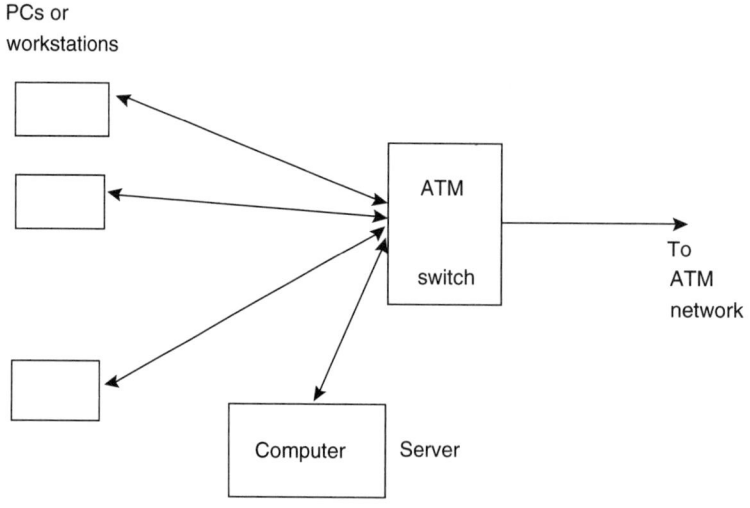

Figure 6.6 *An ATM switch*

drafting. In business, such as in oil prospecting, large amounts of data on simulation have to be moved quickly from graphical workstations to the field offices. Then there are other video applications in the entertainment industry and in education. In the office there is teleconferencing, image archiving, and group work that mixes voice, images and text all in a time-sensitive environment. Also, the traffic load on networks is heavy not just because of the type of application but also because of high volume. Even ordinary applications like file transfer and e-mail have risen exponentially. All must now be handled by LANs and WANs.

In summarizing the discussion of ATM, one can compare it with circuit switching as in Table 6.2. Also, one can safely say that ATM is a proven technology but that its applications lie more in the future (like teleconferencing, films-on-demand) than in current reality. There are many commentators and practitioners who are enthusiastic about ATM and there are some that are more cautious. We shall quote both sides below:

> ATM is often described as the technology that will allow total flexibility and efficiency to be achieved in tomorrow's high speed, mulitiservice, multimedia networks ... its promise of high speed, integrated services and universal connectivity ... the technology that finally enables high-bandwidth, time critical applications to reach the desktop. (Vetter and Du, 1995: p. 29).

> ATM cannot deliver ubiquitously across the network in a failure-proof mode as yet ... Many of the current network management platforms lack the performance features and sophistication required to manage ATM's high speed, highly complex connection-oriented networks ... an enterprise network entirely composed of

ATM will not be possible for some time yet ... most enterprise users will implement hybrid ATM/frame relay and connect to a device that is terminated in the enterprise network on an ATM port ... Interoperability with existing technologies such as frame relay and the X.25 is imperative to a smooth and cost-effective evolution of ATM. (Federline, 1995: pp. 69–70).

The problem that many of us, especially laypersons, have with ATM is not with the structure or its functions but with its name. We tend to confuse ATM in teleporcessing with ATM in banks, Automated Teller Machines.

ATM is one approach to the demands of transmitting data and voice. Another approach is the ISDN, the topic of our next chapter.

Summary and conclusions

A MAN and a WAN are extensions of a LAN. There is more interest in the two extremes, the LAN for short haul and the WAN for long haul telecommunications. They are compared in Table 6.3.

Another way of looking at a LAN and a WAN is that the LAN is an access layer which interfaces other networks at low speed and the WAN is the backbone layer of networks providing high performance connectivity. This relationship is shown in Figure 6.7 where the two layers are represented by 'clouds' as boundaries.

Table 6.2 *Comparison of circuit switching and ATM.*

	Circuit switching	ATM
Traffic type:	Data	Data and voice
Structure:	Variable length	Fixed length
Delays:	Variable	Fixed
	Can be long	Very low
Bandwidth:	Wasted	Efficient
Switching:	Done in software	Done in hardware
Orientation:	Connection	Connection and connection-less

Table 6.3 *Comparison of a LAN and a WAN.*

	LAN	WAN
Distance:	For a few km	For over 1000 km
Speed:	Low speeds (in Mbps)	High speeds (in Gbps)
Error rate:	Low	Can be high
Ownership:	At firm's level	Higher than firm's level
Administration Costs:	Less than WAN	More than LAN
Maintenance:	Less complex	More complex than LAN
Routing algorithm:	Simple	Complex
Switching:	Frame FDDI	Frame FDDI ATM

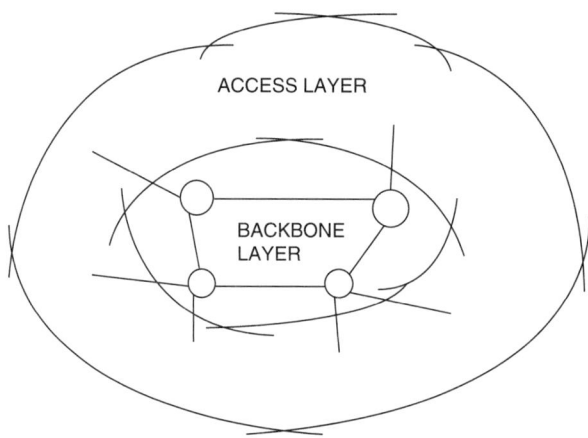

ACCESS LAYER

BACKBONE LAYER

Figure 6.7 *Access and backbone layers*

Smith cautions us about the issues to be addressed by the WAN backbone:

> address capacity planning issues as well as fault reporting and route planning functions. Router management systems focus very heavily on the management of routes/paths and are typically weak in the wide area performance reporting and management ... unless users fully appreciate and are prepared for the consequences of some of the interconnectivity systems that they implement, their WAN resources could become expensive, overloaded, and poorly managed. LAN internet working needs an efficient WAN behind it for users to achieve productivity improvements. (Smith, 1994: p. 55).

To meet the high load demands and variety of traffic we need an ATM. But this does not mean that it cannot be used in a LAN. The LAN of the future will not only have hosts, Internet working devices, interfaces to public networks, but also ATM switches as predicted by Vetter and Wang:

> The bandwidth of traditional LANs is usually on the order of tens of megabits per second, while ATM will support Gbps speeds. Today's LAN also lacks scalability. Tomorrow's LANs must operate in an environment in which computing devices are so inexpensive and readily available that there are hundreds or even thousands in a typical office. With such large numbers of devices any attempt to interconnect them with traditional shared-media LANs would be impossible. The limitations of existing bus and ring LANs, the demand for higher bandwidths, and

larger user populations are major reasons for the growing interest in ATM LANs. (Vetter, 1995: p. 35).

> Another factor supporting the use of ATM in a LAN is that LANs are increasingly managed in a centralized fashion by hubs. Increasing use of twisted-pair and optical fibre media foster centralized hub connections as well, and a switched LAN interconnection is seen as an extension (or a replacement) to existing hubs. As computing power increases, direct connections of high-end workstations to a centralized switch is seen as an attractive, especially for video and other high-end applications. In particular, the expected growth of video-related applications makes the connection-oriented ATM technique a suitable choice for this usage. (Wang, 1995: p. 40).

Budwey and Salameh look at the role of ATM as occupying a niche in networking.

> In the global networks, SMDS and frame relay will prevail in the next five years as the driving technologies. ATM will provide for high performance networks, frame relay for lower end applications, and SMDS will be the LECs switching interface of choice. If priced correctly, SMDS may provide the infrastructure of the future global network. (Budwey and Salameh, 1992: p. 26).

Case 6.1: ATM at sandia

Sandia Labs in Alburquerque, New Mexico, is organizationally related to Livermore Labs, in

California, some 1800 km apart. They were one of the first to have a supercomputer and be on the ARPANET. They both are involved in research and development for the US government defence and energy departments. They serve thousands of scientists and engineers and have some of the most powerful computing power in the world. Sandia has 200 LANs and there are 25 in Livermore. These facilities are to be consolidated for purposes of implementation, security mechanisms, reliability figures, meeting performance requirements, and adherence to networking and communication standards.

Supercomputing at Sandia and Livermore has two computing environments: 'secure' for classified work and 'restricted' for unclassified work. Both environments provide TCP/IP (Transmission Control Protocol/Internet Protocol) access for remote FDDI (Fibre Distributed Data Interface) (also used in the UK Parliament in London), or Ethernet LANs to centralized network resources such as graphic servers and mass data storage. Access to supercomputers is provided at both sites by Ethernet/FDDI and other routers.

ATM switches are used in the consolidation link based on Cisco routers that provide the connection from the LAN technology to the SMDS/ATM (Switched Megabit Data Service/Asynchronous Transfer Mode) to MAN (Metropolitan Area Network) and using the ATM switches to connect the MAN to the WAN (Wide Area Networks). The measured round trip delay time of the end equipment is 7.1 seconds.

Future plans include 'the possibility of migrating from SODS to the emerging standards for transporting IP and other protocols directly over ATM switches'.

Source: Neagle *et al.* (1994). Developing an ATM network at Sandia National Laboratories. *Datamation*, **28** (2), 21–3.

Case 6.2: Navigating LANs/WANs in the UK

LANs, MANs and WANs are being used in the UK for resource sharing and as a mechanism for improving the personal productivity of office workers, as well as for promoting collaboration and work group computing.

Britain's PA Consulting group have found that businesses have increased their sales by 25%;

reduced their administrative staff by 15%; and improved their customer care index by 10%.

In addition to increasing productivity and reducing costs, there is also often an improvement in security and quality levels as well as a reduction in delays. The main factors that influence the LAN/MAN/WAN connectivity solutions are the size of the MAN/WAN, the location of the access points, number of users, patterns of traffic, applications portfolio, inter-LAN software, and hardware compatibility.

The LAN/MAN/WAN connectivity can lead to an enterprise solution that, according to Chris Gahan of the British carrier BT, '... is a blend of services, using individual technologies to their best advantage and having the flexibility to change the blend as business changes'.

Source: *International Herald Tribune*, Oct. 4, 1995, p. 14.

Supplement 6.1: Wan technologies

	Frame relay	ATM	ISDN	SMDS
Transmission mode:	Variable-length packets	Fixed length, 53 byte cells	48-bit packets	Fixed-length, 53 byte cells
Usage:	Data, some voice	Data, voice and video	Data, voice and video	Data
Speed:	56 kbps to 1.5 Mbps	1.5 to 622 Mbps	144 kbps	56 kbps to 34 Mbps

Source: *Computerworld*, Oct. 16, 1995, p. 69.

Supplement 6.2: Survey on WANs

Focus Data, an independent market research firm conducted a survey of users of *Network World* on usage and selection criteria of WANs. The results are shown below:

Analogue dial-up	53.1%
ISDN	33.7%
Switched digital	15.3%
Other	32.7%
Don't know	9.5%

Based on a possible score of 5.0, the scores for the top five selection criteria used are as follows:

Ease of use for remote users	4.56
Throughput performance	4.45
Support for a specific LAN protocol	4.37
Ease of support	4.12
Management tools	3.76

Source: *Network World*, Oct. 30, 1995, p. 62.

Supplement 6.3: Projected pricing of ATM

The price of ATM access is predicted to drop consistently at least for the next three years. The predicted drop is as follows:

1995	$5400
1996	$4000
1997	$3000
1998	$2000

Sources: CIMI Corp., Voorhees, N.J., US; printed in *Computerworld*, Nov. 20, 1995, p. 2.

Bibliography

Alexander, P. (1995). Network management: the road to ATM deployment. *Telecommunications*, **29**(9), 47–50.

Basi, J.S. (1990) Networks of the future. *Telecommunications*, **24**(7), 33–36.

Bryan, J. (1993) LANs make the switch. *Byte*, **18**(6), 113–132.

Budwey, J.N. and Salameh, A. (1992). From LANs to GANs. *Telcommunications*, **26**(7), 23–26.

Fritz, J. (1994). Digital random access. *Byte*, **19**(9), 128–132.

Hurwicz, M. (1997). In search of the ideal WAN. *LAN*, **12**(1), 99–102.

Kim, B.G. and Wang, P. (1995). ATM networks: goals and challenges. *Communications of the ACM*, **38**(2), 39–44.

Lane, J. (1994). ATM knits voice, data on any net. *IEEE Spectrum*, **31**(2), 42–45.

Miller, A. (1994). From here to ATM. *IEEE Spectrum*, **31**(6), 203–204.

Pugh, W. and Boyer, G. (1995). Broadband access: comparing alternatives. *IEEE Communications Magazine*, **33**(7), 34–46.

Richardson, R. (1997). VPNs: Just between us. *LAN*, **12**(2), 9–103.

Smith, B. and Udell, J. (1993). Linking LANs. *Byte*, **18**(12), 66–84.

Smith, P. (1994). Reconciling the LAN vs. WAN bandwidth management mindset. *Telecommunications*, **28**(3), 51–55.

Vetter, R.J. (1995). ATM networks: goals, architectures and protocols. *Communications of the ACM*, **38**(2), 39–44.

Vetter, R.J. and Du, D.H.C. (1995). Issues and challenges in ATM networks. *Communications of the ACM*, **38**(2), 28–29.

Weiss, J. (1990). LAN/WAN internetworking. *Telecommunications*, **24**(7), 57–59.

7

ISDN

...the intrinsic value of a telecommunications system grows combinatorially with the number of subscribers it interconnects.

David Rand Irvin

Introduction

ISDN is the acronym for Integrated Services Digital Networks. It is a digital version of the switched circuit analogue telephone system. 'But,' say the critics, 'we have had switched networks and telephones for a long time. So why the excitement?' 'Well,' reply the proponents of ISDN, 'it is multimedia and handles data, voice, images and video.' 'But,' add the sceptics, 'we have had voice and fax through the modem all these years, and so what is new?' The proponents of ISDN then point out that it is not only a technology for multimedia communication, but it is an enabling technology that will allow computers with applications of integrated multimedia to be as ubiquitous as the telephone is today. The sceptics then say, 'I have been hearing of the ISDN for over a decade and we have seen no results. Maybe we should not upset what we already have. We may not need something that is so complex and difficult to implement.' 'Yes,' counter the proponents of ISDN, 'It has taken a long time and will take longer because we are dealing with a system that is not only integrated but internationally so. And getting international agreement between the carriers, suppliers of telecommunication components, the computer industry, and many governments, does take a long time. It also takes a long time for testing a product carefully and then getting acceptance especially for an advanced concept. Remember, there were only 25% of households with telephones in the US in 1920 and it took 60 years for this percentage to increase to 96%. 'These things take a long time'. So the argument rages.

In this chapter we will examine the myths and realities of ISDN. We will describe ISDN as it is today, look at its objectives for tomorrow and the day after, discuss the implementation of ISDN, and examine the obstacles and future for ISDN.

We start with image processing followed by voice processing. In both cases, we examine the nature of the application and their uses in business and daily life. These applications are constrained by a lack of resources necessary for implementation of the digital technology. These constraints are then examined. For the reader who wishes to read on, there is more on the nature of ISDN and its evolution.

The computing environment

ISDN was designed for an environment where all the needs of computing could be integrated. This included applications of data and voice, as well as images and video. Data and images can be easily digitized while voice and video are basically analogue signals and more appropriate for telephony than for a digital computer. The problem is one of integrating the two types of signals or else pay for the inefficiencies resulting from the interfacing of the two. Either all should be digitized or all be analogued. ISDN takes the approach of all being digitized. However, understanding the nature and magnitude of the digitized and analogue applications is necessary to appreciate ISDN.

All the early processing were computations and transaction processing which were numerical and digital. Later on we added textual processing, but this was digitized by giving each alpha character a unique digital equivalent. Then came graphics which in many cases could be digitized if a curve is viewed as a set of lines, which they are because many discontinuous lines can look continuous.

Likewise an image (like a drawing or even a photo) can be viewed as a set of dots where each dot is digitized. Thus, an image can be represented by an array of numbers. A number code could also represent the intensity (and perhaps colour, if we are not just dealing with black-and-white images). These primitive picture elements are called **pixels**. Thus a computer image is a two dimensional array of numbers, the individual pixel values. For example, we might have a 100×100 array of intensity measurements, each selected from a range of 0 to 100, where 100 represents white and 0 represents black, and the 99 intermediate values represent various shades of grey.

The initial stage of image processing is pixel processing and may involve a 'clean-up' process, i.e. removal of noise (e.g. black pixels that should be white) that is often introduced by the hardware that generates the pixel image. A 'smoothing' operation is then performed in which a small cluster of adjacent pixels are compared, and a single odd-valued one is adjusted to a value that is closer to the 'average' of that of its neighbours. This smoothing operation is computationally trivial but done repeatedly, something like 10 000 times on a 100×100 array of pixels. One can quickly see that the computational needs of image processing can multiply rapidly and become very large very quickly. Digital computers soon became indispensable.

The large set of computations for image processing is worth the price because as the saying goes 'a picture is worth a thousand words'. Also, there are many applications in business that range from simple charting for a report to complicated drawings, blueprints, and CAD/CADD (computer aided design/computer aided drafting design). In medicine, magnetic resonance imaging (MRI) and computerized axial tomography (CAT) scans of parts of the body including the brain may well avoid dangerous surgery and save many lives. In a less dangerous and more entertaining way, images are used in cinema and film-making. In 1982, the film *Tron*, was released and credited as the first feature film that used computer generated imagery as background for live actors. Well, within a decade, this digital technology became so commonplace that it is now being used throughout the entire film industry. We now have digital cameras that take moving and animated pictures and store them for future manipulation by computer. This technology

has applications in business for training, for advertising products and for simulations in decision-making.

Image processing is also important in any office. To give you some idea of the magnitudes involved, consider a study done by Arthur D. Little in 1980 in the US. The study found that an average office worker handles per day: 1 page from files, 5 pages from mail, 4 from catalogues, 11 photocopies, 32 pages of computer printout as input, 14 pages passed along, 5 pages mailed and 8 pages to be filed. Now project this into the future and you can soon see that there is a great potential demand for image processing if office work is to be rationalized and made efficient, especially when office processing shifts from data and text only to their being embedded in graphics and pictures as images.

Even the need for processing data is increasing rapidly as shown in Figure 7.1 where files in the 1970s have grown into very large files in the 1990s and may now occupy 10^{12} bytes. In addition to data, text and graphics, there is the need to process voice which is often indispensable today in the office, in the factory, in the home and in all walks of life. This need is largely met by the telephone and partly by voice processing. The differences between the processing of voice and that of data may be well known, but for the record they are summarized in Table 7.1.

Many applications related to voice are processed by the telephone exchange. An example is video processing. One configuration is shown in Figure 7.2. This can be done through a central facility like a PBX (Private Branch Exchange) carrier with a **BRI** (the basic rate interface) rate for low volume, or through the LEC (Local Exchange Carrier) paying a **PRI** (the primary rate interface) for high volume. The BRI and the PRI are two classes of services to customers of baseband ISDN. The BRI provides up to 144 kbps (two 64 kbps 'B' channels + one 64 kbps 'D' channel for control information). The PRI provides up to 1.54 Mbps which includes twenty-three 64 kbps 'B' channels and one 64 kbps 'D' channel.

The many applications of voice and image processing mentioned above are stand-alone applications. Many can be integrated providing a *raison d'être* for ISDN. Some of these applications are listed in Table 7.2. They vary in bandwidth demand as shown for a sample of applications in Figure 7.3.

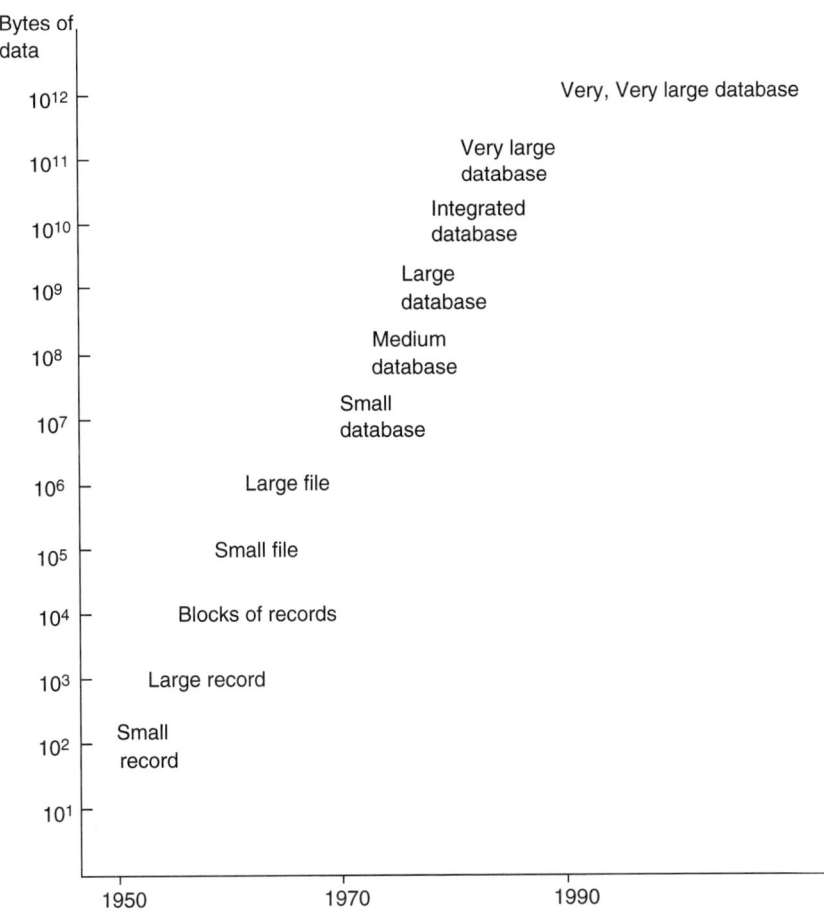

Figure 7.1   *Growth of corporate databases*

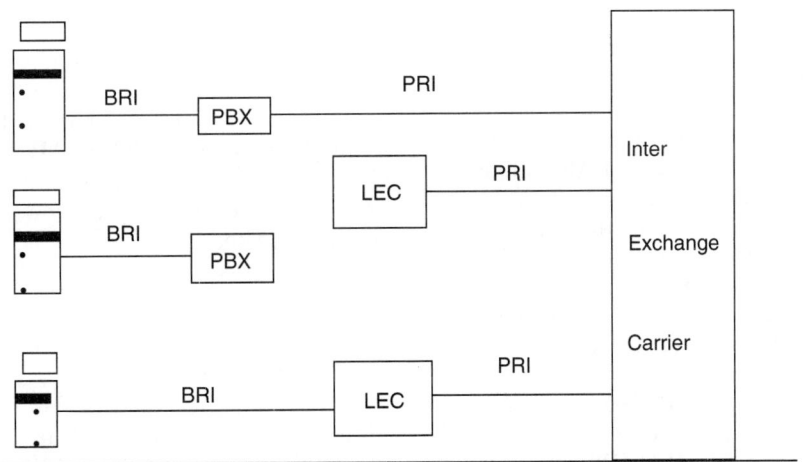

BRI = Basic rate interface; PRI = Prime rate interface; LEC = Local exchange carrier

Figure 7.2   *Networking with video-telephones*

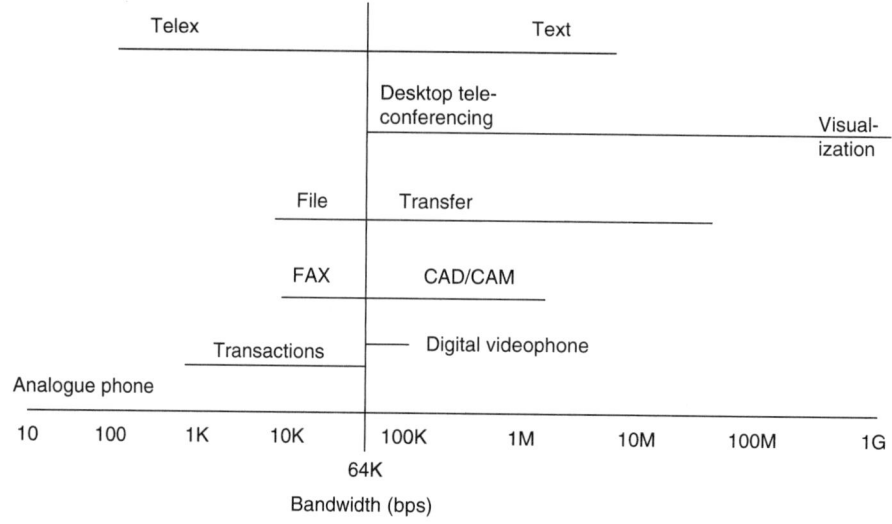

Figure 7.3 *Bandwidth requirements. Not to scale; only approximate*

Table 7.1 *Computer network vs. voice network*

	Computer network	Voice network
Usage:	Digital	Voice
Distance:	Limited	Long
Performance:	High	Limited
Ownership:	Private	Public usually
Charges:	Low, if any	Service charge

The resource environment

The question that could be asked is why these applications cannot be processed by our conventional digital equipment along with modems to do the conversion between digital and analogue signals. The answer is that modems are appropriate for PCs with a 9.6 kbps modem, despite their steady increase in bandwidth from around 100 bps to almost 50 kbps. Up to 100 kbps, the baseband ISDN may well be adequate for connecting LANs as well as for transmitting data faster than before. For example, a facsimile page that took 30 seconds in the pre-ISDN era would now take around 4 seconds. However, for applications beyond 100 kbps, one needs the **B-ISDN** (Broadband ISDN) as shown in Figure 7.4. The B-ISDN is designed for voice and video as well as large volumes of data to be transmitted over long distances. With fibre optics, speeds of the B-ISDN can get data rates between 10 and 600 Mbps. Such bandwidth applications like the high speed workstations, large data repositories interconnected for the purpose of processsing medical images, molecular models, distributed CAD/CAM (computer aided design/computer aided manufacturing), and the like are waiting adoption. Applications of ISDN were listed in Table 7.2. The most spectacular application of ISDN (not even listed in Table 7.2 because it is still a proposition) may well be NASA's Earth Observing System for global-change research, which is expected to transmit more than one trillion bytes of data per day (or equivalently 92 million bits per second) for the duration of a 15 year period. That is to begin in the late 1990s (Irvin, 1993: p. 43). However, we are talking about B-ISDN without really explaining the baseband ISDN or narrow band ISDN. It is time to do so.

What is ISDN?

Voice and data are inputs to an interface hardware equipment which is connected to an ISDN interface. It is connected to an ISDN switch through three channels: two 'B' channels (bearer channels) of 64 kbps and 16 kbps, appropriate for either data or voice transmission, and one 'D' channel, which is a 16 kbps designed to control

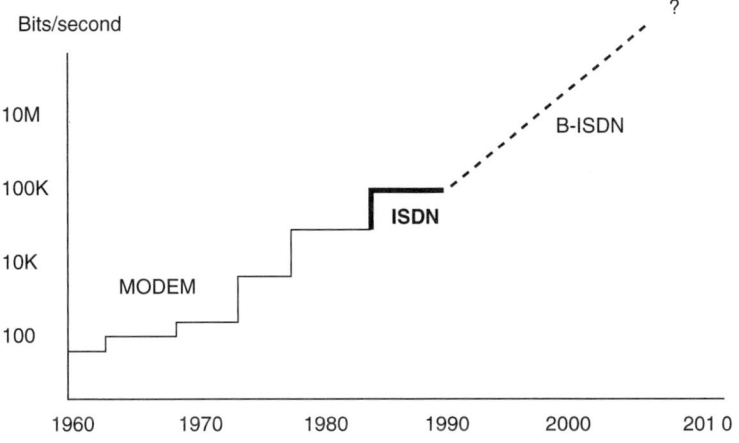

Figure 7.4 *Rise in data rates (adapted from Griffiths, 1990: p. 158)*

Table 7.2 *Applications made feasible by ISDN*

Customer sharing with salesperson the same screen
 on products and conditions of sale

Teacher sharing screen with student

Telecommuter sharing screens of multimedia desktop
 with customer, supervisor or co-worker

Video-conferencing with all parties looking at data,
 text, graphs, pictures and even results of a
 simulation in progress

Access and dialogue with librarian

Medical records and imaging access

Remote medical diagnoses like from an airport or
 form

Security (and identification) and surveillance system

Teleshopping

Telebanking

Telereservations

Telenews

High speed bulk multimedia transfer including
 books from a library

Multidocument image storage and retrieval

Note: All the above applications involve on-line remote processing. Many of the above (e.g. video-conferencing and those involving dialogues with up to the minute updated data) are **isochronous**, that is, they are time dependent and in real time

transmission in the 'B' channel (used to signal the switching system to generate calls, reset calls and receive information on the incoming calls including the identity of the caller). These three channels are sometimes referred to as the 2B+D system. Two of these three channels are of 64 kbps each. They can be multiplexed to form one 128 kbps or by multiplexing four 'B' channels to form one 256 kbps channel. Similarly, one 64 kbps can be submultiplexed into two 32 kbps or eight 8 kbps channels for eight terminals connected in parallel. These channels connect to the ISDN switch which is connected at the other end to networks that may be signalling, non-switched, switched or packet switched networks. This is shown in Figure 7.5.

Implementation of ISDN

ISDN has no system to compete with or to match and copy. ISDN was implemented following all the rules of good development: the system specifications were stated by the users, and the system was designed, implemented and tested. Since the system was to have a world-wide relevance, the development had to be global and this takes a long time.

In 1984, the first specifications of ISDN appeared as the 'Red Book' specs followed by the 'Blue Book' specs, based on experiences with the Red Book. These were made by CCITT, an international organization telecommunications based in Europe.

Suppliers for components were selected: Siemens Stomberg-Carlson in Europe and AT&T and Northern Telecom in the US. The implementation was ahead of schedule and more so in Europe than in the US. This may be because in Europe we have nationalized PT&T

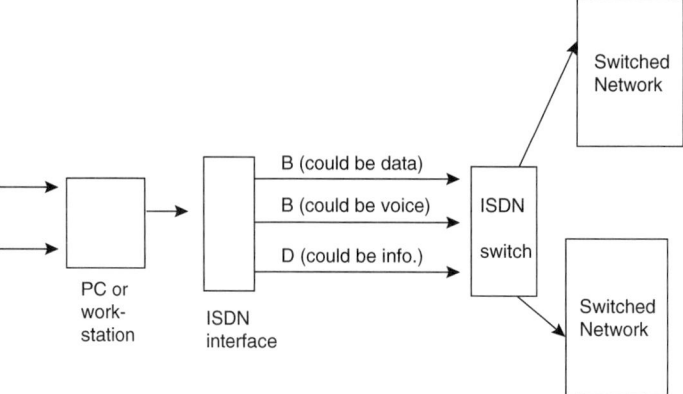

Figure 7.5 *Basic components of an ISDN*

(Post Telephone and Telegraph) organizations, while in the US the telecommunications industry is decentralized. The private carriers had billions of dollars invested in analogue equipment and so there was hesitation and caution but no resistance to a conversion, painful and expensive though it was bound to be. But carriers in the US did anticipate the coming of ISDN and have rapidly added WAN access capabilities to everything ranging from computers to internetworking equipment to video-conferencing equipment. US companies are developing alliances and partnership to redirect their resources to the burgeoning potential market of ISDN. The computer industry is also producing equipment necessary for an ISDN environment such as the ISDN modems, adapters, bridges and bridge/routers (Garris, 1995).

The implementation of ISDN was started in the UK in 1985 by British Telecom followed by Illinois and California in the US in 1986; in Mannheim and Stuttgart, Germany in 1987; and in Brittany, France, also in 1987. By 1988, there were over forty different trials of services of the 'Red Book' standards that were in progress. In many countries, the infrastructure to be required by ISDN was being laid. In the US, the use of fibre grew from 456 000 miles per year to 3 811 000 miles a year by the end of 1991. Japan enjoyed a fibre optic backbone network since 1989. They announced their fibre-to-the-home trials, and a schedule for providing such service to all subscribers by the year 2015. In Europe, the Euro-ISDN (European ISDN standard) was introduced in 1988 and compliance

with the standard was targeted for 95% in 1995. In Australia there is also a 95% availability of ISDN. In the UK, the availability of ISDN has been 100% for several years. In the US, services are expected to be around 90% by 1996 (Galvin and Hauf, 1994: p. 36).

Despite its extensive testing and its international certification, ISDN has not been widely adopted. In a study done in the US, (Lai *et al.*, 1993: p. 49) 17% of the companies surveyed rejected ISDN. The main reasons cited (in order of significance) were:

- other networks can serve the same communications needs equally well;
- nation-wide ISDN not available;
- not able to justify costs;
- not an established technology;
- not available in our area;
- international ISDN is not commonly available;
- not compatible with organization's computing environment.

The same study identified the principal obstacles for the adoption of ISDN (Lai, *et al.*, 1993: p. 50). These are:

national ISDN not available;
unattractive tariff structure;
lack of user awareness;
world-wide ISDN not available;
expense of ISDN equipment;
available only in metropolitan areas;
incompatible equipment;
ISDN services not attractive;
lack of standards for ISDN.

Summary and conclusions

ISDN is a viable technology but has not yet received wide acceptance. Meanwhile it will coexist with other technologies that are often older and better entrenched like the analogue, Switched 56 and SMDS. Another competing technology is the frame relay. This is not older but younger than ISDN; in fact, it is a spin-off of ISDN (Bhushan, 1990). These technologies are compared in Table 7.3.

The 'I' in ISDN stands for the integration of the media of data, images, voice and video into one single digital end-to-end system of seamless transmission and communication. The integration of the two basic media of digital data (from the computer) and analogue signals (from a telephone) are shown in Figure 7.6. Such integration will not only enhance the many applications of integrated data digital and voice as listed in Table 7.2 but will encourage the user and the industry to identify and develop applications that have only been dreamed of. For example, it may well be possible soon to pick up the phone and call someone across the continent or even across the world and download a database or images or video in one window or the screen and see the other party in another window simultaneously. This may well come without any grand opening and fuss but will just creep upon us like a long

Table 7.3 *Comparison of ISDN with other technologies*

	Advantages	*Disadvantages*
Analogue:	Ubiquitous	Slow
	Cheap	Long set-up times
Switched 56:	Faster than analogue	High monthly services charge
	Provides digital service	
SMDS:	Highest bandwidth	Confined to local carrier's region
Frame relay	Best for full time connections such as large and very busy branch offices	Too expensive for low frequency users
ISDN	Fast	Not universally available or compatible
	Flexible configuration	Unfamiliar to many professionals
	Can be cheap	Expensive if packaged wrongly

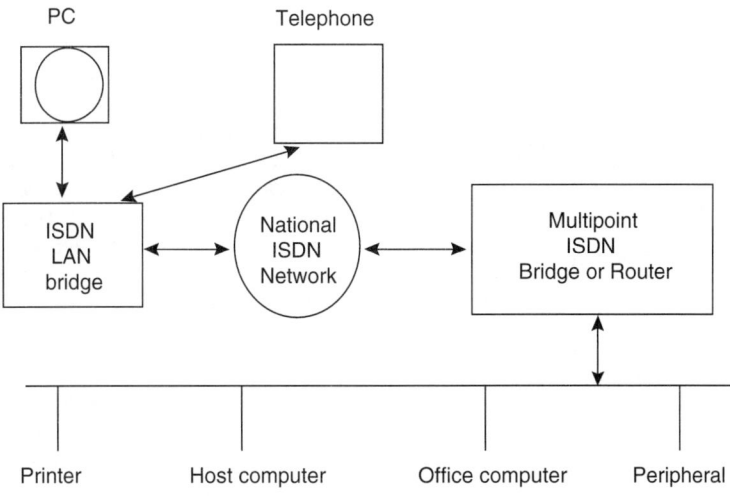

Figure 7.6 *Integration of digital and analogue*

overdue progression of computing applications.

ISDN is not yet a plug-and-play technology, but it is a big leap from the analogue-only communication of the telephone. ISDN is an access technology that allows us access to all data whether in the form of digits or analogue signals simultaneously. It is a vehicle by which business as well as households will cruise on the international information highway.

ISDN is a foundation technology and like the foundation of a house it is fundamental and even invisible. But the infrastructure for the house of ISDN has been carefully planned and well implemented. It is not commonly accepted, not even commonly known. There is still confusion and lack of awareness about availability schedules, rates, standards, and even the benefits and enhancements of ISDN. There are still problems of national standards being in conformity with international standards, the problem of cheap and stable rate structures, as well as interoperability and portability between components and devices using ISDN. We have not yet reached the level of applications that provide an industry with the assurance of economies of scale for the industry to adopt ISDN. Also, ISDN is constrained by bandwidth. A broadband ISDN, the B-ISDN, is now being implemented (a case on the development of its standards is discussed in the chapter on standards). All this will come in due time and as with the telephone, it may take many decades but it will be ubiquitous and indispensable for a worthy style of living that awaits us.

ISDN is consistent with many of the well accepted architectures of networking, which is the subject that we will examine in some detail in our next chapter.

Case 7.1: ISDN at West Virginia University (WVU)

WVU had a FDDI backbone that links its ten buildings but had 90 other buildings that could not be interconnected. Economic analysis indicated low bandwidth connectivity in these satellite buildings dial-up routers that can use modems to link LANs over ordinary phone lines gave only the most casual support of LAN interconnection despite compression being used.

In 1990, WVU decided on a flat-rate ISDN service that connected the satellite buildings to the FDDI backbone using Ethernet bridges. This cost $40 per month for each line and $2100 for a bridge that lashed the two 64 kbps 'B' channels of ISDN to create a 128 kbps pipe. With compression, thruput exceeded 200 kbps. That gave an order of magnitude slower than Ethernet but an order of magnitude faster than an analogue modem. Also being synchronous, it carries a bigger load than does the asynchronous traffic. The alternative was to get a T1 link for $700 per month plus a bridge and a CSU/DSU (channel service unit/data service unit) for $12 000.

Source: *Byte*, Dec. 1993, p. 75.

Case 7.2: ISDN in France

The Numeris ISDN in France has many applications using ISDN that can be classified into four categories as shown below.

DATA	MULTIMEDIA
Remote processing	Radio commentaries
LAN interconnection	Teleconferencing
Software loading	Audiogram Service
	Illustrated video text
IMAGE PROCESSING	**DOCUMENTATION**
Image server	High speed facsimile
Medical Imagery	Electronic mail
Telesurveillance	Document database
Remote teaching	Document exchange
Video Telephony	
Local image stations	

The ISDN implementation was done in three stages. The first stage was to use ISDN for a specific application including image videotext; the second stage was the voice and data integration; and the third stage was the full integration with the architecture of the corporate network. 'The various needs expressed by a corporate network are split amongst the three main bearer networks, which are leased lines, the packet switched network, and the ISDN. The planning and definition of this architecture requires two to three years before becoming available.'

Source: Jean-Pierre Guenin, Therese Morin, Francois Lecrec, Roger Trulent and Pierre Deffin. ISDN in France—1987–1990: from the first commercial offering to the national coverage of numeris ISDN. *IEEE Communication Magazine*, January 1991, pp. 30–35.

Case 7.3: ISDN for competitive bridge across the Atlantic

In December 1996, the author was playing competitive bridge in real time on the Internet. Three of the players were from America and one, with the screen name Atle, was from Norway. Atle was consistently slow in his responses.

In between the plays, the players are able to communicate with each other and the Americans nudged the Norwegian to speed up. Finally Atle responded: 'I apologise for being slow but not for long. Next month I am getting an ISDN connection.' 'Oh' I said, 'How much will that cost you?' $US500. And what does it cost in America?' asked Atle. The American from California responded: 'I do not need an ISDN. I get a fast enough response and unlimited access for less than $20 per month.' 'Wow' said Atle, but soon got immersed in playing bridge and the conversation on ISDN ended.

Bibliography

Bhushan, B. (1990). A user's guide to frame relay. *Telecommunications*, **24**(7), 39−42.

Crouch, P.E., Hicks, J.A. and Jetzt, J.J. (1993). ISDN personal video. *AT& T Technical Journal*, **72**(3), 33−38.

Derfler F.J. Jr. (1994). Betting on the dream. *PC Magazine*, **13**(18), 167−187.

Galvin, M. and Hauf, A. (1994). Expanding the market for ISDN access. *Telecommunications*, **28**(10), 35−38.

Garris, J. (1994). ISDN sleight of hand. *PC Magazine*, **14**(5), NE1−6.

Griffiths, J.M. (1990). *ISDN explained*. Wiley.

Irvin, D.R. (1993). Making broadband ISDN successful. *IEEE Network*, **7**(1), 40−45.

Lai, V.S., Guynes, J.L. and Bordoloi, B. (1994). ISDN: adoption and diffusion issues. *Information Systems Management*, **10**(4), 46−52.

Sankar, C.S., Carr, H. and Dent, W.D. (1994). ISDN may be here to stay ... But it's not plug-and-play. *Telecommunications*, **28**(10), 27−33.

Viola, A.J. (1995). ISDN solutions: ready for prime time. *Telecommunications*, **29**(6), 55−57.

Walters, S.M. (1991). A new direction for broadband ISDN. *IEEE Communication Magazine*, 39−42.

8

NETWORK SYSTEMS ARCHITECTURE

OSI is a lovely dream; SNA has a lot of clout; TCP/IP is current reality; but using multiple protocols is the trend

Anon, 1995

Introduction

Architecture is concerned with structure, and style and design. The architecture of network systems is the style and design of the structure of networks that enables electronic communication. It provides a framework for all the components and protocols discussed in earlier chapters and provides a frame of reference for much of the discussion to follow. It is an overview of the structure of networking. Typically, an overview is done early in any text in order to facilitate placing the components in their relative position. We have done the opposite. It is like travelling around much of the world and then looking at the world map. But, in the case of networking, we make the exception on pedagogical grounds. An early overview of network systems architecture could be very intimidating. Networks and telecommunications are very rich in the names and acronyms of the many components, devices and protocols involved. Using them in an overview without explaining them first could be difficult for both the reader and the author. So we discussed important components and protocols, and now we are ready to see how they all fit together and provide a basis for the remaining discussion.

The earliest architecture of networks was SNA by IBM. There were many others developed in America, including DNA, DCA, OSA and TCP/IP. Across the Atlantic, there was IPA and XBM developed by ICL in the UK, as well as the OSI model developed by the international organization CCITT. It was expected that all these models would fold up or merge into one internationally accepted standard. And there was a shakedown as expected, but three models seem to have emerged; SNA, TCP/IP and OSI. In

this chapter we will examine each. First, we will examine SNA and OSI and compare them. We will then look at APPN, the updated version of SNA, and compare it with its competitor in the US, the TCP/IP. Finally, we will compare the OSI models with APPN and TCP/IP. We conclude with observations on how this rivalry between the three models will affect the end-user as well as the computer industry and telecommunications in general.

Systems network architecture (SNA)

SNA was developed by IBM beginning in 1972. It is a layered architecture where each layer is a group of services that is complete from a conceptual point of view. Each layer has one or more entities, where an entity is an active element within a layer. Each entity provides services to entities in the layer above them, and in turn receive services from the layer below them (except for layer 1). In addition, each layer provides a set of functions so that communication can take place not only for file transfer and transactional processing, but for the management of on-line and real-time dialogues that may take place between two parties. The traffic expected was not only data but also voice and video.

The SNA architecture consists of sets of layers within groups of the physical unit (PU) and the logical unit (LU). Each layer is numbered sequentially from 1 to 7 starting from the bottom. Thus, layer 1 is the physical layer that provides the mechanical and electrical level interconnections for the two stations (sending and receiving). Layer 2 is the datalink layer that provides rules for transmission on the physical medium such as packet formats, access rights, and error detection and correction. It is responsible for the transmission between two nodes over a physical link and is serviced through bridges. Layer 3 is the network layer that provides routing of messages between two transport entities. This layer is sometimes called the path control layer because it is responsible for the path that a message takes through the network which could include more than one node. This network layer 3 is serviced through routers.

The physical layer in SNA is addressed in the architecture, but SNA does not actually define specifications for protocols in this layer. Instead, SNA assumes the use of different approaches including national and international standards already in place.

In the logical unit, there is layer 4 which is for transmission and provides functions for error-free delivery of messages such as the flow control, error recovery and acknowledgment. This layer also provides an optional data encryption/decryption facility (to be discussed in a later chapter). Layer 5 is for data flow control. It initiates and establishes connections and keeps track of the status of sessions and connections. This layer controls the pacing of data within a session; arbitrates users' rights and services when there is a conflict; synchronizes data transfers; and controls the mode of sending, receiving and response. Layer 6 is the presentation layer which

provides the necessary services for formatting different data formats used in each session and manages the sessions dialogues. Finally, there is the upper most layer, layer 7, the application layer. It provides the application service elements for the end-user. This includes the control of exchanged information, operator control over sessions, and resource sharing, file transfer, database management, and document distribution and interchange.

The LU (logical unit) is between the PU (physical unit) and the applications of the end-user who enters into sessions for communication. The sessions' medium will vary in mode and hence there are seven LUs that correspond to each type of session mode. For example, the LU2 supports sessions with a single display terminal of the 3270 type while LU6 supports peer-to-peer connection with the application subsystem or application program. The PU, LU, end-user and the seven layers of network architecture are portrayed in Figure 8.1.

For communicating in the SNA schema, the message must first go to the top-most layer (closest to the application and end-user), layer 7 as shown in Figure 8.2. The message then goes down to the physical unit (path OA — origin to the point A), across the transmission path AB, and finally up the seven layers to the destination D, through path BD. Both paths OA and BD are controlled by network software which is additional to the OS (operating system) software

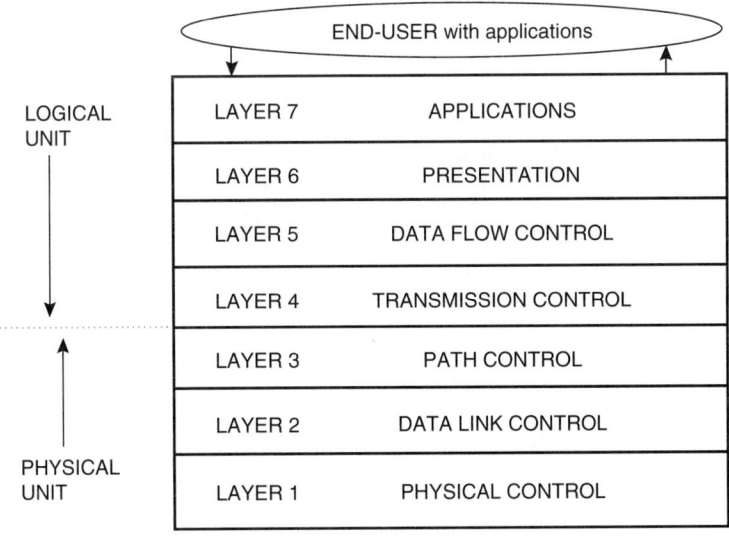

Figure 8.1 *Seven layers of SNA*

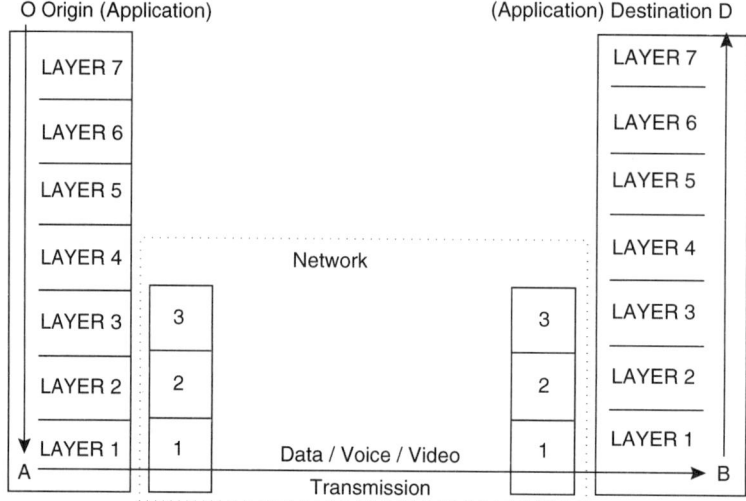

Figure 8.2 *Communication through the layers*

and the applications software. The transmission is through the network. The communication path is shown in Figure 8.2.

Corresponding to the use and development of the SNA layers, there were protocols being developed that addressed the individual or group of layers as shown in Figure 8.3. Thus, the IEEE 802 standards of the Institute of Electrical and Electronics Engineers in the USA developed protocols for the physical layer as well as for medium access control in the datalink layer and for logic control for most of layer 3. The IEEE 802 included standards and definitions that became *de facto* standards for the industry for the Ethernet and token rings, as well as providing definitions for concepts like connection and connectionless service. Other protocols developed were for other sets of layers such as the IP for layer 3

SNA	IEEE 802	DOD	CCITT	ISO
Layer 1				
Layer 2				
Layer 5		TCP		Sessions
Layer 4		TCP		TP
Layer 3	Logic control	IP	X-25	
Layer 2	Medium access control		LAP-B	
Layer 1	Physical		X-21	

Figure 8.3 *Early sources of protocols*

and the TCP for layers 4 and 5. These protocols were developed by the DOD (Department of Defense) in the US. We mentioned them in an earlier chapter and will come back to them later in this chapter. Other early protocols were developed by the international organizations like CCITT for layers 1–3 and by the ISO for layers 4 and 5. The protocols by the CCITT and the ISO corresponded directly to the OSI model, a subject that we will now address.

The OSI model

OSI stands for the Open Systems Interconnection reference model by the ISO, International Standards Organization. The **open** refers to a specification made openly in the public domain in order to encourage third-party vendors to develop add-on products to it. The interconnection refers to procedures for exchange of information between computers, terminal devices, people, processes and networks.

OSI was announced in 1977 as a response to the need of international standards for communications and networking. The main objectives of the OSI model were:

1. to provide an architectural reference point for network design;
2. to serve as a common framework for protocols and services consistent with the OSI model; and

3. to facilitate the offerings of interoperable multivendor services and products.

In design, the OSI model was not very different from the SNA model, as is clear from the conceptual comparison in Figure 8.4, except at the lowest level and physical layer. While there is an almost one-to-one mapping for most of the upper layers, there is no direct counterpart in the SNA model for the lowest layer of the OSI model. The SNA model left the definition of the communications devices outside its model: the OSI did not have any such reservations.

For a summary comparison on SNA and OSI in general terms see Table 8.1.

The OSI model never did catch on in the USA despite the fact that it was developed by an international organization. OSI, however, was (and is) very popular in Europe. The European Common Market Commission had made the OSI and ISO its standards for connecting systems products and networks of different

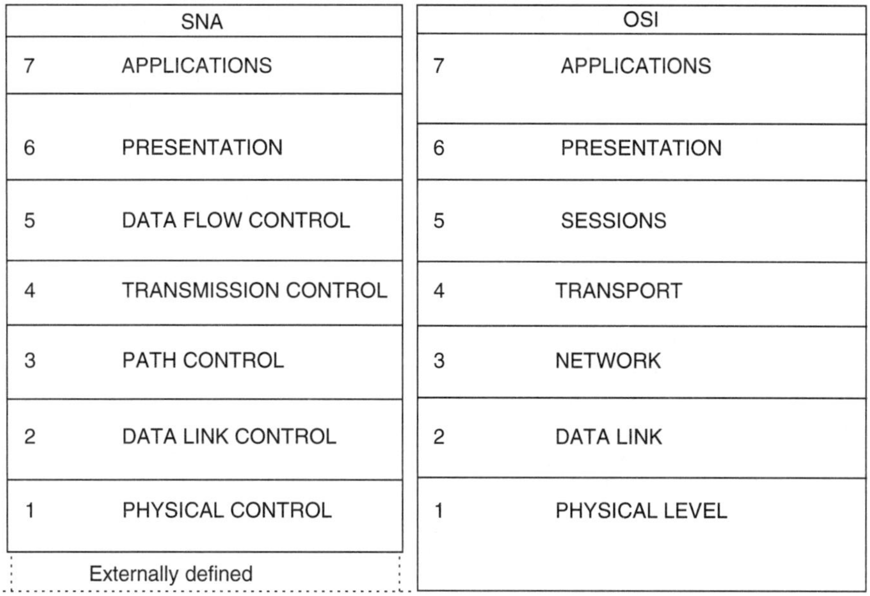

SNA		OSI	
7	APPLICATIONS	7	APPLICATIONS
6	PRESENTATION	6	PRESENTATION
5	DATA FLOW CONTROL	5	SESSIONS
4	TRANSMISSION CONTROL	4	TRANSPORT
3	PATH CONTROL	3	NETWORK
2	DATA LINK CONTROL	2	DATA LINK
1	PHYSICAL CONTROL	1	PHYSICAL LEVEL
Externally defined			

Figure 8.4 *SNA versus OSI*

Table 8.1 *A comparison of SNA and OSI*

	SNA	*OSI*
Source:	IBM	ISO (International Standards Organization)
When initiated:	1972−74	1977
Architectural design:	*De facto* standard in the US and IBM mainframe users	Largely like SNA but has more for the physical layer 1
Levels implemented in 1985:	All seven layers	Bottom three layers
Acceptance:	Welcomed by IBM mainframe users and many in the US	Largely in Europe with SNA/OSI interfaces negotiated with IBM
Future:	Updated by APPN and other products being steadily introduced	Work being done at all seven levels in addition to the network management level (see Deirtsbacher *et al.*, 1995)

manufacturers. The Europeans used its leverage as a large customer of IBM equipment to persuade IBM to agree to the OSI model. In the USA, the DOD (sponsor of TCP/IP) had executive directives to use international protocols. It seemed that the model was bound to become the international standard. But IBM did not give in to the OSI model. It did, however, agree to a gateway between the SNA and the OSI model. There are three ways to integrate the two models: one is directly at each of the seven levels; the second is indirectly through the three bottom levels of the physical unit; and the third is to have an SNA/OSI interface in between (Tillman and Yen, 1990: pp. 219—220). There are many interconnection devices that include repeaters, bridges, routers, bridge/routers, gateways, device emulation, Internet transmission, and interstation transmission (Tillman and Yen, 1990: pp. 216—218).

There are many reasons for IBM's hesitation to give up SNA for international harmony. One was that at the time of the most pressure, 1985, IBM had all its seven layers implemented while OSI had only three bottom layers implemented. Also, IBM was no small firm to be pushed around. IBM not only had a dominance in all segments of the computing market but also a sales presence in all the world, including Europe. In 1985, IBM had a revenue of $48.554 billion, more than all the next 12 world-wide competitors combined. (This revenue was more than the GNP of all the countries in Africa outside South Africa and two of the oil rich countries of North Africa.) In 1985, IBM employed 405 535 people (*Datamation*, June 15, 1986, p. 56). IBM continued with its proprietary network architecture and had a 'captured' market from all the users of IBM mainframe computers.

The APPN

The problem facing IBM was that its networking architecture in SNA was designed for a mainframe host serving a lot of dumb terminals in a hierarchical master—slave configuration. This was appropriate for the 1970s when SNA was first implemented, but the world of computing changed in the 1980s. The dumb terminals were replaced by PCs and workstations; centralized management was replaced by decentralization of computing; allocation of resources by the host was to be replaced by the sharing of information

and computing resources without host interventions; the mainframe host was being replaced by computers that were more powerful than many mainframes as servers of computing resources including databases, programs and peripheral services; and the single host was now being replaced by multiple hosts and servers. This new computing paradigm was for peer-to-peer computing and client—server computing. We shall discuss these configurations and the downloading and downsizing of the mainframe to the end-user client and the server in a later chapter. In this chapter we will examine how IBM responded to the changing environment with its APPN and the competition it faced from TCP/IP.

APPN is the acronym for Advanced Peer-to-Peer Networking. A **peer** is a functional unit on the same protocol level as another. In peer-to-peer communication, both sides have equal responsibility for initiating a session. This is in contrast to the master—slave relationship, where only the master unit initiates and the slave responds.

The hierarchical host to terminal configuration of the SNA is now replaced by networking that handles peer communications among hosts, departmental computers and desktop PCs, and is compatible with the hierarchical SNA structure while maintaining connectivity with dumb terminals. Because APPN is an update of SNA, the new traffic can still be handled without much conversion or **encapsulation** (in telecommunications, it signifies the headers for transfer from a high protocol level to a lower protocol level).

APPN was originally targeted in 1986 of midrange systems. Since then there have been many network announcements to enhance the APPN. An important enhancement was the SAA, the Systems Applications Architecture, that had a sophisticated and fundamental routing technology. The routing was now possible without host intervention. APPN now kept track of network topology making it considerably easier to connect and reconfigure nodes. Such enhancements to the APPN added more functionality and made the APPN more **robust** (a program that works properly under all normal but not all abnormal conditions).

In addition, performance increased. The ten messages required under SNA now required only two messages. But despite all the enhancements to APPN, there still is strong competition from TCP/IP.

TCP/IP

TCP/IP stands for transmission control protocol over an Internet protocol. We visited TCP/IP earlier as a set of protocols. They were developed by the DOD in the mid-1980s and was concerned with the middle layers of the seven layer cake architecture. The TCP/IP was expected to retire, unable to resist the pressure from IBM'S SNA and OSI supported by the international community. But, TCP/IP took on a life of its own. Since it was based on the UNIX operating system, it had the support of many UNIX users who liked the multitasking and multithreading features which are ideal for multiplatform networks. UNIX was becoming the system of choice for many mission critical applications in IS (information sytems). In the early 1990s, it was generally accepted that TCP/IP had made many inroads into the SNA market, perhaps close to 20% of the over 50 000 SNA networks across the world serving some 300 000 nodes (Kerr, 1992: p. 28).

The success of the TCP/IP approach was that it relied for its critical backbone on the multiple protocol routers and the OSPF: the 'open shortest path first' approach that was a new routing algorithm. The routing was adaptive giving the ability to route around failed circuits. 'These routers evolved out of the TCP/IP and Ethernet environments; as LANs expanded, the routers were enhanced to accommodate more protocols and to provide critical segmentation capabilities. This enables IP routers to become the key to controlling broadcast storms and ensure reliability in large meshed internet backbones connecting local and remote LAN users'. (Rosen and Fromme, 1993: p. 79).

It is very tempting to marry the SNA and the TCP/IP. One way is to run TCP/IP on a IBM mainframe using SNA. The other approach is to implement SNA gateways on UNIX machines. What is emerging though is the use of TCP/IP to create what is called the SNA/IP networks that competed directly with the APPN, sharing responsibility for the different layers in the network architecture with SNA still specializing in the upper layers. Incidentally, even IBM approved of the SNA/IP alternative to internetwork, for IBM offered the system as an alternative to their pure APPN. The two are independent enterprise architectures but they can coexist on the same physical network. There are several vendors that offer routers that handle both IP and APPN simultaneously. This allows enterprises to deploy pure APPN or IP backbones, or both. The backbone is connected to access networks through access devices, as shown in Figure 8.5. The access network is within the sphere of influence and control of the end-user

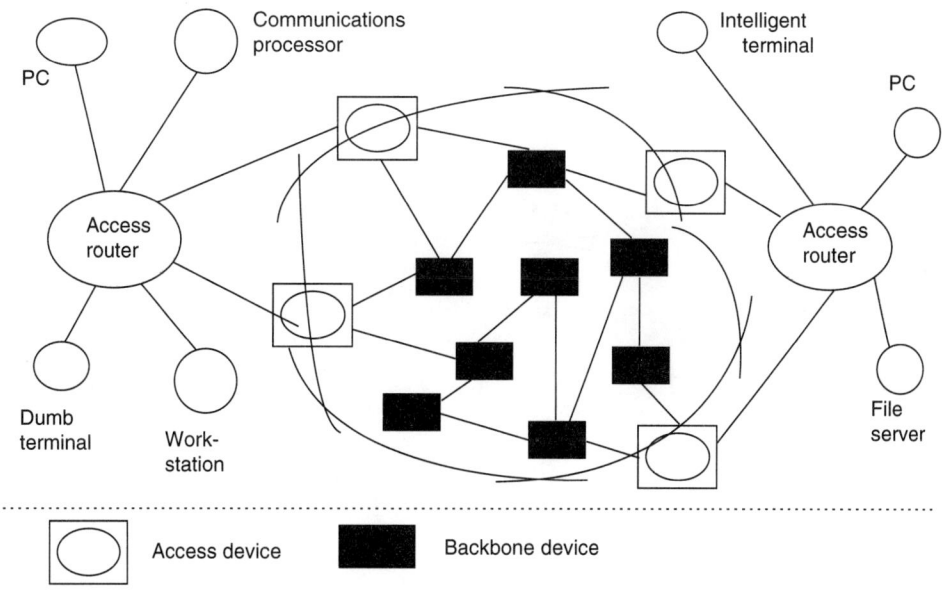

Figure 8.5 *Access and backbone networks*

and could include PCs, workstations, terminal (dumb or intelligent), communication controllers, and fileservers. The access devices are routers or communications processors. In APPN networks, the access devices can be routers, FEPs (Front End Processors), or LAN-attached communication controllers. These access devices channel traffic to the backbone routers where the data fields for the backbone include addresses of source and destination, start and end delimiters, and frame check. The backbone may be a MAN or a WAN with wide-area infrastructure for communications across the region or country, or even globally across the world, linking to all points of the enterprise network.

One way to link SNA terminal traffic and a LAN is to put a token ring LAN adapter in the SNA device. As defined by the IEEE 802.5 standard, the token ring frame contains fields for the addresses of source and destination, start and end delimiters, and frame check sequence.

One or more token rings could be used with an IP backbone. This is shown in Figure 8.6, where there are two token rings on the sender's side with a source routing bridge in between and one token ring on the receiver's side. There are two routers accessing the backbone.

Multiple protocols

There are many protocols under almost continuous development, like the X.400. The X designation tells us that it has been developed by CCITT. The set of recommendations from X.400 to X.430 defines standards for a general purpose system and recommendations that may solve many of the interconnectivity problems like those of e-mail and EDI (Electronic Data Exchange) so important for financial institutions.

Achieving a homogeneous system may not be as easy as one would like. In the absence of internationally agreed architectural model and protocols we must mix protocols within the seven-layer architectural model. This is shown in Figure 8.7 where there is often a choice in each layer. Another view, that from the other side of the Atlantic, is the use of the OSI model and all its protocols. This view is shown in Figure 8.8.

Multiple protocols give flexibility to the network manager, but the dream of the network manager may well be of having the ability to mix and match not just topologies and protocols but also network management strategies. The alternative strategies available to the network manager are the subject of the next part of this book.

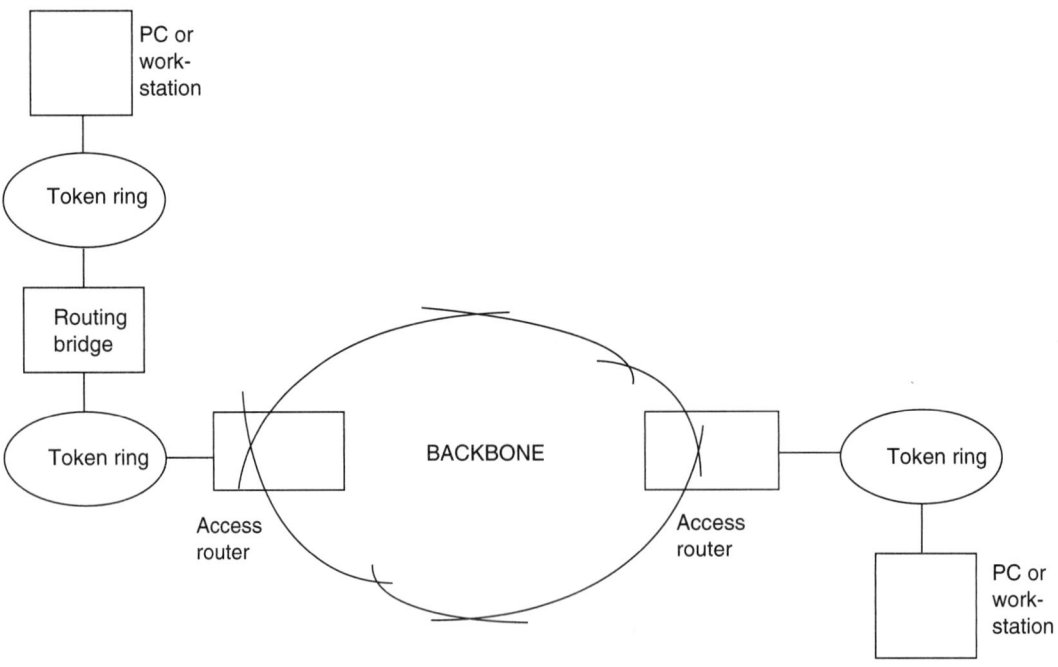

Figure 8.6 *Token rings and backbones*

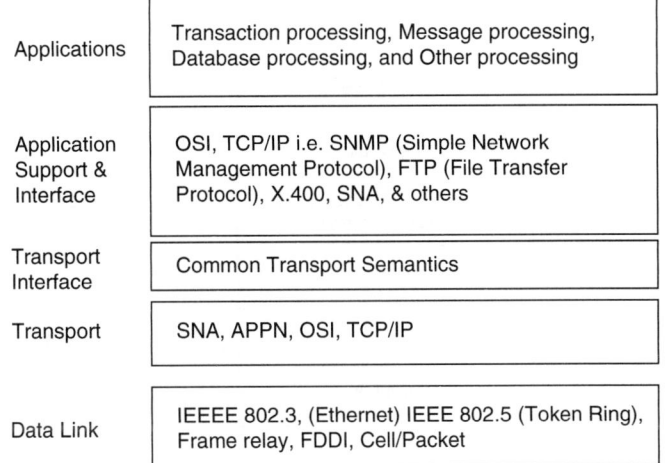

Figure 8.7   *Protocols at different layers*

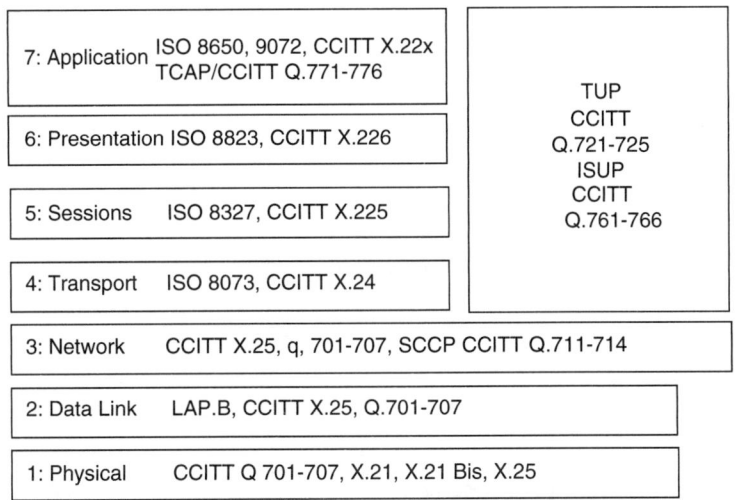

Figure 8.8   *International protocols*

Summary and conclusions

Network architecture is the design of a communication system including hardware, software, access methods and protocols. We have discussed access methods in a previous chapter; in this chapter, we discuss protocols. They are the basis of hardware and software products that are interoperable, a topic which will be discussed in a later chapter.

Network architecture is a complex subject, with many books being written on each of the main models (see Meijer and Peeters, 1982;

Martin, 1987; Meijer, 1988; and Tang 1992 for but a small sample). In this chapter we have taken an overview of some of the models. We examined the SNA along with its updated APPN, the international OSI model, and TCP/IP protocols. A summary comparison of APPN, TCP/IP and OSI are shown in Figure 8.9. The different architectural models being accessed through a gateway network are shown in Figure 8.10.

There is no general acceptance for a single model for network architecture like there is one international standard for ISDN. The main reason is that ISDN had no competition. In

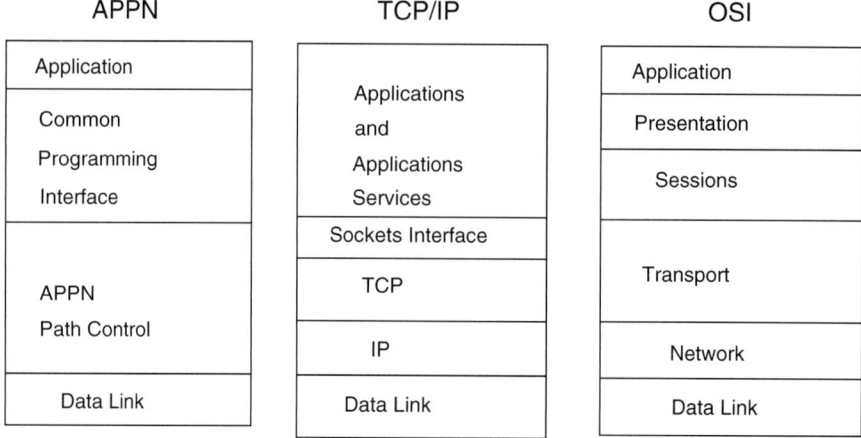

Figure 8.9 *Appn vs. TCP/IP vs. OSI (adapted from Friedman, 1993: p. 90)*

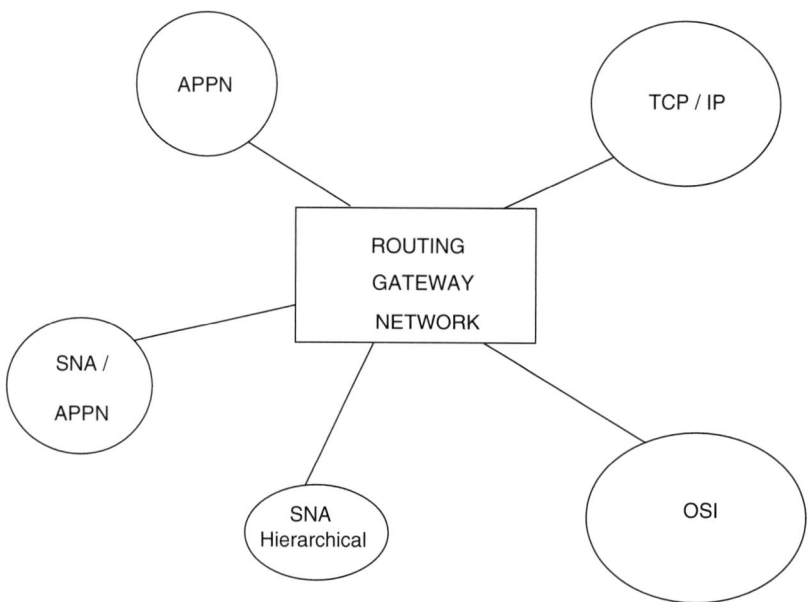

Figure 8.10 *Many architectures being supported*

contrast, the international model OSI had competitors that were well established with some 3–5 years' experience. Also, the IS (information systems) managers were in no hurry for the ISDN. They were preoccupied with expanding their applications portfolio; logically integrating their applications; integrating vertically with their DSSs and EISs and ESs (expert systems); and of integrating their databases with their knowledge bases and making their systems intelligent. In contrast, in the mid-1980s, when network managers had to make decisions on their architectural designs, they were actively looking for network performance and a network architecture that worked. They had no ideological loyalty to international standards. Those that had IBM hardware stuck with their SNAs and the upgraded APPN, while those with UNIX systems preferred the TCP/IP partly because these protocols had a built-in integration with their operating systems. The alternative would have been to wait for the international standards to stabilize and be fully implemented. This would also have required the network systems and components manufacturers to guarantee the offerings of OSI compatible products. The manufacturers (at least in the US) showed no inclination for wanting to offer guarantees, whilst the network managers did not want to wait in a shroud of uncertainty about which network architecture was to be universally accepted. Fortunately for the network managers, they did not have to wait and adopt an architecture that was universally accepted. The competition in the world market gave the consumer and network manager greater choice and pushed the industry into offering a menu of (albeit only three) cost-effective and high performance architectures, the OSI model being just one of them.

The fact that the international marketplace did not accept the OSI model is a source of great sorrow not just from the point of view of network architecture but also from the viewpoint of the development of standards. Robert Amy, who has observed the rise and fall of OSI for over a dozen years, has this to say on the failure and success of OSI:

> ... the market-place has not accepted OSI ... because the world moved out from under OSI before OSI could complete its tortuous way through the standards process ... a further deterrent to the acceptance of OSI was the industry's lethargy in developing new services and products that were visible to end-users ... The acceptance of the layered model served to focus attention on the need to disentangle applications from communications protocols. As a result, data communications equipment has become in effect multilingual, and multiprotocol encapsulation has emerged in today's router products ... the original goals of OSI have been met, but without reducing the number of protocol options available to each network user. (Amy, 1994: pp. 52–3)

One lesson that comes out of this rivalry of architectural models is that SNA prospered and survived because IBM managers adapted to the changing environment of peer-to-peer and client–server paradigms. Could the OSI model have done just as well if it had a head start like SNA? Do international standards have an inherent sluggishness built into them that prevents them from adapting quickly to changing

environments? This is a subject that we shall return to in the chapter on standards. It is the part of the book that deals with strategies of network management: the set of chapters that is the next subject of our discussion.

Case 8.1: AAPN in HFC Bank

HFC is a retail bank in Windsor, UK. It has over 200 AS/400 terminals and is perhaps the largest user of APPN in the world. John Hogan, director of Information Technology at HFC, has this to say about APPN: 'It has limited design options for disaster and recovery for us. We had to put in extra switches in the event of a failure, so it would be better if APPN had a way to deal with it'. Hogan thinks that the basic advantage of APPN is not speed, it's that it takes a fundamentally different approach from hierarchical SNA to making applications run across networks. 'More traditional mainframe sites will say "Why do I need it?"... But a paradigm like APPN will be important for client–server applications, and that's the way the world is going. Even the most hardened glass house has got to recognize it'.

Source: *Datamation*, Oct. 1, 1992, p. 31.

Case 8.2: Hidden costs of APPN

Kathryn Korostoff, principal consultant at' Strategic Networks Consulting Inc. of Rockland, Mass., says that 'the cost of adding AAPN to a front-end processor (FEP) is really huge. Usually, you have to increase the hardware capacities – memory and processing power'. Korostoff says that upgrading a base-configured 3745 front-end processor to support APPN may cost as much as $26 800 per FEP. She says that it will cost an additional $11 800 per FEP in hardware and somewhere around $11 500 in software.

Source: *Datamation*, Oct. 1, 1992, p. 31.

Case 8.3: Networking at SKF, Sweden

SKF is the world's largest manufacturer of bearings. SKF has a workforce of about 41 000 working in 150 companies in 40 countries. For its 3000 end-users it had an SNA backbone in 1978 with five centres in Europe. In 1992, SKF went from an SNA backbone to a router backbone with 26 routers. Its backbone is still 90% SNA, 9% DECnet and 1% LAN traffic. The backbone also links Ethernets that support CAD/CAM file transfers, over 100 VAX minicomputers and a link to the company's data centre (King of Prussia) in the USA. In addition, SKF supports over 11 000 dumb terminals, more than 4000 PCs, and leases more than 300 private circuits. SKF uses frame relay to link small sites largely because of its ability to handle bursts.

Routers and more specifically the IP backbone had a distinct advantage 'because they divert traffic back onto its original path once a line is restored'. But the router backbone cost more that the earlier SDLC (Synchronous Data Link Control) network. 'That's partly because it employs higher bandwidth lines to move more traffic and hold down delays. Redundant links, used to ensure reliability, also contribute to the steeper charge'. Skr 6.5 million a year was budgeted a year 'for the leased lines in the backbone's meshed core and another Skr 250 000 a year for the routers'.

SKF consolidated the five computing centres into two, one in Sweden and one in Germany, both using IBM 3090 mainframes. It closed down the mainframe operations in Clamart (France), Turin (Italy) and Luton (UK). The consolidation saved SKF around 30 million Swedish kronor (approximately £3 million) per year while still maintaining reliability and performance on its corporate network. Most of the savings came from staff reductions and 'elimination of IBM's hefty license fees for mainframe software'.

Source: Peter Heywood (1994). Shifting SNA onto a global router backbone: SKF shows how it's done. *Data Communications*, **23**, (5), 58–72.

Bibliography

Amy, R.M. (1994). Standards by concensus. *IEEE Network*, **8**, (1), 51–55.

Coover, E.R. (1992). *Systems Network Architecture*. IEEE Computer Society Press.

Deirtsbacher, K.-H., Gremmeimaier, U., Kippe, J., Marabini, R., Rössler, G. and Waeselynck., F. (1995). CNMA: a European initiative for OSI network management. *IEEE Network*, **9**, (1), 44–52.

Friedman, N. (1993). APPN rises to the enterprise architectural challenge. *Data Communications*, **22**, (2) 87–98.

Kerr, S. (1991). How IBM is rebuilding SNA. *Datamation*, **37**, (20), 28–31.

Martin, J. (1987). *SNA: IBM's Networking Solution*. Prentice-Hall.

Meijer, A. (1988). *Systems Network Architecture*. Wiley.

Meijer, A. and Peeters, P. (1982). *Computer Network Architectures*. Pitman.

Rosen, B. and Fromme, B. (1993). Toppling the SNA internetworking language barrier. *Data Communications*, **22**, (6), 79–86.

Tang, A. and Scoggins, S. (1992). *Open Networking with OSI*. Prentice-Hall.

Tillman, M.A. and Yen, D. (Chi-Chung) (1990). SNA and OSI: three strategies of interconnection. *Communications of the ACM*, **33**, (2), 214–224.

Part 2

ORGANIZATION FOR TELECOMMUNICATIONS AND NETWORKS

9

ORGANIZATION FOR NETWORKING

The illusion of autonomy causes us to ignore the connection with other systems ... The illusion of control causes us to deny the nature of the system in which we are a part.

Peter Stein

When you fail to plan, you are planning to fail.

Robert Schuller

Introduction

Let us suppose that on the basis of the knowledge you have gained in the previous chapters and with your experience in computing, you have been offered a job as Director of Networking. The firm has just made mergers and acquisitions of manufacturing facilities both at home and abroad that anticipate a heavy use of networking including applications in multimedia, video-conferencing and e-mail, as well as global transfer of data and information for an employee base of 3000 employees spread around the country and abroad. You will be responsible for offering network services that are of high quality and performance, reliable, robust, and accessible to end-users. You may make three assumptions. One is that there are no budgetary constraints provided that you can justify the costs. Two, you have a free hand in organizing your department. And three, you have been made an offer that you (and your family) cannot afford to turn down. So what do you do?

Well, first you read the rest of this book quickly and carefully. In the last part it will discuss all the applications of networking. In this part you will learn about management of computing. The many functions of managing a network system will be delegated to other staff members and their functions will be identified and discussed in later chapters in the book with forward references to these chapters. The one function that you may not wish to delegate (at least at this point) is planning for networking. But before one does that one must have an organization. This can be discussed at two levels: at the corporate level and at the departmental level. Here one must identify the different positions needed for network management with the functions that they must perform. We shall do this first, followed by a discussion of the important function of planning for telecommunications and networking.

Location and organization of network management

In the early days, network management was part of IT (information Technology) and was considered as part of support services, much like the librarian, information centre, security officer, standards officers, training officers, etc. They all reported to a Director of IT, also known as the **CIO**, Chief Information Officer. This is shown in Figure 9.1. The location of network management and of IT depending largely on the organization culture of the organization and also on the importance and complexity of networking to the organization. However, as the need for telecommunications and networking increased and was recognized, it became part of communications and was lumped together with other communications related departments like mail, telephone and telegraph; voice and facsimile; teleconferencing and text processing. This configuration is shown in Figure 9.2. Network personnel now reported to a Vice-President of Information Services, also known as **Director of MIS** (Management Information Services) or **MOT** (Manager of Technology). This organizational grouping recognized the importance of telecommunications and networking to all communications and the desire

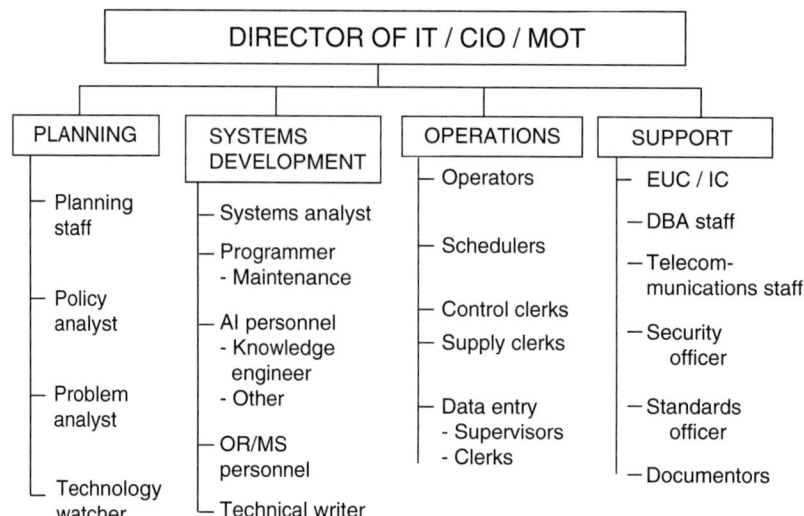

Figure 9.1 *Organizational structure of an IT department*

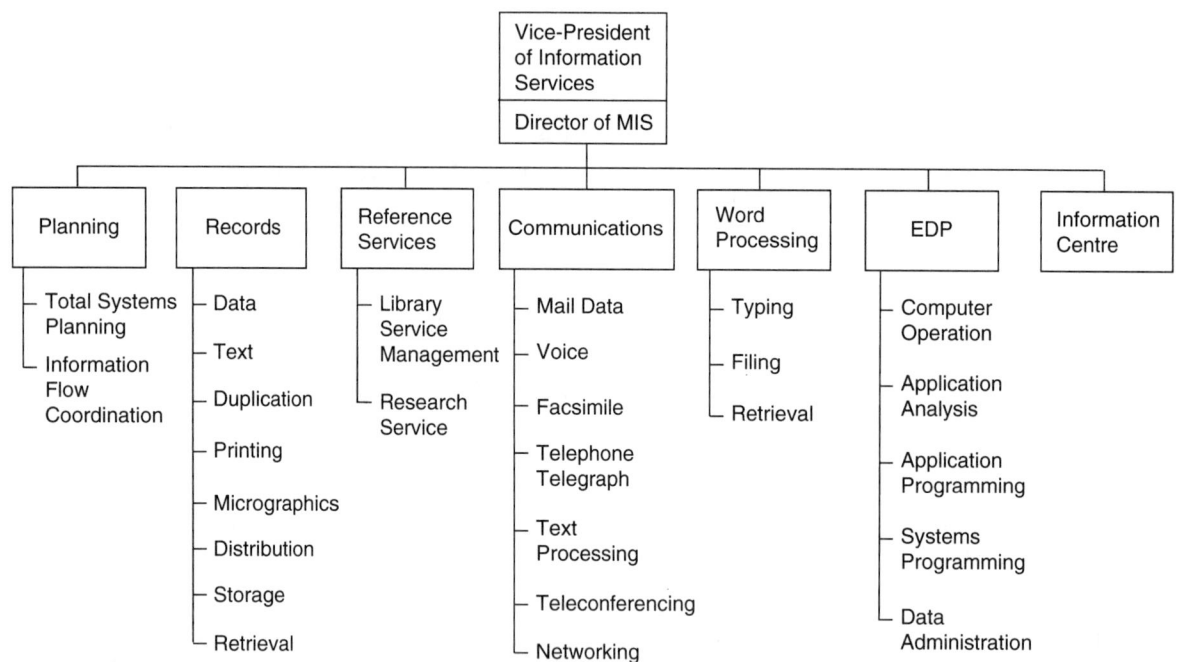

Figure 9.2 *Horizontal integration under Vice-President for information services*

to rationalize these related activities and integrate them not only for higher efficiency and lower costs but for greater effectiveness. This importance grew as PCs and workstations proliferated and the desire for downsizing increased. Organizations become decentralized and distributed with end-users demanding their own computer access from their desktop to other databases and computer peripherals that they did not have or could not afford. The end-users became clients of an interrelated computing system served not by central hosts but by servers dispersed not just in the organization but remotely. This led to the client—server systems and downsizing connected by LANs to be discussed in Chapter 10.

In addition to an increase in the volume of processing, there has been an increase in its complexity. Computer processing and its remote communication is no longer confined to data and text but has now expanded to voice and image processing. A further extension would be their integration into multiprocessing.

Another change in the computing environment is that trade and other daily interactions are no longer confined to the organization but extend across the country and the world. Communications are no longer just national or regional but international and global. Thus telecommunications and network systems have to be global, which requires large bandwidths and large capacities for wide area networking. They ideally require a national telecommunications infrastructure (to be discussed in Chapter 15) and an international infrastructure (to be discussed in Chapter 16).

The importance of telecommunications and networking can be found in the demand for networking personnel. Salaries is one indication of the demand for personnel. A look at the top ten positions in IT in 1994 will reveal that two of the top four and four of the top ten are network personnel. This is shown in Table 9.1 and is well collaborated by other studies (Chiaramonte and Budwey, 1991: p. 56).

The high salaries of network personnel is not just a reflection of the high demand for such personnel but also a reflection of low supply. The low supply is partly the result of the fact that these personnel have not only to be technical but also to have a very broad based education. This is especially true for senior personnel in a public regulatory organization in telecommunications. The mix of professionals required by Oftel (Office of Telecommunications) in the UK is shown in Table 9.2.

It shows that only a little less than 40% of all personnel were technical personnel. In a firm, the ratio of non-technical personnel will be lower but a point that can be made is that telecommunications and networking is a multidiscipline field. This makes the recruitment of technical personnel for networking a little difficult especially at the higher levels of network management where the non-technical and managerial skills are in high demand. At the lower levels, skills

Table 9.1 *Ranking of the top ten positions in IT by Salaries*

IS Director
Telecommunications Director
Data Centre Manager
LAN/WAN Network Manager
Senior Software Engineer
Systems Analyst/Programmer/Project Leader
Object-Oriented Developer
Client—Server Administrator
Communications Specialist
Programmer Analyst

Source: *Data Communications*, May 1994, p. 17

Table 9.2 *Mix of personnel in a regulatory agency*

Profession	Number
Technical experts	8
Information Officers	3
Economists	2
Internal auditors	2
Librarians	2
Lawyers	2
Accountants	1
Technicians	1
Total	21

Source: *Telecommunications*.

are mostly learned on the job. At all levels, personnel have to be attracted, trained and retained in the highly competitive world of telecommunications personnel.

What are these skills? What are the professions in telecommunications and networking and how are they organized in a department? What specialists are required? What is the role of consultants in networking administration? Are they on a retainer or hired *ad hoc*, or both? What is the relationship of telephone and mail management to network management? We shall now answer these and other related questions.

Structure of network administration

The structure of network administration will be greatly affected by existing organizational culture, personalities in top management, and the volume and complexity of the network applications portfolio. It also faces jurisdictional problems that could be very contentious since telecommunications and networking will replace or threaten to replace many existing departments that are large and well entrenched, because they are very labour-intensive. Take the example of internal mail which has to be sorted and delivered. Can that be done by e-mail? E-mail has experienced an exponential growth in the last decade and is the subject of Chapter 19. It may replace much of surface and air-mail, referred to as 'snail-mail' because it moves at the relative pace of a snail when compared to the speed of e-mail. This will cause much displacement and some unemployment even though there still will be some traditional mail. Much will depend on the competitiveness of the Post Office and the PT&T as well as the speed and effectiveness with which ISDN is implemented. Another area of contention with networking will be in telephone communications management which is still quite labour-intensive despite much automation of their services. The analogue signals of telephony may soon be replaced partially by ISDN. Yet another area is image processing. This will be partly replaced by ISDN but this is not so labour-intensive and will not cause as many contentious jurisdictional problems as with telephones and the mailing room.

It is very likely that all these departments will coexist for at least some time in the future but the battle lines are drawn and the battle is soon to come. Whatever the battle, there will be some organizational changes that will be necessary for networking. One configuration is shown in Figure 9.3.

Teleprocessing and networking will report to the Director of Information Processing, or Director of IT, or the CIO (Chief Information Officer), who is often a Vice-President. Sometimes an alternative title to a CIO is the MOT, Manager of Technology. This emphasizes the high-tech aspect of the job and is sometimes a preferred title.

There are many personnel involved with operations. We will discuss operations in the chapter on network administration, Chapter 13. In this chapter, we will briefly discuss, or at least identify, the other personnel required for telecommunications and network administration.

The core personnel would be those involved in development. They will help operational personnel especially on the systems and subsystems that they developed. These personnel would be the network hardware engineer, the network software engineer, and the network engineer, who would most likely be a senior person with a knowledge of both network hardware and software. They will be supported by specialists, like those on voice processing, image processing. In

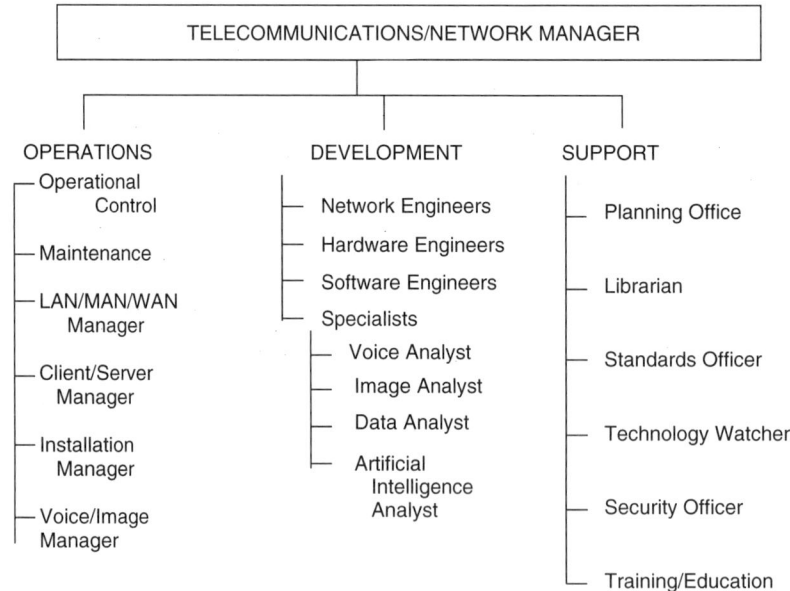

Figure 9.3 *Organization of telecom and networking*

small organizations, the voice and image processing specialist may be rolled into one person specializing in pattern recognition, a subdiscipline of AI, Artificial Intelligence. Alternatively, there may be an AI specialist.

In organizations where such specialists cannot be supported for lack of substantial projects, a consultant is engaged for the purpose. The consultant could either be on a retainer or be engaged on an *ad hoc* basis for consulting on important, complex and expensive decisions. Otherwise the consultants and full-time professionals could be organized on a project basis. Organization by project is not unique for IT and is used for systems development especially in large projects that use the SDLC, Systems Development Life Cycle. Even with prototyping used for development, the project approach is desirable because it involves end-users and management as participants and has a better chance of good problem specification, project implementation and eventually project acceptance.

The development staff as well as the operational staff have supporting personnel that include the Planning Officer, which in some cases may well be the Network Manger or Assistant Manager. Then there is the Librarian, a person not unique to IT. So also with the Standards Officer, who unlike duties in, say, Systems Development is not mostly generating standards but enforcing them. Telecommunications and networking is a service that is shared by many others and so standards have to be national or regional if not international. This subject is important enough to deserve a separate chapter for itself, Chapter 11.

Security is also a problem known to all of IT, but security in telecommunications is very unique and very important. In these days of LANs, MANs and WANs, the danger of security breaches through communication channels is both serious and complex. It is the subject of Chapter 12. They are related to the operation and management of networks, the subject of Chapter 13.

Another support function of any department of networking (and telecommunications) is that of training and education. All these functions have to be planned. The Director/Manager of Networking will participate in this task and in small organizations may be totally responsible for it. This task will be the focus of the remaining part of this chapter. In it we will examine the nature of planning, the dynamics of planning, its process and its implementation.

Planning for networking

Planning is the visualization of the future and taking steps to achieve and strive towards this vision. But is planning for networks any different from corporate planning or even planning in IT? The answer is yes, and no. Yes in terms of the concept and process, but no in terms of the inputs and outputs, dynamics and uncertainty involved. Take for example the IS department. We select this department since it is the highest paid in IT for many corporations and is responsible for all if not most of the mission critical corporate applications. For example, the payroll is critical and affects everyone but this criticality occurs only once a month or perhaps once a week. But for some firms, networking is critical during all working hours as for example the reservation system for airlines or hotels or car rentals. For a firm with world-wide sales, production or suppliers, the working hours may be most of the day, and, for teleworking, it is all hours of the day all year round. And this criticality may be even more extensive as e-mail (and voice mail) and ISDN become more ubiquitous, not just in the corporate world but in all our daily lives. We cannot do much without communications. And so networking is very important and critical.

One problem with planning for networking is that it is so dependent on technology which is so unpredictable and unstable. In IS, we have a stable technology and a fairly predictable response from end-users. The telecommunications and networking world is far more volatile and dependent on international standards and governmental actions. The dynamics are different.

Planning variables

A problem in planning is the transformation of long-range goals into strategic and operational objectives. This is especially difficult with a telecommunications and networking (and computing) technology that is often unpredictable. For example, consider the applications environment postulated in the case stated early in this chapter. Does that include image processing? Not explicitly, but if you do know that the firm is in the manufacturing business with factories and suppliers spread around the world, you should predict that the firm will want to send images in the form of drawings and blueprints on the network. Should you plan for image processing?

Why not ask management? The answer is that IS has taught us that management does not always know, and if it does know it cannot always articulate its needs in operational terms that a planner can use. There are numerous cases when management has been asked about an application and has responded, 'No, we will not need that application.' A little later the same managers demand that application, and if you remind them of their denial of the need of that application, they will quickly reply, 'Ah, but you should have known my needs. That is why you are paid such a high salary.' And so it becomes incumbent on the planner to predict the future and do so correctly, or at least close to being correct.

Dynamics of network planning

The main players in networking are the governmental agencies, standards organizations, the telecommunications and computer industry, and, of course, the corporate world and its end-users. Their interrelationship is shown in Figure 9.4

The important interactions are between the corporate entities (with their end-uses) and the industry (computer and telecommunications). The industry offers a product and the firms and end-users accept or reject it. Historical

examples include the picture telephone that was demonstrated in the 1965 World's Fair, but the public was not ready for it. Even today, the integration of picture and voice as well as voice along with data is not fully available or accepted in the market-place.

A more recent example is with PCs. The industry offered large mainframes but the end-user wanted the PC. When small firms responded, there was demand for them to be integrated. And so increased the demand for LANs. Meanwhile the end-user became more computer literate and more demanding. They wanted decentralization and a distribution of computing power and wanted it to gravitate to the desktop with them being the clients with access to computers acting as servers of databases and other computing services. And so came about the client–server paradigm which the industry eventually recognized and then offered the necessary technology for servers and clients.

Technology is often driven by the industry and responds to market forces and demands from the industry. Thus the forces of change are both upstream and downstream. But often, and in many countries, technology is driven by the availability of international standards without which there is less incentive to produce products that will have international acceptance. Technology

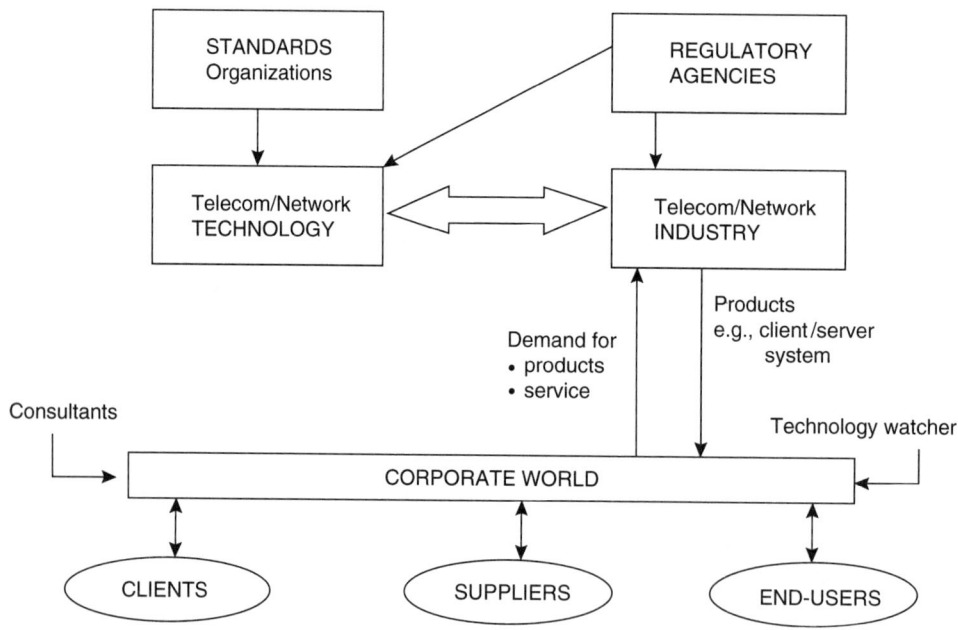

Figure 9.4 *Main players in network planning*

is also driven and influenced by governmental agencies depending on the country. In the US, the intervention is not direct but indirect through funding for academia and industry especially in the area of space exploration. In Japan the intervention is more direct with seed-money to industry and the selection of products to be produced and market-share to be achieved. In other countries including Europe there are the nationalized PT&Ts and intergovernmental agencies that have considerable influence. We shall discuss these dynamics in a later chapter, Chapter 15, but for now it is sufficient to say that the interrelationship exists and does influence the external variables in the planning process.

Planning process

The planning process for networking is conceptually much the same as for IT and corporate planning. Planning is done for varying planning horizons each feeding into the other. Thus the long range plans provide objectives and goals that are to be achieved by strategic medium range plans which in turn dictate the short term plans through instruments like budgets. This process is shown in Figure 9.5. The inputs to this process are many and summarized in Figure 9.6. They include data on external events as much as on the internal environment of goals, objectives and constraints. There is a dependence on advice

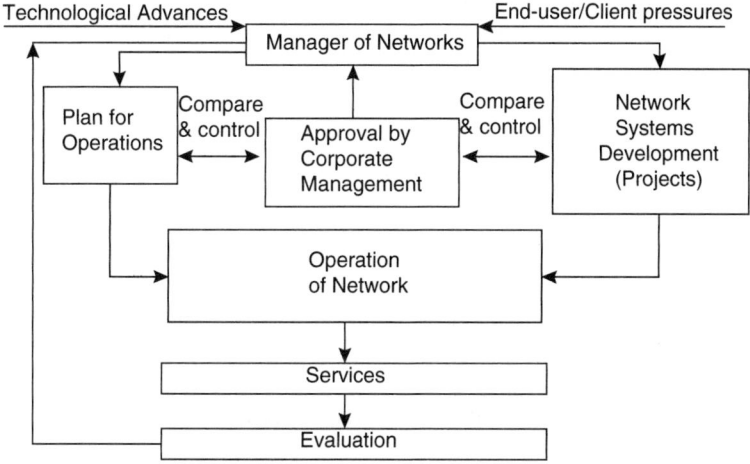

Figure 9.5   *Implementation of telecom/network plan*

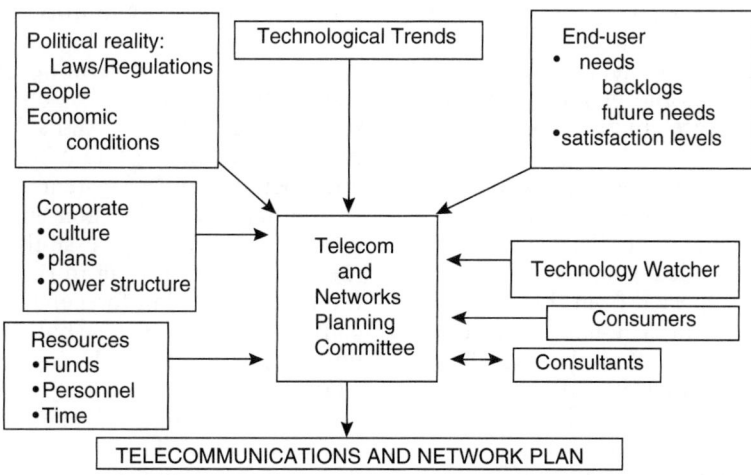

Figure 9.6   *Inputs to network planning*

from **technology watchers** and **consultants**. The consultants can be a source not only of objectivity but also of a *Weltanschauung* and global view not often found within an organization. The consultants need to be experts not just on technology but also on the acceptance behaviour of the end-users and customers of networking.

Implementing a plan

There is often a problem in going from a long range set of goals and objectives to a medium term and short term set of operational objectives. This problem occurs because the manager tends to make statements in rather general terms such as 'very reliable and accessible' and 'high quality and performance'. For one thing the variables are not too operational. What is meant by 'reliable' and 'accessible', and what are 'quality' and 'performance'? The other problem is to specify what is meant by 'very' (accessible) and 'high' (quality and performance). Lofti Zadeh calls these identifiers **fuzzy** and has an elaborate procedure for converting these 'fuzzy' variables into numerical values (Partridge and Hussain: 1994: pp. 242—53). It is unlikely that the typical manager will have the time to convert networking goals into numerical values, but the planner could state these goals in operational terms. For example, reliability could be stated in terms of delays or response times, like an average delay of no more than 30 seconds or an average response time of 1 minute from request completed to start of response. The values could be in ranges like the average response time of 1—2 minutes. And this may be the start of a dialogue till convergence is reached. This may take a few iterations and one or two plans and planning sessions before the fuzzy values are converted into more operational terms.

An example of operational objectives for networking is shown in Table 9.3. It incorporates the long range goals postulated in the Case Study stated earlier. It has added some applications based on what can be expected from the clientele and technology. For example e-mail and voice mail may be desired applications by the clientele but may well be dictated by technology and available whether you want it or not.

The performance variables are mostly what is known in IT or have been mentioned in an earlier discussion. One variable **survivability** may need

Table 9.3 *Possible objectives of network administration*

Systems applications
Should allow for the possibility
  of the following applications:
Distributed processing (e.g. client—server
  system)
    E-mail
    Telecommuting
    Image processing
    Voice processing
    Video processing

Desirable systems characteristics
High speed access to LAN/MAN/WAN
Easy access to LAN/MAN/WAN
Access should be for asynchronous
  system (i.e. real-time systems)
Masking of complexity for end-user
  (end-user friendliness of system)
Delays and average accesses per
  successful attempts should be
  below a specified threshold
Low down-time
System should be cost-effective
Systems should allow for:
    expansion
    varying traffic pattern
    maintainability
    security
    survivability

a definition. It concerns the ability to operate despite a given probability of the presence of disruptive and dysfunctional influences. The concept is related to **reliability**. If the system is a real-time system, then high reliability is important and this will determine the topology to be used. You may recall that one topology is more reliable than another. And so the objectives of a plan do have an important bearing on the design and operation of the system.

Another design decision will relate to accessibility. If, say, high accessibility is required and there is a new building in the design stage, then the long range plan for networking should be to consider making the new building an 'intelligent' building with network connections in rooms where professionals would be working. But that is in the long run. How about the medium term and short term? The answer will depend on the distribution of end-users and resources and the computer literacy and attitude of the end-users.

If there is a potential for many end-user clients and many servers of databases and computing resources that are distributed, then we may have the potential of a client—server computing paradigm. It is such computing paradigm that we will consider in our next chapter.

Summary and conclusions

This chapter concerns the front-end of network management. We use the term network management for telecommunications management, which includes not only the network but also the hardware, software and all the necessary interfaces required.

In this chapter we examined the options for the location of the network administration in the organization chart and discussed some of the alternative organizational structures for the network administration department. We also identified many of the functions of network administration and gave forward references to chapters in this book where these functions are discussed in great detail.

One of the functions discussed in this chapter was that of planning. Planning has many inputs. These are shown in Figure 9.6.

We recognize that telecommunications is strongly linked to market forces and the industry of telecommunications and computers. They in turn under pressures and influence of governmental agencies as well as standard organizations, both national and international. These are top-down and are in addition to the bottom-up pressures from the end-user and corporate management to mask the complexities of the systems and make it end-user friendly without any loss of efficiency or effectiveness. This is a difficult act for the Network Administrator.

Case 9.1: Headaches for network management

A survey of 427 corporate networkers at large global companies were asked to list the 'risks and threats' they felt. The percentage responses are as follows:

Unauthorized systems access	28%
Viruses and malicious code	24%
Password exposure	24%
Internet access	23%
Disgruntled employees	21%
Information leaks	18%
Use of laptops	13%
Natural disasters	13%
PBX fraud	8%
Hackers	6%
Terrorism	5%

Source: Datapro Information Services Group (Datran, NJ, USA), printed in *Data Communications*, Jan. 1995, p. 14

Bibliography

Adler, J. (1989). The case for centralized LAN management. *Telecommunications*, **23**(6), 58—62.

Chiaramonte, J. and Budwey, J. (1991). Finding and keeping good people. *Telecommunications*, **25**(2), 55—59.

Davies, J.E. (1990). Professional development and the institute of information scientists. *Journal of Information Science*, **16**(6), 369—379.

Desmond, C. and Templeton, R.S. (1989). Planning a business communications network. *IEEE Network*, **3**(6), 8—10.

Doll, D.R. (1992). The spirit of networking: Past, present, and future. *Data Communications*, **21**(12), 25—28.

Flanagan, P. (1994). Taming the network: how are telecom managers coping with change? *Telecommunications*, **28**(8), 31—32.

Healy, P. (1991). Good advice, bad advice. *Which Computer?* **14**(6), 76—88.

Mingay, S. and Peattie, K. (1992). IT consultants — source of expertise or expense. *Information and Software Technology*, **4**(5), 341—350.

McLean, E.R., Smits, S.J. and Tanner, J.R. (1991). Managing new MIS professionals. *Information and Management*, **20**(4), 257—263.

Partridge, D. and Hussain, K.M. (1994). *Knowledge-Based Information Systems*. McGraw-Hill.

Torkzadeh, G. and Xia, W.-D. (1992). Managing telecommunications by steering committee. *MIS Quarterly*, **16**(2), 187—196.

THE CLIENT–SERVER PARADIGM

In the 21st century, it will no longer be sufficient to put computers into environments. They must be part of the environment.

Bill Joy, Vice-President of Micro Systems

Introduction

In the early days of computing, computers were expensive and computer personnel were scarce. Development and operations were centralized. Soon, there was a backlog of applications and a lot of 'noise' between the end-user and the computing providers. A reaction was towards decentralization into distributed processing where computing was decentralized while much of the centralized computing shifted to the nodes and later to the end-user. Meanwhile, there were two important technological developments. One was the arrival of the PCs in the early 1980s with the promise of performance/price ratios higher than the mighty mainframe and yet small enough to stay on the desktop. But, the PCs were isolated from other computing resources like expensive peripherals and the corporate database. In the mid-1980s came the next computing development which was the LAN (Local Area Network). This enabled PCs (and other computers) to 'talk' to each other and share the scarce resources of peripherals and databases. The LAN provided interconnectivity. What was also needed, however, were computer processors that would facilitate sharing and serve out these resources. Thus came about the servers. The file server was for data sharing and the printer servers were for sharing printer resources. But, soon this was not adequate. The number of PCs rose dramatically even within one organization. The number of end-users and clients rose too. Processors ranged widely in power and price. Peripherals became more versatile, requiring not just printer servers, but also image processing servers, and *voice* processing servers. These servers, with the database and application programs residing on them, were accessible to any PC, mini or mainframe through a bus or a LAN. Such a client–server system is shown in Figure 10.1.

There were many limitations to the PC-centric approach. One was the lack of adequate administrative control and systems management tools. Also, there was a high cost of data swappage and network traffic as well as the inefficient use of the processors both at the end-user end and at the server end. A solution was to distribute the load and responsibilities between the processors at the front-end, (the client) and the back-end (the server). The client–server paradigm that was thus born is the focus of this chapter.

We start with an operational definition of the client–server system, identify some of its functions and advantages, and list its many implementations and success stories. With this background as motivation, we shall examine the components of such a system: the hardware (processor and user interfaces); software; communications facilities; the human resources need; and, finally, the organizational environment necessary for the client–server to be successful. We shall also examine the concern of corporate management with 'downsizing' and 'right-sizing', and consider the advantages of the client–server paradigm as well as the obstacles involved in its implementation. We conclude with a discussion of the client–server paradigm used by many end-users for cooperative processing.

Components and functions of a client–server system

An overview of the client–server approach is shown in Figure 10.2. It is compared with the

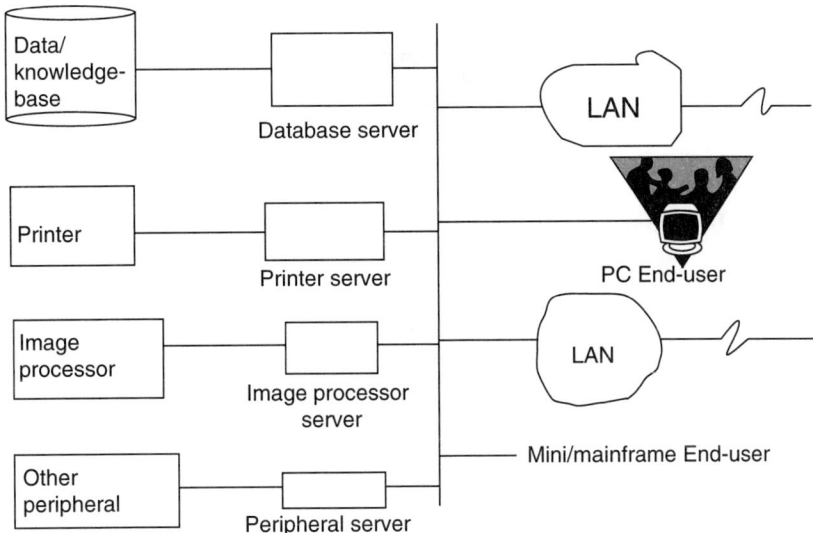

Figure 10.1   *Schema for client—server system*

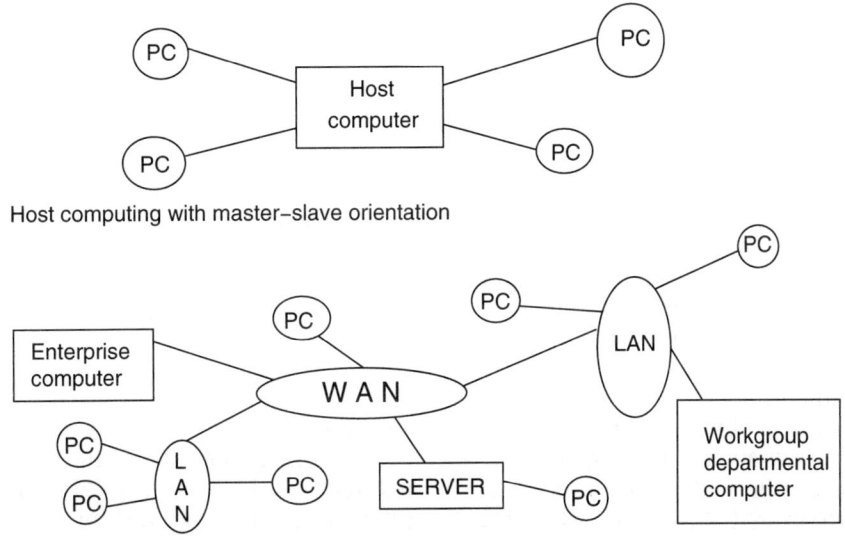

Host computing with master–slave orientation

Distributed networking with client–server orientation

Figure 10.2   *Host computing vs. the client—server system*

earlier configuration of a host acting as 'master' and the connected PCs (or minis) acting as 'slaves'. With the client—server approach, each PC is independent for local processing and sharing the centralized enterprise equipment (especially servers) through a LAN. In the case of widely dispersed clients and servers, the access could be through a MAN (Metropolitan Area Network) or a WAN (Wide Area Network). The corporate database and enterprise-wide application programs typically reside with the server and are accessed by the end-user through the client processor, the LAN and the server, as illustrated in Figure 10.3. Each of the components of the system will now be discussed in turn. (The section numbering will correspond to the numbers in Figure 10.3.)

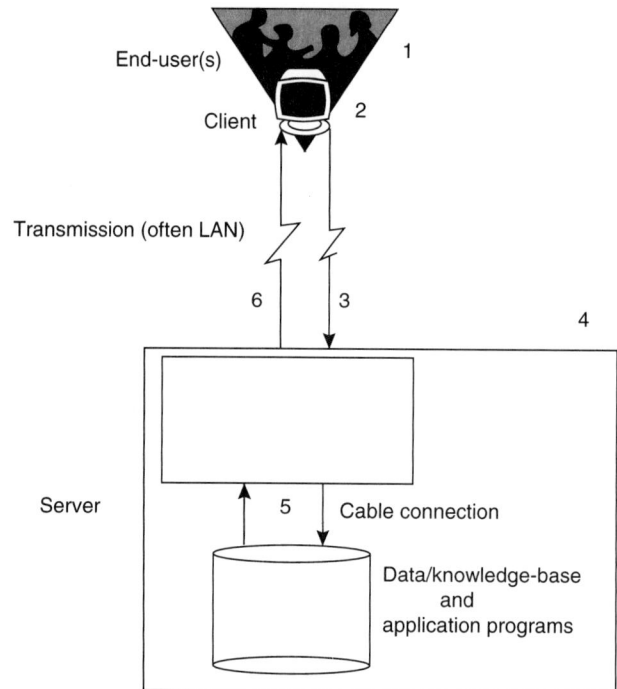

Figure 10.3 *Components of a client—server system*

THE USER

The user is typically the end-user, who accesses the client for service. The end-user could be a corporate manager, a professional or employee of the corporation, or it could be customer. Here is where some confusion can arise. The customer in business and commerce is called a client, but this client is a human, not to be confused with the client in computing which is a processor. We shall have more to say about the user and end-user when we reach the end of the process of using a client—server system, that is, when the user or end-user receives the output and results from the system. First, however, we discuss the client.

CLIENT

The client could be a powerful processor or a dumb terminal with no processing capability. Typically, it is at least a PC with its own operating system. Most of the processing is often done by the server and the division of work is determined by a computer program, which is why programs for a client—server system are different from those of a traditional transactional system. Thus, the client—server system is an enabling technology with applications written specifically for it. Sometimes this means rewriting existing applications. Whether this means all applications or selected applications or even 'mission critical' applications will depend on the confidence in the client—server system and the propensity for risk on the part of IT management.

The applications may include message processing like e-mail; accessing local files and databases (with or without a DBMS) for browsing and local computing; computing on server-supplied data; and (through the server) the sharing of computing resources like the corporate image processing system, an optical character recognition system, advanced graphic processing systems, a colour plotter, or just a fast laser printer. Furthermore, these peripherals may belong to a variety of vendors. Given a well designed application, a powerful server and a well tuned fast network, one can have quick results with the entire system behind the client being totally invisible and transparent to the end-user who does not need (or care) to know where or how the processing is done.

To facilitate the query processing from a client, most client–server systems use a Structured Query Language (SQL) which is a high level dialogue language competing in its end-user friendliness with many a 4GL. SQL is compared with its equivalent in English in Figure 10.4.

SQL, along with its relational database (residing with the server), is almost a *de facto* standard for client–server systems, though more recent semantics of the access language is increasingly object-oriented. Meanwhile, SQL does allow access to multiple database servers providing a wide range of decision-making information and knowledge residing in remote and dispersed locations.

An important component of a client processor is its **UI**, User Interface, through which the user communicates. For a user such as a programmer, the UI need not be very user-friendly, but for an end-user it had better be friendly or else it may not be used at all. A **GUI**, Graphical User Interface, is the more preferable for the end-user because it enables access through graphical icons rather than through programming commands. In the future, a GUI may well be able to accept and deliver information not just in terms of numbers and even text but also as graphics, voice, video, images and animation, thereby becoming a multimedia terminal. However, the more facilities available, the more powerful the client processor needs to be. For simple data manipulation, even a laptop would suffice, but for more complex and large amounts of processing, a workstation may be appropriate. However, a workstation can be three or six times as costly as a PC depending on the 'bells and whistles' (features) attached. At the least, the workstation is faster and has more memory and access to more programming languages depending on what functions have to be performed. If it is a general purpose workstation, the most popular business programs would be the word processor, the spreadsheet, e-mail, DBMS processing and business graphics. For a DSS (Decision Support System), greater speeds, computing power for, say, business simulation computing and access to programming languages, including simulation languages like SIMSCRIPT, are required. For a KBS (Knowledge-Based System) like an expert system, AI languages like Prolog and LISP will be necessary. For an engineering workstation one needs not only computing power but also powerful graphics capability for applications like CAD (Computer Aided Design) and systems simulation.

A solution to the expensive workstation is to have an **X terminal**, which is essentially a workstation without a disk. It is designed specifically for the type of processing and functions of a client–server system with the focus on network communications, graphics performance and an end-user friendly interface. The X terminal is actually an 'application-specific' workstation optimized for running the X protocol. The X windows protocol 'can support any number of windows with any type of font or window size. Also, the X terminal's bitmapped screen allows the application to display all sorts of data formats (text, images, drawings, and so on) simultaneously, thereby enhancing the user's productivity'. (Socarras, *et al.* 1991: p. 52).

An X terminal can be connected directly to a LAN, through a workstation by cable, or with more than one X terminal attached to one workstation.

The X Windows system is the *de facto* standard for the larger machines in the UNIX environment. For the PC, the *de facto* standard is Microsoft Windows. These *de facto* standards provide network transparency so that an application run on a server thousands of miles away will appear on the screen of the client as if it were running on the local client processor.

'Ironically, the X terminal's greatest virtue is the lack of functionality. By having no programmability, no local diskette drive, it is impossible for anyone to introduce unqualified software at the desktop. As a result, all software on the application server is installed under the quality assurance of the IS organization. In addition, the inability to download files and copy them to a diskette, improves the data security of the network.' (Connor, 1993: p. 52)

It is about time to discuss a network.

```
English: Find the number of employees
working for Mr. Smith and making at
least $40,000 annually
Query for Client processor:


    SELECT NAME
    FROM PERSONNEL
    WHERE MANAGER='SMITH'
    AND SALARY<40000
```

Figure 10.4 *English and its Query language equivalent*

Network and transmission

The server and the client can be connected together by hardwire. However, when they are widely dispersed, they must be connected to a LAN, which in turn is part of a MAN or WAN.

This enables a corporation to have an enterprise network that may be strung around the country and yet operate individually, in a workgroup, or at a department or local level. For this to occur, it is important that there is interoperability, i.e. the operation and exchange of information in a heterogeneous mix of equipment and software using the network. The essence of openness is that there is interchangeably of components of the system and therefore vendors have to compete with better products and better service. Both customers and vendors benefit from interoperability and the resulting competition.

Interoperability also implies an open network architecture. The earliest architecture was the SNA but this was proprietary and not open to other vendors. Some, especially those in Europe, objected and supported the OSI proposed by the ISO, the International Standards Organization. But ISO did not get much acceptance in the US where SNA was accepted by IBM users whilst TCP/IP had many supporters in the UNIX User's Group. Thus we do not have one universal open system but the systems available are not closed either.

One very desirable feature of a client (and a server) is that it has an open architecture, which ensures interoperability, that is the hardware and software can operate interchangeably on each other's equipment despite their being non-homogeneous. This is not the vendors' preference because they would like you to acquire all your equipment and software from them and have a captured market. In conflict with the vendor, it is in the interest of the customer to have the flexibility to mix and match components of the client—server system.

SERVERS

Connectivity, though very important, is not all that is needed for efficient and effective sharing of computing resources. This was recognized with the increase in the number of PCs when each manager of a PC could not afford the database and peripheral resources needed. The importance of resource sharing became obvious. This was achieved in the mid-1980s with servers, which is software that enables and controls fast and easy access to databases and application programs.

Databases evolved into knowledge-bases and increased not just in number but greatly in types of models and complexity. Also, there was often the need to share software, not just applications software, but also compilers for programming languages, such as the one for SIMSCRIPT 11 for the occasional user of simulation. And there was also the need to download some of the applications from the minis and mainframes to the client PCs. All this included an increased recognition of the importance of the client and the end-user, who demanded not just fast and easy access, but also a user-friendly environment that facilitates and even encourages the sharing of resources.

The new paradigm of sharing computing resources has its own demand on specialized supporting resources: it requires a network server OS (Operating System), multiple user interface, sometimes a GUI (Graphical User Interface), a dialogue oriented client—server language (such as a version of SQL) and a database architecture. Since the resources are spread out spatially, not just in one country but also abroad, there is the need of fast and reliable telecommunications and interconnectivity. And all this must be implemented while the hardware/software platform and the communications technology being used is transparent to users. The user must pay too by following the procedures and protocols required by the systems. The users must also learn and observe some of the rigidities for using the standard user interfaces and interface languages.

Much of the software required is available from vendors and software houses and the client—server systems vary greatly in emphasis and capabilities. They may be mainframe-centric (centred), PC-server-centric, or may have an emphasis on data/knowledge-base distinct from being communication-based. But, despite the availability of software packages, there is often a need for in-house software development. There is also the need to integrate the client—server systems with existing information systems and to use the system not only independently as an end-user but also to work cooperatively among groups of end-users.

The server is a processor, conceptually very much like the client processor. Yet, it is very different. For one thing it does not have a UI, let alone a GUI. It is designed for networking, database processing and applications processing. It is different from a general purpose processor and is sold as such: a server processor. The server may differ as to their responsibilities and functions. For example, the server may act as a repository and storage of information, in which case it is a **file server**; or, it may perform data retrieval, in which case it is a **database server**.

Which type of server is desirable depends on the needs and objectives of the system. In any case, the server must be able to do multitasking (perform multiple functions simultaneously), use multiple operating systems, be portable, have **scalability** (the ability to upgrade upwards without loss of software performance), and have a fast response time despite the time required for teleprocessing. Because of these capabilities, servers are much more expensive than a client processor, in the range of $20 000 to $50 000 at 1995 prices.

Why are servers so expensive? Because they perform many functions. These functions include:

- Network management;
- Gateway functions, including access to outside access and public e-mail;
- Storage, retrieval and management of documents;
- File sharing;
- Batch processing;
- Bulletin Board access;
- Facsimile transmission.

The platforms for sever processors are PC-LAN servers with mainframes and minicomputers as alternatives. The server platform and issues relating to the servers are hidden from the end-user, but there are issues which include: internetworking; disk space utilization; data management; gateway and other access control; backup and recovery; and fault tolerance. Another important issue is the management of data (and knowledge if any).

Database processing

By processing data at a server instead of at a mainframe there are some principles of processing data that are applicable. These include the issues of integrity, security and recovery of data. The enterprise data that is a corporate resource needs to be unified and integrated; access to data must be controlled to maintain security; and recovery must be possible after systems failures. This should be done without affecting the access for legitimate sharing and without an uncontrolled proliferation of islands of data cropping up everywhere. Access optimization, security utilities and I/O (Input/Output) handling would vary with servers manufactured by different vendors, but should not impact on the consistency of data and the responsiveness for the end-user. This improved responsiveness is important, especially for a query which could now be a fraction of a second instead of 3—5 seconds under a mainframe, thus allowing the processing of customized 'wild' queries. Most of the processing, however, is data entry and could still be done in batch with updating done in the day and processing done at night and transmitted by LAN for use in the early morning at the local client end. This enables multiple users (such as, say, small businesses or professionals like doctors) who cannot afford their own processing to use a client processor.

Much of data management (and resource sharing) is automated. Some of this is done through a DBMS that resides with the server which controls access between multiple processing systems (and even multiple distributed databases) and integrates data access with network management.

Applications processing

Data is used by application programs and most of these reside with the server. There are some off-the-shelf client—server applications now available and these are increasing in number and scope. Still, however, many applications must be painfully developed from scratch. Application tools and lower prices have made client—server systems development more competitive in terms of cost performance when compared with mainframe and minicomputer development. But the client—server application development does differ from traditional software development in some significant ways that include:

1. Processing functions are distributed between client and server. The front-end client portion is run by end-users using languages like SQL that have simplified data request protocols and extract data from whereever it might be located, whatever computer stores it, and which ever operating system controls it.

105

2. UI and more often GUI are used because end-user friendliness is still very important, if not most important to the end-user.
3. Advanced networking, mostly LANs, are used.
4. 4GLs and code generators are used extensively, though OO methodology is being increasingly used.
5. Development tools like SQL Windows, FLOWMARK, Progress, ObjectView, and Uniface for OS/2 are emerging.
6. CASE tools are being used with Rapid Prototyping.

SERVER BACK TO CLIENT

Once the applications are processed and data retrieved as per request by the client, the results are sent back the way they came, through the LAN and the client processor. There the application results may have to be formatted for better display or the data retrieved has to be processed. All this is done by the programs residing at the client. This is then given to the user through the UI, User Interface. Thus far we have examined simplified line diagrams of the client−server approach. A formal schematic diagram of a real-world application is shown in Figure 10.5.

Organizational impact

A user who may be a computer programmer, unlike the end-user, may well be less than enthusiastic with the client−server paradigm. Why? Because the programmer is concerned (justifiably or not) about the loss of control over development, processing and use of information. Also, like the DBA (Database Administrator), the programmer is concerned about the loss of integrity and security of data. The programmers and the DBAs are supported in their concerns by the Manager of IT, but for an additional reason. The manager is concerned about a loss of power through a reduction of span of control and a loss of control over the acquisition of resources, powers and responsibilities that now gravitate to the department administrating the client node.

While the professional computing personnel (including IT management) are not enthusiastically in favour of the client−server system, the corporate management is enthusiastically in favour. Why? Because there is an opportunity of reducing costs at the centre. Some cost components increase and some decrease, but the net is a decrease of overhead costs of computing at the centre. There are other organizational consequences of the client−server systems: downsizing.

The client−server paradigm is largely minicomputer- or microcomputer-centric and is competing well with the mainframe-centric systems where most of the 'mission critical' applications reside. However, the costs for the client−server system is higher than the mainframe in total costs though the initial costs are lower. The annual costs per user in the fifth year (with 3584 mainframe users) in one study came to $1484 for the mainframe and $2107 for the client−server system (Semich, 1994: p. 37).

The downloading of computer processing from the mainframes (and minis) is referred to as downsizing by corporate management. It is the decreasing of their overhead costs and is downsizing for them. But, downsizing in management parlance is the reduction of the labour force and the firing or displacement of personnel. This is popular in recessions and bad economic times while upsizing and expansion is popular in good economic times. The right size and right timing is referred to as right-sizing. But, in terms of computing resources there need not be any change in labour force. There is a reduction of overhead computing costs at the centre, but not necessarily a reduction of total costs. What we have instead is a shift of costs (and responsibilities) to the client nodes. This shift is welcome to corporate management at the centre though they do share the concerns of losing control over the development and processing of information.

Support for the client−server system comes from the end-user who can see the reduction of time and 'noise' and an increase in responsiveness in the development of systems. The end-user gains greater control over the system. End-users are now more computer literate and experienced and no longer fear the computer which is becoming more robust and friendly. They are willing to take the responsibilities of controlling the client-end and leave the server-end to the centralized processing centre. This coalition of corporate management and end-user with help from the computer industry (in the delivery of appropriate systems) will overcome any resistance to the client−server system which will be increasingly implemented throughout business and industry.

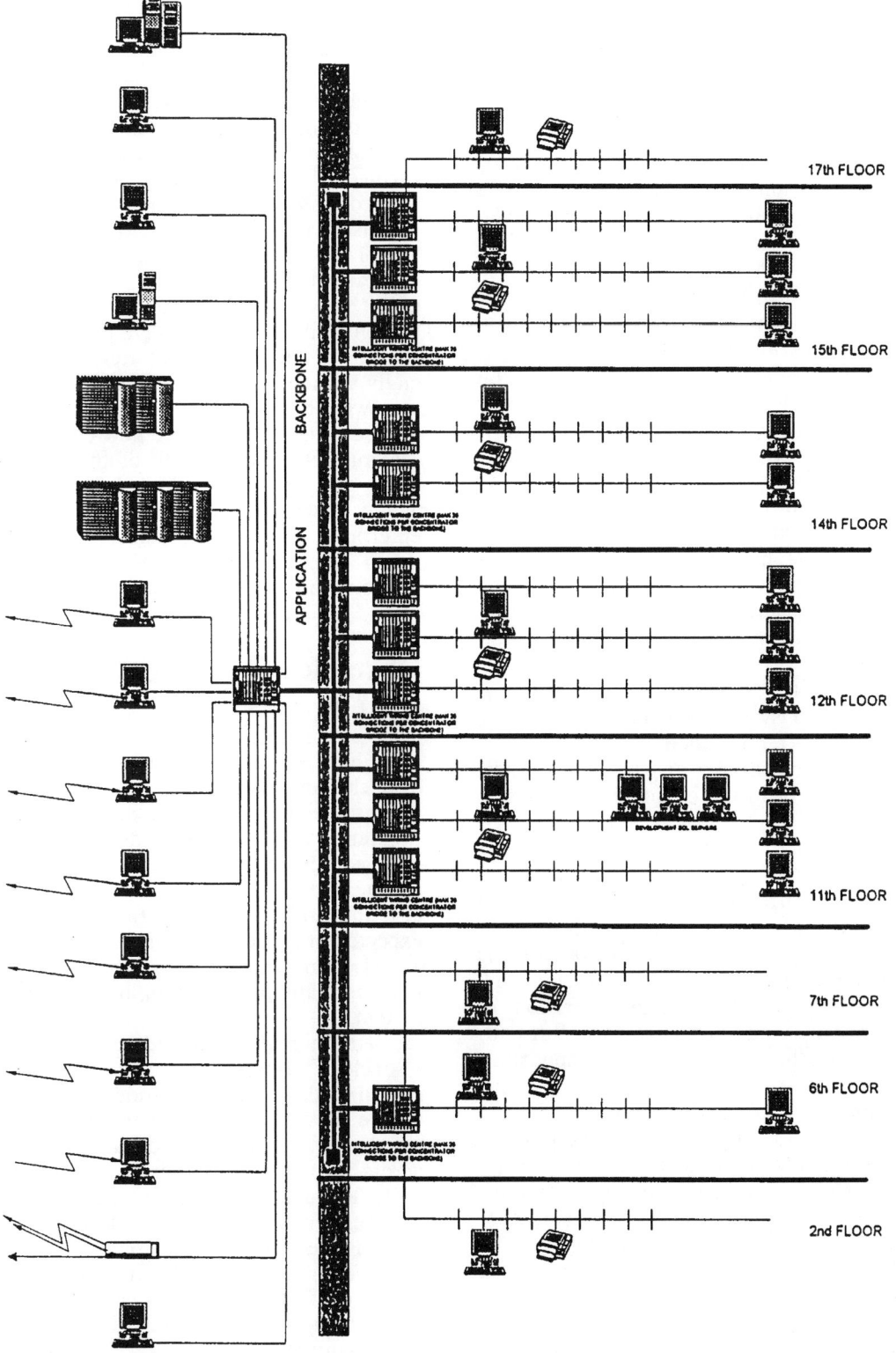

Figure 10.5   *A real-world application of the client–server system*

There are three alternatives in implementing a client–server system. One is to go vertically and implement all the levels of implementation, but for only one business at a time. However, implementing, for example, a LAN for one business does not make economic sense because once implemented a LAN can be used by other applications. The other approach is to implement horizontally, i.e. the LAN, the server, the clients, applications and the end-users (their training). This approach is also flawed in that some resources lie idle while others are being developed. The third alternative is to do all at once, horizontally and vertically.

It is likely that the third approach is often not feasible because of the high investment necessary in money and skilled personnel. It is also highly risky. The less risky approach is to implement vertically and do so on a pilot study basis. The first implementation in one case cost $2 million, but then the next implementation of similar scope was almost half the cost. So, there is a learning curve involved. Also, in implementing a pilot study, it is desirable to select an application that is less risky and has low costs and low visibility.

Bad economic times and recession are allies for the client–server system. Management wanting to downsize and reduce their centralized overhead will be more willing to give up some of their centralized control to the client, the server and the end-user.

It will be sometime before IT personnel and corporate management are fully comfortable giving up their centralized processing, which in many cases is fairly well stabilized with its security regime, its back-up and recovery practices, its fault tolerant mechanism, and an experienced IT staff running the development and operations of their information and knowledge base. But end-users are taking the initiative and the responsibility. They want more control over their operations and the time required for development.

Downloading to the end-user's desktop is perhaps inevitable, especially with increasing open architectures in hardware, open systems in telecommunications, and the refinement of multimedia equipment. The technology will work in favour of the end-users and the corporation and make the computer industry more competitive and responsive to the end-user's needs.

Advantages of the client–server system

The advantages of a client–server system are context-sensitive and depends on your vantage point. If you are a corporate manager you will be thrilled with the advantages of downsizing and the reduction of costs at the centre. If you are an end-user you will be pleased with getting control of the system even though it may mean that you cannot blame computer personnel for things going wrong and instead have to take certain responsibilities for the system (at the client-end). If you are one of the computer personnel, especially at the management level, then you will be somewhat happy at your reduced responsibilities for the system but you are now concerned about the integrity and security of the system, especially that of the data/knowledge-base. On the whole, viewing the system from the point of the enterprise, there are more advantages overriding the disadvantages, risks and limitations involved. These advantages are summarized in Table 10.1.

Obstacles for a client–server system

In achieving the advantages, there are some obstacles and risks. Some have been implied in the discussion of the organizational implications: problems associated with downsizing and the need for greater coordination and cooperation among the end-users of the system. There is also the problem of resistance, which must be expected of all new technological changes. There are also problems of conversion, especially the need to train personnel in the use of the new system.

Training may be considered a technological obstacle. End-users have to be trained not only on using the client and knowing the functions of the server, but end-users need to be educated about networking and trained in navigating across the LAN and perhaps even the Internet. Other technological obstacles are the lack of tools of development and products of the client–server system; the lack of methodologies; the shortage of experience in the planning and implementation of a client–server system; and the lack of national and international standards relating to the equipment and operations of the system.

Table 10.1 *Advantages of a client—server system*

Reduction of responsibilities and cost overhead

Better local cost control of operations and development (original and modifications)

Faster response time to requests for processing

Greater access to corporate data/knowledge otherwise maintained in a highly protected and centralized data structure. The client—server system strips data off transactional systems and stores it in the server to be shared for analysis and even manipulation locally

Enables distribution of processing from centralized to desktop computing

Offers cooperative processing between individuals and group departments across organizational boundaries, geographies and time zones

Rewriting systems for the client—server system is often an opportunity to purge obsolete software from the application portfolio and to consolidate and integrate the system, and make it more efficient

Offers more friendly interfaces for end-users, especially knowledge workers and customers

Greater involvement of end-users in IT implementations

The open architecture and open systems offer flexibility in choosing different configurations of hardware, network and DBMS from multiple vendors

There are greater possibilities for expansion by adding hardware (even laptop computers) to networks without replacing existing hardware. The plug-and-play possibility applies at least in theory when parts of a system can be replaced without impact on the rest of the system

Table 10.2 *Obstacles in the way of a client—server system*

Organizational

Lack of personnel skilled in the client—server system and in networking

Resistance to change and new technology

Risks of downsizing

Costs of conversion

Need for greater coordination and control of more end-users

Technological

Need for LAN/WAN infrastructure

Lack of skills and equipment resources

Lack of methodology and experience in planning for a client—server system

Lack of client—server products and tools of development

Lack of client—server applications

Lack of national and international standards for the client—server paradigm.

The resources and infrastructure needed for the client—server paradigm is displayed in Figure 10.6, where we see that an end-user navigates through the client facility, the network infrastructure, the server facility, and then to the applications and data/knowledge-base. Having done the desired computing, the reverse path is taken back to the end-user.

The end-user can do some local processing on a local database at the client facility and also use the many software productivity tools such as word processing, spreadsheets and e-mail. Some of these productivity tools, like e-mail, will be used more intensively in cooperative processing which connects the clients horizontally and improves horizontal communications. Productivity tools extend communication to different times but the same place, as well as different times and different places. However, cooperative processing does increase the cross-currents between end-users and thus raises new problems of collaboration and cooperation in intercommunications and may lead to adhocracy.

We conclude with a view of future paradigms of client—server systems moving from the Ethernet era (of file servers, to database servers, groupware and transactional processing monitors) to the intergalactic era of the late 1990s and early 21st century with object-orientation in distributed processing (Orlafi *et al.*, 1995:

These obstacles and others have been summarized in Table 10.2 and can be overcome only by close cooperation between IT personnel, corporate management and end-users.

Summary and conclusions

Alok Sinha has described the client—server system well:

> …one or more clients, and one or more servers, along with the underlying operating system and interprocess communication systems, form a composite system allowing distributed computation, analysis, and presentation.

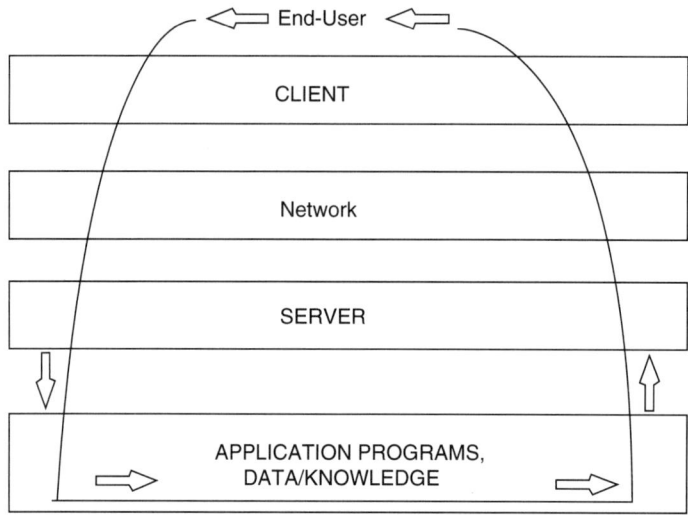

Figure 10.6 *Navigation in a client—server system*

Table 10.3 *Paradigms of client—server systems*

1982—1986	FIRST WAVE (Ethernet Era)
	File Servers
1986—1995	SECOND WAVE (Ethernet Era)
	Database Servers
	Groupware
	Transactional Processing Monitors
1995—20xx	THIRD WAVE (Intergalactic Era)
	Object-Orientation in Distributed
	Processing

p. 122). This evolution of the client—server system is summarized in Table 10.3.

Case 10.1: Client—server at the 1994 Winter Olympics

IBM installed a client—server system for the Winter Olympics in 1994. It was designed to serve 100 000 potential users that included 50 000 accredited personnel, 2000 athletes, 8000 media representatives and over 100 000 visitors per day.

The application was 'mission-critical' and reliability had to be absolute since there would be no second chance to capture say a ski slalom race or a lug run. And what if the results were inaccurate? That would not only cause a flurry but bring many an athlete and relatives to tears!

There was token ring LAN connecting PS/2 (PC computers) at 16 sites with clients and servers at over 3000 sites. The network architecture used was IBM's SNA whilst the PS/2 handled all the accreditation and games management. The OS/2 graphical user interface (GUI) offered easy access to all users accessing a client. There was also an IBM ES/9000 mainframe which served as a central database with over 250 000 files available for the network. In addition, there was the RISC System 6000 that was used for the design and planning of the games.

Case 10.2: Citibank's overseas operations in Europe

Citibank started moving workload from its 176 mainframes in 17 countries to a client—server system. The strategy was one of re-engineering to have a more open and flexible platform that would support an object-oriented network which would provide information for enhanced decision-making and lead to higher productivity.

The client—server platform handled 30 000 transactions per day with a value of approximately $200 million. The objectives of the system were to have a very high level of fault tolerance, good contingency and a platform that would be scalable to the required processing power.

Citibank runs Windows NT on Compaq servers. They deliver transactions to other banks, do bookkeeping, accounting, MIS, reporting, drafts, DSS support, and check writing. By 1995, plans are for all locations in Europe to be linked by a wide area network and document imaging system tied directly into the new platform, delivering transactions in an electronically structured format.

Source: George Black (1994). Citibank's big gamble. *Which Computer?* May, pp. 42—44.

Case 10.3: Applications of client—server systems

The Morning Star group used the client—server approach to downsize from a 15 year old mainframe system and achieved a faster time-to-market response and greater productivity. It used the Hewlett-Packard HP 9000 Business Server.

Heinz Pet Products used the client—server paradigm to increase productivity and improve customer service.

Electronics Distributor built its client—server system incrementally for tracking marketing information, increasing productivity, enhancing decision-making support, improving customer service and bringing the product faster to the market.

The Bank of Montreal in Canada, with assets of $116 billion, re-engineered its operations to provide innovative and timely solutions. The bank worked with Digital Equipment using a integrated suite of software products for enterprise information delivery developed by the SAS Institute.

Motorola, a $3.5 billion manufacturing enterprise, reports its right-sizing effort to result in halving its costs under a mainframe (Connor, 1993: p. 56).

GE in one application showed a one-year payback for its start-up cost through its downsizing effort using a $6000 PC.

Unisys, a large manufacturer of computer systems, cut its information system's costs by

a third and improved service to end-users by implementing the client—server technology.

Bibliography

Cameron, K.S. (ed.) (1994). Special Issues on Downsizing. *Human Resource Management*, **33** (2), 181—298.

Canning, (1992). Plans and policies for client/server technology. *I/S Analyzer*, **30** (4) 1—16.

Connor, W.D. (1993). The right way to rightsize. *UNIX Review*, May 45—55.

Datamation, **37** (20) (1993), 7—24. Cover story on 'Client/Server Computing'.

Levis, J. and von Schilling, P. (1994). Lessons from three implementations: knocking down the barriers to client/server. *Information Systems Management*, **11** (2), 15—22.

Liang, T.-P., Hsiangchu, Chen, N.-S., Wei, H.-S. and Chen. M.-C. (1994). When client/server isn't enough. *Computer* **27** (5), 73—79.

Miranda, M.H. and Tellerman, N.A. (1993). Corporate downsizing and new technology. *Information Systems Management*, **10** (2), 1993, 32—38.

Muller, N.J. (1994). Application development tools: client/server, OOP, and CASE. *Information Systems Management*, **11** (2), 23—27.

Orfali, R., Harkey, D. and Edwards, J. (1995). Intergalactic client/server computing. *Byte*, **20** (4) 108—122.

Pinella, P. (1992). The Race for client/server CASE. *Datamation*, **38** (5), 51—54.

Semich, J.W. (1994). Can you orchestrate client/server computing? *Datamation*, **10** (16), 36—43.

Seymore, J. (1994). Application server at your service. *PC Magazine*, **13** (3), 277—302.

Sinha, A. (1992). Client—server computing. *Communications of the ACM*, **35** (7), 77—98.

Socarras, A.E., Cooper, R.S. and Stonecypher, W.F. (1991). Anatomy of an X terminal. *IEEE Spectrum*, **28** (3), 52—55.

Schultheis, R.A. and Bock, D.B. (1994). Benefits and barriers to client/server computing. *Journal of Systems Mangement*, **45** (2), 12—15.

Watterson, K. (1997). Client—server meets the Web generation. **2**(2), 93—96.

11

STANDARDS

If you think of 'standardization' as the best that you know today, but which is to be improved tomorrow — you get somewhere.

Henry Ford

Introduction

We discussed standards when we described the OSI model. It was to be the international standard of network architecture, but it did not get public acceptance, largely because it did not 'complete its tortuous way through the standards process' (Amy, 1994: p. 52). So what is so tortuous about international standards? Are national or local standards any less tortuous? What are standards anyway? Are they necessary for telecommunications and networking? If so, what are their advantages and limitations? What does it cost? What is the process of standardization? Who formulates them and who enforces them?

It is these and related questions that we will examine in this chapter. We will also look at the national and regional organizations that have done most for the development of international standards. These are the Europeans, the Japanese and the Americans. In each case, though, we will look at a different perspective to give us a balanced and 'total' picture of the process. In the European case we examine a regional organization and its standards-making process; in the Japanese case we look at the national organization and its decision-making process; and in the American case we look at a case study in which an international standard was developed, that of the B-ISDN.

For a view of decision-making at the international organization itself, we look at the development of standards for the OSI model at the ISO, International Standards Organization. First, however, we look at standards, their need and their rationale.

What are standards?

When measuring quantity, time or distance, we adhere to standards — such as 12 to a dozen, 60 minutes to an hour. These are standards we all unconsciously acknowledge. **Standards** are accepted authorities or established measures of behaviour, operations or performance.

Sometimes things in life are not as standard as we would want, like the day of rest in the week, or even the number of days in a year. But these are not very crucial. Some standards are crucial like when lives could be lost because we do not drive on the same side of the road in all countries of the world. Sometimes, a lack of standards is just a big pain in the neck, like when we travel with electrical appliances and our plug does not fit the socket in the wall or we do not have the right frequency of electrical current. Why can we not have one shape of plug that fits in all sockets and have just one frequency all around the world? This was not necessary in early times when people did not travel with electrical appliances that were crucial for shaving or drying hair. But as the world gets interrelated we need standards that are not only national but international. This is very true in telecommunications where, without international standards, equipment will not be interoperable, protocols will be different, transmission media may vary and international communications will be impossible. This will affect the community at large but definitely the business community doing international trade. Telecommunications are also needed in our world today because we are in an age of multivendors and multicarriers where interconnectivity and interoperability is necessary and this can be achieved only through international standards.

Lack of international standards will hamper the marketing of products beyond their national borders; reduce information exchange between customers, manufacturers and retailers; and slow technological progress because producers will create duplicate products instead of building on previous work. The existence of standards not only improves communication, but also provides portability, helps planning and control of operations, reduces uncertainty, stimulates demand, spurs innovation, encourages competitiveness, and provides an organized way of sharing and transferring technology.

Standards in telecommunications is part of the problem of standards for the computing industry where standards are needed for programming languages, operating systems, interface devices, chips and wafers, database and knowledge-base design, systems development, data representation — the list can go on and on. Without such standards we would not have the interoperability we now have. These standards are part of the work done by international standards organizations like the **ISO**, International Standards Organization. In the early 1980s, they were active in many areas like those shown in Figure 11.1. Note that there was some interest

and acknowledgment of interconnections and communications. Since then, telecommunications has become more ubiquitous and standards have been developed in areas including specifications for wiring, modems, connectivity devices, transmission, ISDN, OSI network architecture and protocols, voice and video processing, and so on. We look later at just one of these, the OSI.

We need international standards if we are to utilize the potential of telecommunications and increase world trade and international interactions. But international standards are far more difficult to achieve than national standards. In many countries there is a nationalized communications PT&T and a government that can dictate national standards. In countries like the US, there is no ministry that can dictate standards. Instead there is a National Institute of Standards that must react to the needs of standards. But even in the US, it is easier to agree on standards than to have international standards because the players involved are global and greater in number and diverse in interests. Theoretically it is all the countries in the United Nations that must agree to an international standard, but in practice it is a few industrialized countries that do the work and have the expertise. The rest follow. However, getting agreement among these developed and interested countries is still slow and difficult

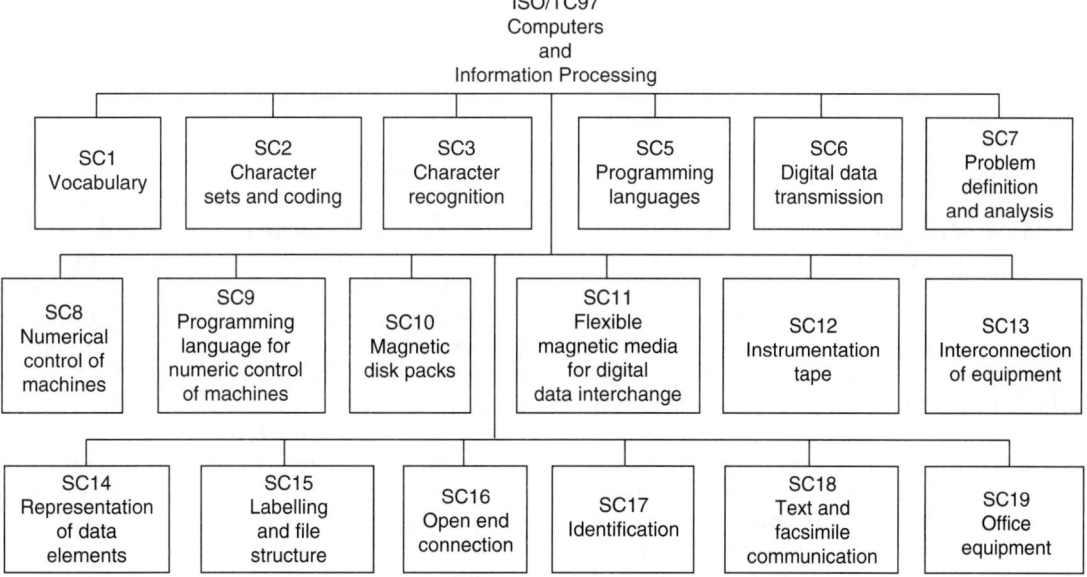

Figure 11.1 *Topics under consideration by subcommittees of the International Standards Organization, Technical Committee 97*

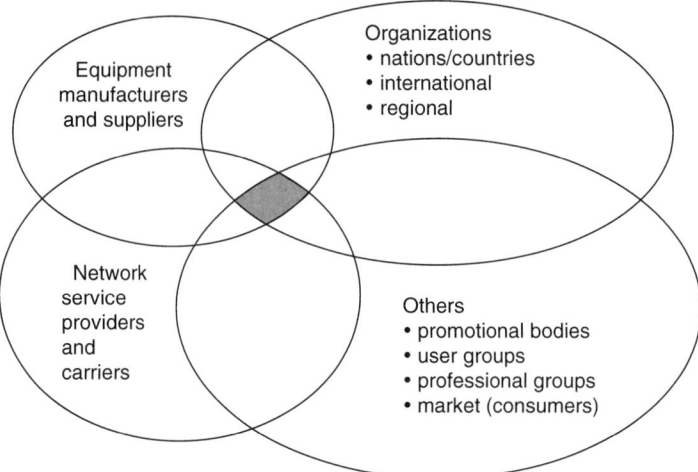

Figure 11.2 *Players in Developing International Standards*

because they may have conflicting objectives and interests.

The standardization players are shown in Figure 11.2. One set of players is the computing and telecommunications industry. This includes not only carriers but also manufacturers of equipment that are directly part of telecommunications or use telecommunications like the computer manufacturers which are far from being a homogeneous group. In some countries the carriers and manufacturers are the same. This was true in the US till recently when Western Electric produced equipment for its parent company that was a carrier: AT&T. With deregulation laws, the two have split and now compete with different vested interests. Another set of players includes the consumer, consumer groups, promotion groups and user groups. They represent the market forces.

The relative strength and influence of players in telecommunication standards (shown in Figure 11.2) is changing as shown in Figure 11.3. The organizations that reigned supreme in the early years are now giving way to the technological forces which in turn are giving way to the market forces that are largely the consumer. This consumer driven sector is also changing. From what was businesses using telecommunications for national trade and commerce, it is now a global market-place also being used by the household and individual who uses telecommunications for e-mail and entertainment. Hence international standards have to be increasingly conscious and sensitive to these needs which

makes the mix of the market more heterogeneous and difficult to predict and to cover every aspect of suture intended applications.

Not all parties and all players are involved all the times. Different players and parties are involved in different stages in the development of standards. We can see this in Figure 11.4, where the **base standards**, containing variant and alternative methods that are implement-defined facilities, are proposed by an international organization (with participation by other players at a high policy level). These base standards develop into **functional standards** also called **profiles**, and contain a limited subset of the permissible variants. After feedback and participation by regional and national standards bodies, trade associations and professional organizations, the functional standards are then tested. Testing could be done by independent organizations where the emphasis is on conformity with the design specifications of the standards. In parallel, the technical personnel from countries that are members of the international organization also perform testing but with emphasis on interoperability and performance first and conformance second.

Testing is also concerned with compatibility and at different levels: multi-vendor compatibility, upgrading and multi-vintage compatibility, and product-line compatibility.

The process of decision-making for standards is both upstream, from the corporation and user to the international organization, and downstream, from the top down to the user and

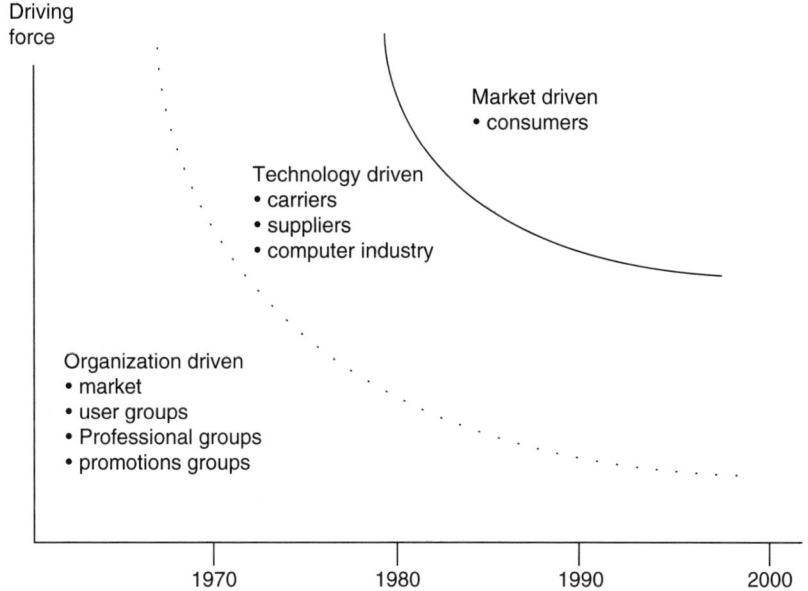

Figure 11.3 *Forces Driving Standards*

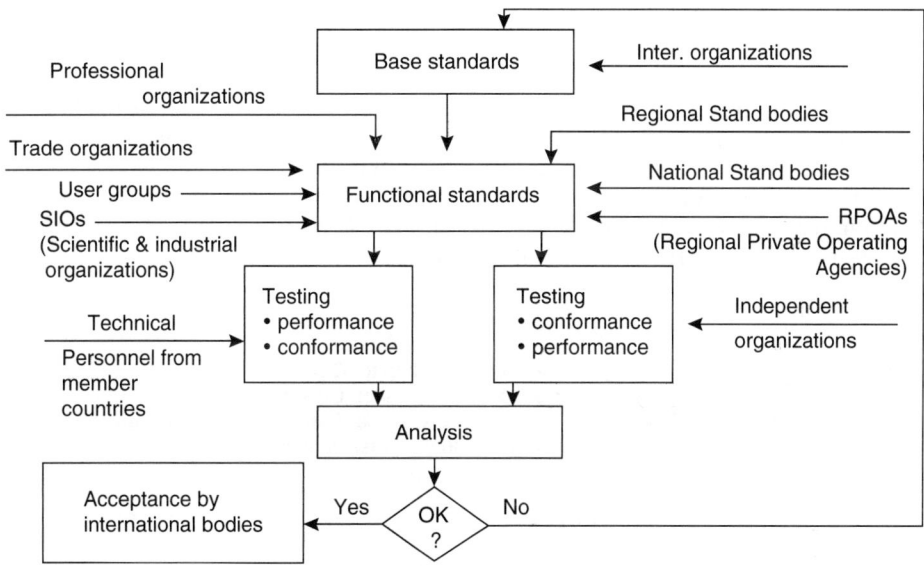

Figure 11.4 *Decision-making process by international organizations*

corporation. Ideally, there is much feedback all along the way, both upwards and downwards.

It is not always possible or desirable to have complete agreement by all parties (the innermost and shaded part of Figure 11.2). Sometimes, standards proposed by ISO are accepted by the globe. At other times, an international standard is not adopted by all countries as in the case of the OSI model. Fortunately, not all players have to agree on everything all the time, not even in the negotiating process where only a few sets of players are involved. For example, in the development of standards for a modem or say a cable connector, the players involved are the suppliers and carriers and the international standards organization represents the remaining parties. But there are some situations where many players are involved and in a very serious way with high stakes. An example would be the standards for network architecture and protocols, the OSI model. It is perhaps appropriate to discuss the process of producing this standard for it is a technology that we have discussed in previous chapter. Also, the standard can have a far-reaching and important impact on global telecommunications as well as being an interesting and controversial case from the organizational point of view.

The development of OSI

It took ten years to develop standards for all the seven layers of the OSI. This delay is partly due to the fact that the OSI was so comprehensive and important, and partly due to the thoroughness of the process of developing an international standard. The standards were developed in workshops with feedback from numerous groups and tested by yet another set of personnel. The organizational structure and personnel involved are shown in Figure 11.5. The names chosen for Figure 11.5 are somewhat different from the generic names used in earlier figures but are not inconsistent. The profile bodies had representation from regional bodies around the world as well as from international and national

governmental bodies. Most of the government participation came from Australia, Canada, Sweden, UK and the USA. They developed the base standards which were then passed on for feedback from numerous promotional bodies. The testing was done in parallel; one for conformity and one for interoperability. Again, this is not conceptually inconsistent with Figure 11.4.

As mentioned earlier, the OSI has been adopted in Europe but not in the US. There are many reasons for this. One is that OSI had to compete with models that were operational in the US. One of them the SNA was developed by IBM and it was difficult to persuade IBM that their design was not the best. Another set of protocols, TCP/IP, was also popular and was based on UNIX, a very popular operating system. Many of their concepts were adopted by OSI but it is more difficult to build on an old structure than it is to start afresh from the bottom. Starting from scratch was what happened with ISDN and it is one of the many success stories for ISO. The second reason perhaps is that it took too long for the international standard to come out in what was a highly volatile technological environment. Timing is highly critical for the success of international standards. In the case of the OSI, the ISO came on to the scene too late and took too long. By the time that the standards were ready, the environment had changed: PCs had proliferated, distributed processing had become popular and LANs had emerged. The assumptions underlying the start of the standards specifications

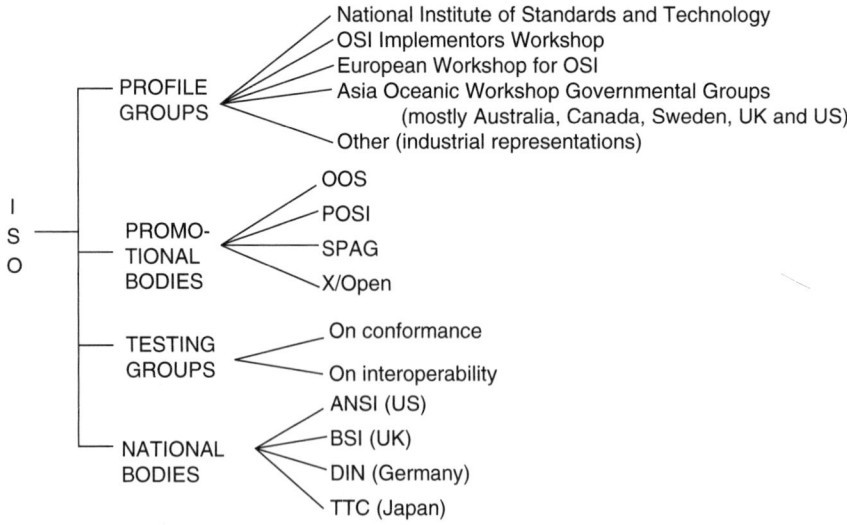

Figure 11.5 *Players for work on OSI*

had mostly changed during the long period of gestation. This long period was partly due to the fact that it was an international standard prepared by international organization and this process just takes a long time. Why? To answer this we must look at the organizational structure and process of international standards organizations.

The ISO

The OSI model may not have been too successful (in the US, that is), but it is only one of many ISO products. One indication of the work of the ISO and the growth of ISO standards can be seen from Figure 11.6. It shows the maintenance of 40 international standards in 1992. Not all these are related to telecommunications but many are, some directly and some indirectly.

	1989	1990	1991	1992
Drafts for Publication	33	116	90	106
Drafts for voting	0	21	104	103
Published standards and technical reports	0	18	47	184
Standards being maintained	0	0	0	40

Figure 11.6 *Growth of standards and technical reports (source: International Herald Tribune, Oct. 14, 1993, p. 14)*

The ISO (and its successor organization) have a vertical hierarchy of decision-making involving its international membership at the regional and national levels. The national levels have their own hierarchy. One configuration of the hierarchy of effort in the US is shown in Figure 11.7. The international body has its own internal hierarchy which includes councils, directors, advisory groups, study groups, as well as regional conferences and international conference (Irma, 1994).

Working parallel to the ISO is the **CCITT** (Comite Consultatif Internationale Telephonique et Telegraphic). CCITT is part of **ITU** (International Union Telegraphique) that was founded in 1934 and in 1947 it became an agency of the UN. In 1991, the ITU had 165 members, each exercising one vote. Mostly bodies like governments that were concerned with telecommunications or state owned PT&Ts were members. The UK is represented by the Ministry of Trade and Industry and the US by a mix of representatives from both government agencies and suppliers but coordinated by the Department of State (Foreign Ministry).

In 1993, a structural reform of the ITU saw the demise of CCITT and its resurrection as the standardization sector of ITU (ITU-T). It has 15 study groups including those on transmission equipment and service systems, modems, switching, voice network operations and maintenance, languages for telecommunications, and open systems.

International organizations have a difficult task of determining when to start on standards.

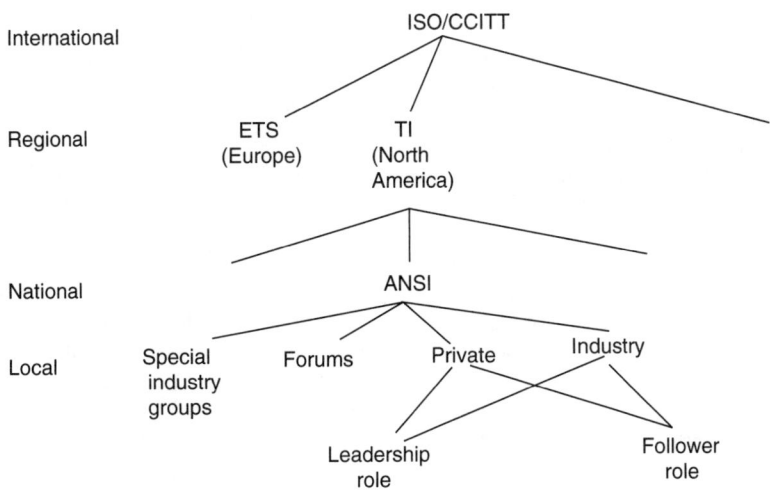

Figure 11.7 *Heirarchy of effort at ISO*

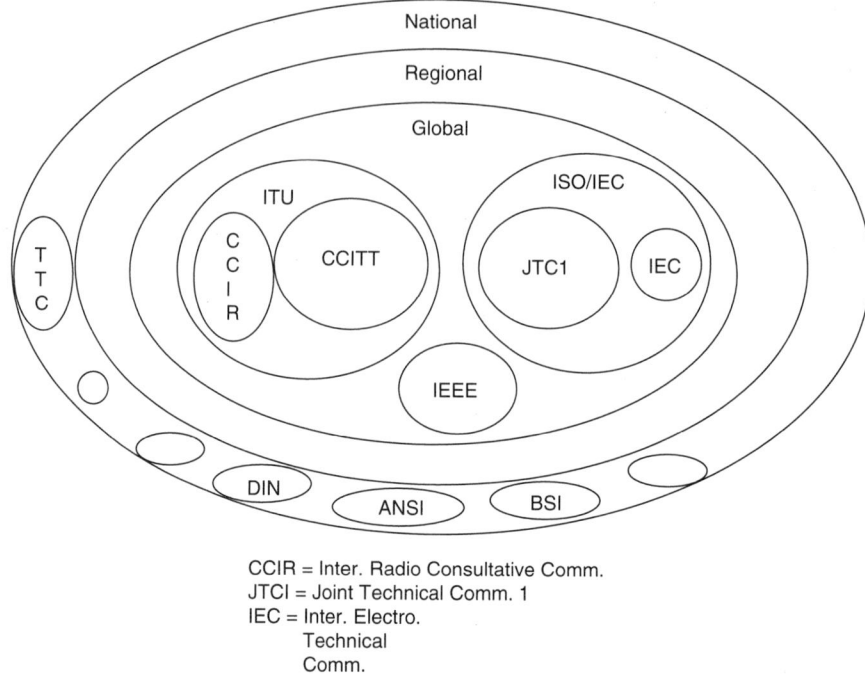

CCIR = Inter. Radio Consultative Comm.
JTCI = Joint Technical Comm. 1
IEC = Inter. Electro.
Technical
Comm.

Figure 11.8   *Organizations involved in International Standards*

You cannot develop too early before the technology has stabilized or the standards will not be relevant because they are obsolete. You cannot start too late because then vested interests dig-in with a loyal following and universal acceptance becomes difficult.

ISO and the CCITT have numerous organizations with which it has a lateral and horizontal relationship. These are summarized in Figure 11.8 to give a picture of the many organizations involved which may explain the sluggishness of international organizations. The diagram also emphasizes that each one of them has a potential input through national organizations. If you wish to influence an international contract you will have to contribute your time and effort. The choice is yours.

You may still be in the dark as to the type of interactions that take place at the national and regional level. To throw some light on these processes, we discuss one regional (in Europe) and one national (in Japan) organization and see its structure and decision-making process. We make the choice largely on geographic distribution. Coincidentally, these are the organizations and countries that are most active in the formulation of standards.

European standards organizations

The European standards organizations are shown in Figure 11.9 with the important relationship down under being identified. Not identified are the many interrelationships with the CCITT, the ISO and other international organizations, though some national bodies like the national standards bodies and national telecommunications administration do have additional direct links to international bodies as indicated in Figure 11.9.

At the national and corporate level, the connections are through trade associations and user groups at the same level as our government. Thus both the private sector and the public sector as well are represented. Other inputs to the European Standards Bodies are from National Standards Bodies and the National Telecommunication Administration.

The **ETSI** (European Telecommunications Standards Institute) is the body that issues the standards. It has 269 members with much of the detailed work being done in project teams appointed for a defined task and for a limited period of time. The proposals of the project teams

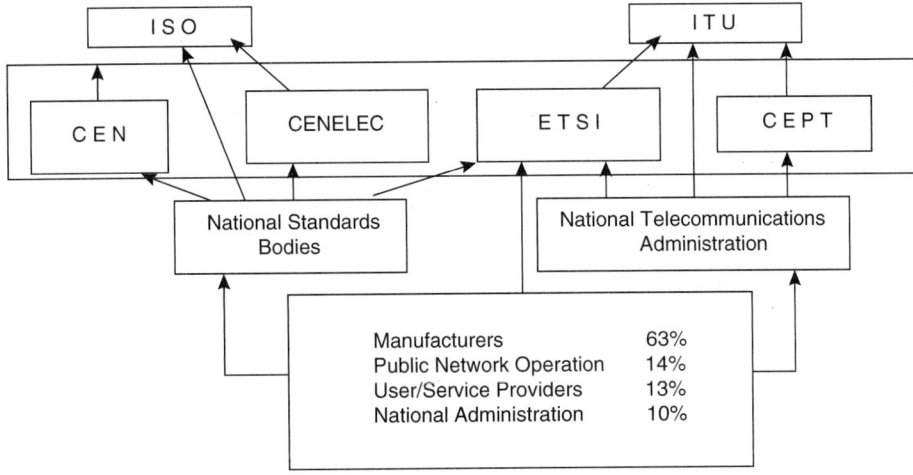

KEY:

CEN = European Committee for Standardization
CENELAC = European Comm. for Electrochemical Standardization
ETSI = European Telecommunications Standards Institute
CEPT = Conference European des Administrations des Poste et des Telecommunications

Figure 11.9   *Regional standards body in Europe*

goes to the technical committees and is then passed on to the technical assembly for approval where a weighted voting method is used.

The ETSI uses the 'working assumption' process as shown in Figure 11.10. Proposals are base on the 'best' estimate of technology. 'In this way, concepts can be put together quickly and adjusted to take account of changes, while still providing a measure of stability so important for implementors.' (Mazda, 1992: p. 61).

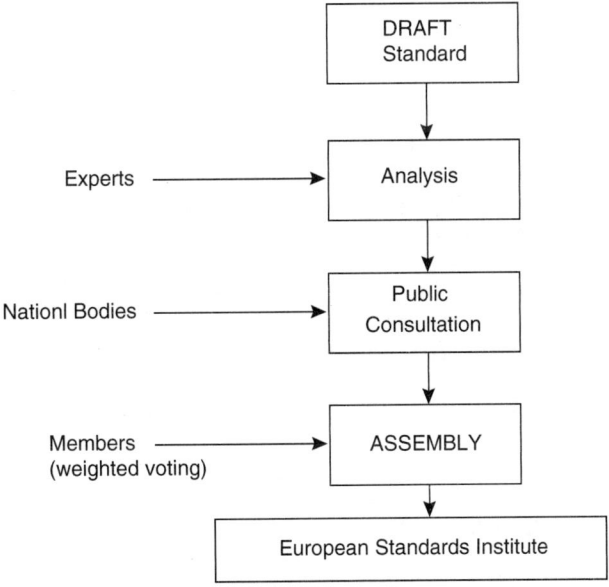

Figure 11.10   *Standards-making process in Europe*

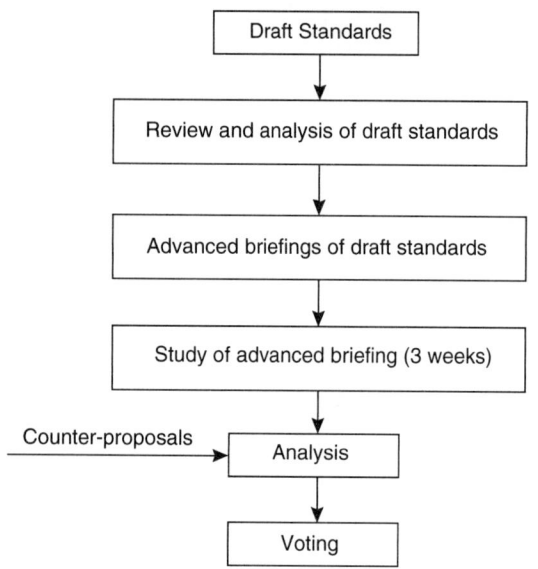

Figure 11.11 *Standardization process in Japan (adapted from Iida, 1994: p. 49)*

TTC in JAPAN

TTC stands for Telecommunications Technology Committee. It is organized with a Board of Directors, Council and Technical Assembly to which one Research and Planning Committee and five technical subcommittees (TCS) report. Each committee and the five TCS has a set of between two and eight working groups each, 27 in all in 1993. Their process of decision-making is shown in Figure 11.11. It is quite different from the other side of the Pacific, in the US, and is discussed in the case study for this chapter.

The TTC is the dominant organization in the Pacific and the Far East. It is perhaps the *de facto* regional organization though Australia and Singapore are becoming increasingly active in telecommunications. TTC is also a member of the regional INTUG which includes memberships from Australia, UK and the US. The US is also a member of the North American regional organization on Telecommunications, T1 (Telecommunications 1).

Summary and conclusions

International standards in telecommunications is at the confluence of two sets of dynamic forces: the telecommunications industry and the telecommunications environment. The telecommunications industry is changing rapidly with the widespread installation of fibre optics, digitization of networks, intelligent networks, EDH (Electronic Data Handling), B-ISDN, EDI (Electronic Data Interchange), evolving switching technology, as well as mobile and personal communications.

The telecommunications environment is changing because of the large proliferation of PCs, the demand for distributed processing and the client—server system, the popularity of e-mail and multimedia, the persuasiveness of wireless technology, and the increasing desire for high speed access. These are made largely feasible because of liberalization, privatization and deregulation of the computing and telecommunications industry in some countries, and the globalization of these industries.

To respond to these changes and to meet these diverse demands we need global standards that would provide a platform for compatibility and interoperability of telecommunication equipment and services. The standards not only have to be feasible but they must also be acceptable to the computing industry and the important governments involved. Standards must also be flexible in the ever-changing environment. Also important, the standards must be in place at the proper time: not too early lest an undeveloped technology gets 'frozen' by the standard; nor too late lest vested interests of equipment manufacturers and carriers become too entrenched for them to compromise and cause them to resist their products becoming outdated and replaced by products following international standards.

Once the right standards are in place at the right time, there are many beneficiaries. Business can expand their markets abroad, thereby increasing sales and achieving economies of scale, enabling them to offer a more competitive and better product for a lower price.

International standards cost millions of dollars every year. Little of this cost is borne by the sale of standards and most of the cost is borne by a few countries, many corporations and some individuals dedicated to working on standards. The coordination of this international effort is done by international organizations line the ISO and the CCITT, now the ITU. They produce many standards, one every 18 months on the average, but some can taken up to 10 years in some cases.

They are slow in delivery because of the inherently slow process of downstream and upstream feedback and consensus building between corporations, industry and trade associations, as well as national and regional bodies that often have vested commercial interests if not national interests to support. Self interests are often at stake. Manufacturers of equipment want their designs to be internationally accepted thereby increasing their market share, whilst carriers do not want to change what they already have any more than necessary, and that if they must change then the change should be easy and inexpensive. They agree to temporary alliances for specific projects among national and international competitors for they all stand to gain from the participation with peers. The ultimate beneficiaries are the public at large who benefit from the many applications that telecommunications made possible, especially for the rural and remote areas all around the world. This is covered more fully in Part 3 of the book.

Despite this laudable effort, there are still some unresolved matters that relate to international standards in telecommunications. One is the transfer of proprietary knowledge and technology to the public domain. This raises the delicate problems of extracting proprietary information for sharing, despite the actual (or perceived) loss for those who hold the proprietary knowledge, patents and experience, and of making the parties concerned participate in the development of international standards. They must be brought into the consensus-building and cooperation process of developing international standards.

Case 11.1: Development of international standards for the B-ISDN in the US

The development of international standards in the US is different from that in other countries largely because it is greatly influenced by its experience in the development of the 802 standards by the IEEE (Institute of Electrical and Electronic Engineers) and the development of the TCP/IP (Transmission Control Protocol/Internet Protocol) standards by IETF (Internet Engineering Task Force). Encouraged by the success of the IETF and the IEEE 802 standards, there evolved another layer to the traditional standards bodies called the Forum. The Forum is a consortium representing a very wide spectrum of industrial players highly focused on working on a tentative standard developed by the traditional standards organizations.

> The new group then builds rapidly on the base to reach an agreement among commercial interests on the details of implementation. Because these agreements are driven by the impetus of commercial products and services, they tend to deal with short-term considerations. Such agreements, however, are quite important to mobilize many segments to develop actual products and services based on these technologies. (Amy, 1994: p. 53).

The Forum focuses primarily on customer premises interests and produces a consensus-based standard. This was true of the Forum on the ATM (Asynchronous Transfer Mode) in the development of consensus on standards for the B-ISDN standards. The Forum along with IETF interact with the four technical subcommittees which are accredited to the ANSI (American National Standards Institute) via their parent committees. The Forum differs with the ANSI committees in that the ATM focuses primarily on the interests of the customer premises whilst the ANSI committees emphasize the interests of the public networks.

T1 is the North American standards organization. It has many working groups, like the T1A1 group which focuses on performance issues; T1E1 is concerned with physical transmission interface; T1M1 Focuses on network management issues; and T1S1 in concerned with network signalling protocols and network service definition. There are also interrelationships with international organizations such as with the international organization ITU (successor of CCITT) through the US National Committee which is coordinated by the Department of State (Foreign Ministry).

The effort on the B-ISDN is contributed by over 200 individuals coming from more than 60 organizations. Members who have voting rights pay a membership fee and every sponsoring organizations pays for the expenses and the time allocated to the project. This contribution is considerable for it is estimated that over 200 000 impressions are made for a normal meeting.

Source: Amy, Robert M. (1994). 'Standards by consensus'. *IEEE Communications Magazine*, **32**(1), 52−55.

Case 11.2: Networking standards in Europe

There are four organizations in Europe that are concerned with standards of networking. These are:
ITU-T, the International Telecommunications — Telecommunications sector; the ETSI, the European Telecommunications Standards Institute; the NMF, the Network Management Forum; and the Tina-C, the Telecommunications Intelligent Network Architecture Committee.

The ITU-T is mainly concerned with TMN, Telecommunications Management Network. The TMN solution has its critics including the spokesperson of the German giant Siemens for having 'always been divided into four layers — element, network, service and business management. This certainly does not facilitate the rapid provision of comprehensive solutions because the standard interfaces are required to link the individual layers.'

Working against these criticisms of the TMN solution is the ETSI which has helped to flesh out the TMN framework along with customer administration and fault and traffic management. ETSI has also made recommendations to the ITU on UPT, Universal Communications System.

NMF is working on information module definitions that are available for transmission, switching, broadband and access management.

Tina-C embraces the TMN as well as IN, Intelligent Networks. It attempts to increase the increase the IQ of the IN 'with focus on distributed processing and object communications ... also addresses service management more specifically than TMN.'

Source: *International Herald Tribune*, Oct. 11, 1995, p. 12.

Bibliography

Burton, J. (1995). Standard issue. *Byte*, **20**(9), 201–205.

David, J. (1992). LAN security standards. *Computers and Security*, **11**(7), 607–619.

Duran, J.M. and Visser, J. (1992). International standards for intelligent networks. *IEEE Communications Magazine*, **30**(2), 34–36.

Iida, T. (1992). Domestic standards in a changing world. *IEEE Communications Magazine*, **32**(1), 46–49.

Irmer, T. (1994). Shaping Future telecommunications: the challenge of global standardization. *IEEE Communication Magazine*, **32**(1), 20–28.

Knight, I. (1991). Telecommunications standards development. *Telecommunications*, **25**(1), 38–42.

Mazda, F. (1992). Standardization on standards. *Telecommunications*, **26**(9), 54–61.

Mossotto, C. (1993). Pathways for telecommunications: a European look. *IEEE Communications Magazine*, **26**(8), 52–57.

Mostafa, H. and Sparell, D.K. (1992). Standards and innovation in telecommunications. *IEEE Communications Magazine*, **30**(7), 22–28.

Nak, D. (1994). Coordinating global standards and market demands. *IEEE Communications Magazine*, **32**(1), 72–75.

Peterson, G.H. and Dvorak, C.A. (1994). Global standards. *IEEE Communications Magazine*, **32**(1), 68–70.

Trauth, E.M. and Thomas, R.S. (1993). Electronic data interchange: a new frontier for global standards policy. *Journal of Global Information Management*, **1**(4), 6–17.

Tristam, C. (1995). Do you really need ISO 9000? *Open Computing*, **12**(5), 65–66.

12

SECURITY FOR TELECOMMUNICATION

Every new technology carries with it the opportunity to invent a new crime.

Laurence Urgenson

Introduction

Problems of computer security go back to days long before telecommunications and LANs became ubiquitous. A study for the period of 1964—73 identified 148 cases of computer crime that were made public. Telecommunications has only increased the exposure to computer crime.

> Anyone who can scrounge up a computer, a modem and $20 a month in connection fees can have a direct link to the Internet and be subject to break-ins — or launch attacks on others ... In European countries such as the Netherlands, for instance, computer intrusion is not necessarily a crime ... some of the most respected domains on the Internet contain computers that are effectively wide open to all-comers — the equivalent of a car left unattended with the engine running ... Computer science professors ... assign their students sites on the Internet to break into and files to bring back as proof that they understand the protocols involved.   (Wallich, 1994: p. 94).

The US Secret Service has estimated the annual cost of fraud by telecommunications at around $2.5 billion; industry numbers range from $1 billion to $9 billion. One reason for this high cost is that PCs are now very common and the population that know how to use PCs is so large. The opportunity is great and 'opportunity makes a thief'. There are also many people who are either using computer systems or have access to computer systems. One author has identified 55 sources for potential computer intrusion. (Wallich, 1994: p. 90—91). Some intrusions are the unintended consequences of corporate action when in their enthusiasm

for downsizing, they pushed applications (and sometimes mission critical applications) on to remote servers which had to use a LAN or some other telecommunications system. This increased the points of vulnerability to intrusion and unauthorized access. With every change in technology comes an opportunity to violate the system and the newer intrusions are always getting to be more creative and effective. In one case in 1993, a long distance telephone company cardholder compromised his card and 600 unauthorized international calls were placed on that card before the network specialists detected the problem and disconnected the violator. All this happened in less than 2 minutes. We have to be faster and smarter. We cannot continue the traditional approach of 'security through obscurity' which is the keeping of vulnerable data secret. We need to use technology and devise security policies that not only catch the violator but dissuade the violator from breaking into the system. It is such technologies and policies that are the subject for this chapter. In it we examine the threats and responses to access which is a new twist to an old problem of computer centre security. We also examine the unique threats to telecommunications including the threat of computer viruses.

We will not discuss organizational considerations of security such as administrative controls, the appointment of a security officer and operational security, because these considerations are common to corporate security (and not entirely unique to telecommunications). Hence these considerations lie outside the scope of this book. The one exception is the assessment of risk and the determination of how much security is needed, because this is somewhat different and often more critical in telecommunications.

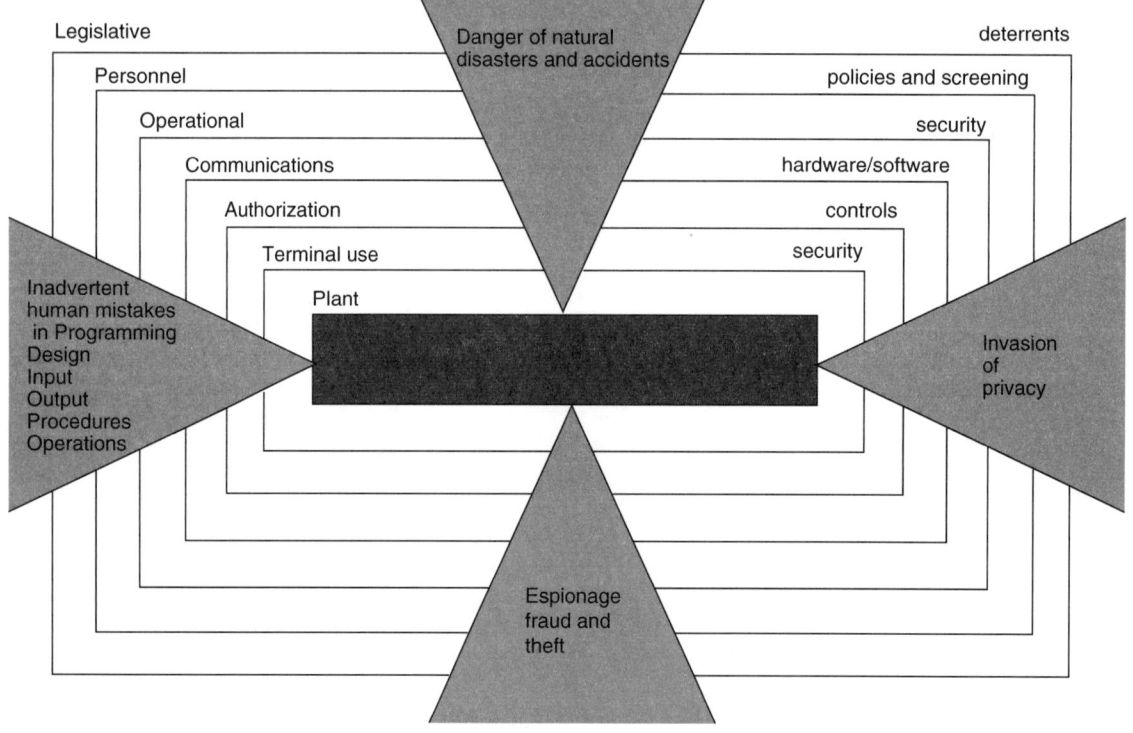

Figure 12.1 *Layers of control*

Security

Security is where data (and information) are protected against unauthorized modification, capture, and destruction or disclosure. Personal data are not the only vulnerable data. Confidential data on market strategies and product development must be kept from the eyes of competitors. Large sums of money transferred daily by electronic fund transfer must be protected against theft. The very high volume of business information processed by computers today means that the rewards of industrial espionage and fraud are of a much higher magnitude than in the past and are ever increasing.

Records must also be protected from accidents and natural disasters. For example, a breakdown in air-conditioning may cause some computers to overheat, resulting in a loss of computing facilities. Fire, flood, hurricanes, and even a heavy snowfall causing a roof to collapse can cause the destruction of data and valuable computer equipment.

Security measures described below are designed to guard information systems from all the

above threats. These measures can be envisioned as providing layers of protection as shown in Figure 12.1. Some controls guard against infiltration for purposes of data manipulation, alteration of computer programs, pillage or unauthorized use of the computer itself. Other measures guard against physical plant, monitor operations and telecommunications, and regulate personnel. These controls are now discussed below.

Terminal use controls

Badge systems, physical barriers (locked doors, window bars, electric fences), a buffer zone, guard dogs and security check stations are procedures common to restricted areas of manufacturing plants and government installations where work with secret or classified materials takes place. A vault for storage of files and programs and a librarian responsible for their checkout provide additional control.

With on-line systems using telecommunications, security is a greater problem, since stringent access controls to terminals may not exist at remote sites. The computer itself must, therefore, ascertain the identity of persons who wish to log on and must determine whether they are entitled to use the system. Identification can be based on:

- what the user has, such as an ID card or key;
- who the user is, as determined by some biometric measure or physical characteristic;
- what the user knows, such as a password.

Keys and cards

Locks on terminals that require a **key** before they can be operated are one way to restrict access to a computer. Another way is to require users to carry a **card** identifier that is inserted in a card reader when they want to use the computer.

A microprocessor in the reader makes an accept or reject decision based on the card.

Many types of card system are on the market. Some use plastic cards, similar to credit cards, with a strip of magnetically encoded data on the front or back. Some have a core of magnetized spots of encoded data. Proximity cards contain electronic circuitry sandwiched in the card; the reader for this card must include a transmitter and receiver. Optical cards encode data as a pattern of light spots that can be 'read' or illuminated by specific light sources, such as infrared. In addition, there are smart ID cards that have an integrated circuit chip embedded in the plastic. The chip has both coded memory, where personal identification codes can be stored, and microprocessor intelligence.

The disadvantage of both keys and cards is that they can be lost, stolen or counterfeited. In other words, their possession does not absolutely identify the holder as an authorized system user. For this reason, the use of passwords is often an added security feature of key and card systems.

Biometric systems

Some terminal control systems base identification on the physical attributes of system users. For example, an electronic scan may be made of the hand of the person requesting terminal access. This scan is then measured and compared by computer to scans previously made of authorized system users and stored in the computer's memory. Only a positive match will permit system access.

Fingerprints or palm prints can likewise be used to identify bona fide system users. Such security systems use electro-optical recognition and file matching of fingerprint or palm print minutiae. Signature verification of the person wishing to log onto the computer is yet another security option. Such systems are based on the dynamics of pen motion related to time when the signer writes with a wired pen or on a sensitized pad. A biometric system can also be based on voiceprints. In this case, a voice profile of each authorized user is recorded as an analogue signal, then converted into digital from which a set of measurements are derived that identify the voice pattern of each speaker. Again, identification depends on matching: the voice pattern of the person wishing computer access is compared with voice profiles in computer memory.

Biometric control systems, of special interest to defence industries and the police, have been under development for many years. Although technological breakthroughs that enable discrimination of complex patterns have been made recently, pattern recognition systems are still not problem free. Many have difficulty recognizing patterns under less than optimal conditions. For example, a blister, inflammation, cut, even sweat on hands, can interfere with a fingerprint match. Health or mood that changes one's voice can prevent a voiceprint match. A combination of devices, such as voice plus hand analysers, might ensure positive identification; but such equipment is too expensive at the present time to be cost effective for most operations in business.

Passwords

The use of **passwords** is one of the more popular methods of restricting terminal access. One example of a password system is the required use of a personal identification number to gain access to an automated teller machine at a bank.

The problem with passwords is that they are subject to careless handling by users. Some users write the code on a sheet of paper that they carry in their wallet, or they tape the paper to the terminal itself. When given a choice, users frequently select a password that they can easily remember, such as their birth date, house number or names of pets, wives or children. Top of the list in Britain seems to be 'Fred', 'God', 'Pass' and 'Genius'.

Someone determined to access the computer will make guesses, trying such obvious passwords first. Even passwords as complex as algebraic transformations of a random number generated by the computer have been broken with the assistance of readily available microcomputers. Of course, the longer a password is in use, the greater the likelihood of its being compromised.

One-time passwords are a viable alternative. But systems of this nature are difficult to administer. First of all, each authorized user must be given a list of randomly selected passwords. Then there must be agreement on the method of selecting the next valid password from the list, a method that is synchronized between computer and user. Finally, storage of the list must be secure, a challenge when portable terminals are used by personnel in remote sites where security may be lax.

Recently, a number of password systems have been put on the market that generate a new password unique to each user each time access is attempted. This is done with a central intelligent controller at the host site and a random **password generator** for each user. Typically, the system works as follows: To gain mainframe access, the user enters his or her name (or ID code) on a terminal keyboard. The computer responds with a 'challenge number'. This is input to the user's password generator. By applying a cryptographic algorithm and a secret key (a set of data unique to each password generator) to this challenge 'seed', a one-time password is generated. The user then enters this password into the computer. The central controller simultaneously calculates the correct password and will grant access if a match occurs.

Such password management systems are difficult to compromise, because passwords are constantly changed. Only a short period of time is allowed for entry of the correct password. Furthermore, the control system is protocol dependent. This compounds the problems of a person trying to breach the system in a network having a variety of protocols. The advantage to the user is that the password generator is portable, usually a handheld device, and easy to use.

In recent years, much publicity has been given to **hackers**, usually youths, who often derive malicious pleasure in circumventing computer access controls.

Authorization controls

In addition to the identification systems outlined in the preceding sections, control systems can be installed to verify whether a user is authorized to access files and databases, and to ascertain what type of access is permitted (read, write or update).

Data directory

A computer can be programmed to reference a stored **data directory security matrix** to determine the security code needed to access specific data elements in files before processing a user's job. When the user lacks the proper security clearance, access will be denied. In a similar manner, the computer might be programmed to reference a table that specific the type of access permitted or the time of day when access is permitted.

The data elements accessible from each terminal can likewise be regulated. For example, according to a programmed rule, the terminal in the database administrator's office might be the only terminal permitted access to all files and programs and the only terminal with access to the security matrix itself. A sample printout from an access director, sorted by user identification number, is shown in Table 12.1.

Assigning access levels to individuals within an organization can be a difficult task. Information is power, and the right to access it is

Table 12.1 *Access directory*

User identification: 076-835-5623
Access limitation: 13 hours (of CPU time for current fiscal year)
Account Number: AS5842

Data Elements	Type of access	Security level	Terminal number	Time lock
Customer number	Read	10	04	08.00−17.00
Invoice number	Read	10	04	08.00−17.00
Cash receipt	Read/write	12	06	08.00−12.00

a status symbol. Employees may vie for clearance even when they do not require such clearance for their jobs. Manager should recognize that security measures designed to protect confidential data and valuable computing resources may antagonize loyal employees. It is important that the need for security be understood by workers and that security controls be administrated with tact.

Security kernel

Unfortunately, the use of a security matrix does not provide foolproof security. In a multiuser system, data in a file can be raided by installing a 'Trojan horse' program. Figure 12.2 shows how this is done. Although the data directory does not authorize Brown to access File A, confidential data from that file is copied into another file that Brown is entitled to access, on the direction of a secret program, thereby circumventing system security.

The concept of a **security kernel** addresses the Trojan horse issue. A kernel is a hardware/software mechanism that implements a **reference monitor**, a systems component that checks each reference by a subject (user or program) to each object (file, device or program) and determines whether the access is valid according to the system's security policy. Figure 12.3 shows how Brown is foiled by a reference monitor in his attempt to raid File A.

A security kernel represents new technology still in the developmental stage. Although a number of projects have attempted to demonstrate the practicality of this security approach, results thus far have been mixed.

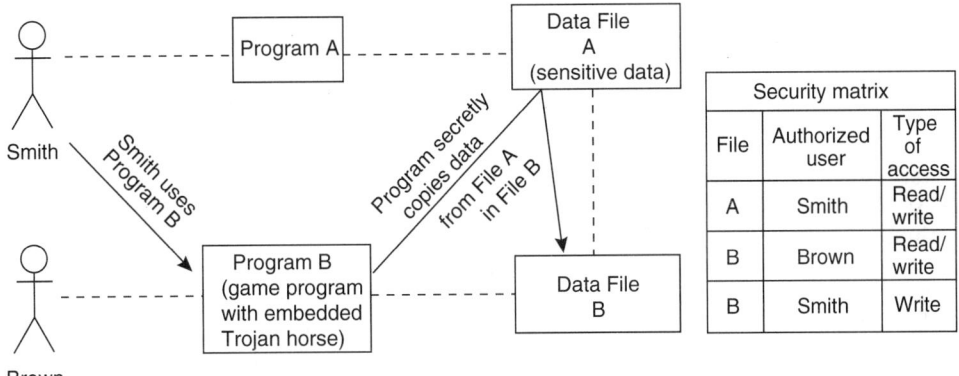

Figure 12.2 *Raiding files: a Trojan horse program*

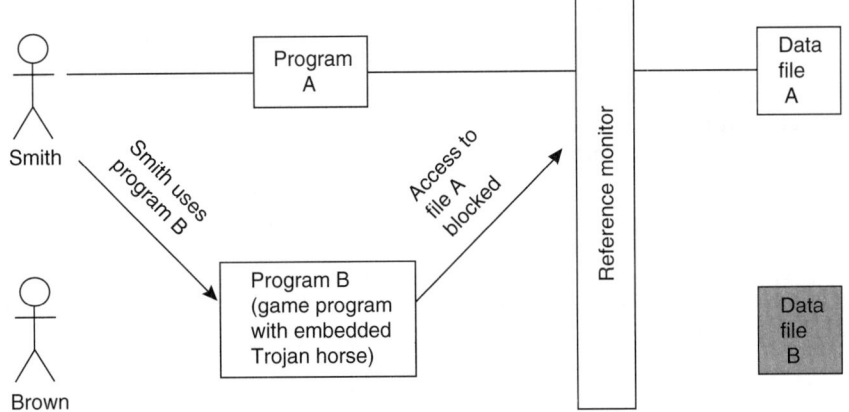

Figure 12.3 *How a reference monitor blocks a Trojan horse raid*

127

Virtual machine

An entirely different approach to security in a multiuser environment is a **virtual machine**. With this systems structure, each user loads and runs his or her own copy of an operating system. In effect, this isolates one user from another although they use the same physical machine, because each virtual machine can be operated at a separate security level. With a virtual memory structure, several user programs can reside in computer memory simultaneously without interference.

Communications security

Computer processing is today closely linked with telecommunications, which allows the transference of computer data between remote points. Protecting the confidentiality of this data at the initiating terminal, during transmission itself or when transmission is received, has required the development of sophisticated security techniques. For example, a **handshake**, a predetermined signal that the computer must recognize before initiating transmission, is one way to control communications. This prevents individuals from masquerading, pretending to be legitimate users of the system. Most companies use **callback boxes** that phone would-be users at preauthorized number to verify the access request before allowing the user to log on. A hacker who has learned the handshake code would be denied access with such a system. Protocols, conventions, procedures for user identification (described earlier in this chapter) and dialogue termination also help maintain the confidentiality of data.

During transmission, messages are vulnerable to wiretapping, the electro-magnetic pickup of messages on communication lines. This may be eavesdropping, passive listening or active wiretapping involving alteration of data, such as piggybacking (the selective interception, modification or substitution of messages). Another type of infiltration is reading between the lines. An illicit user taps the computer when a bona fide user is connected to the system and is paying for computer time but is 'thinking', so the computer is idle. This and other uses of unauthorized time can be quite costly to a business firm.

One method of preventing message interception is to encode, or encrypt, data in order to render it incomprehensible or useless if intercepted. **Encryption**, from the Greek root 'crypt' meaning

to hide, can be done by either transposition or substitution.

In transposition, characters are exchanged by a set of rules. For example, the third and fourth characters might be switched so that 5289 becomes 5298. In substitution, characters are replaced. The number 1 may become a 3, so that 514 reads 534. Or the substitution may be more complex. A specified number might be added to a digit, such as a 2 added to the third digit, making 514 read 516. Decryption restores the data to its original value. Although the principles of encryption are relatively simple, most schemas are highly complex. Understanding them may require mathematical knowledge and technical expertise.

An illustration of encryption appears in Figure 12.4. A key is used to code the message, a key that both sender and receiver possess. It could be a random-number key or a key based on a formula or algorithm. As in all codes, the key must be difficult to break. Frequent changing of the key adds to the security of data, which explains why many systems have a key base with a large number of alternate keys.

In the past, the transportation of the encryption key to authorized users has been as Achilles' heel to systems security. An additional problem is that there sometimes is insufficient time to pass the key to a legitimate receiver. One solution is a multiple-access cipher in a public key cryptosystem. This system has two keys, an E public encryption key used by the sender, and a D secret decryption key used by the receiver. Each sender/receiver has a set of D and E keys. To code data to send to Firm X, for example, a business looks up Firm X's E key, published in a public directory, and then transmits a message in code over a public or insecure transmission line. Firm X alone has the secret D key for decryption. This system can be breached but not easily, since a tremendous number of computations would be needed to derive the secret of D. The code's security lies as much in the time required to crack the algorithm as in the computational complexity of the cipher, because the value of much data resides in timeliness. Often, there is no longer need for secrecy once a deal is made, the stock market closed, or a patent application is filed.

Cryptography, in effect, serves three purposes: identification (helps identify bona fide senders and receivers); control (prevents alteration of a message); and privacy (impedes eavesdropping). With the increased reliance of businesses on

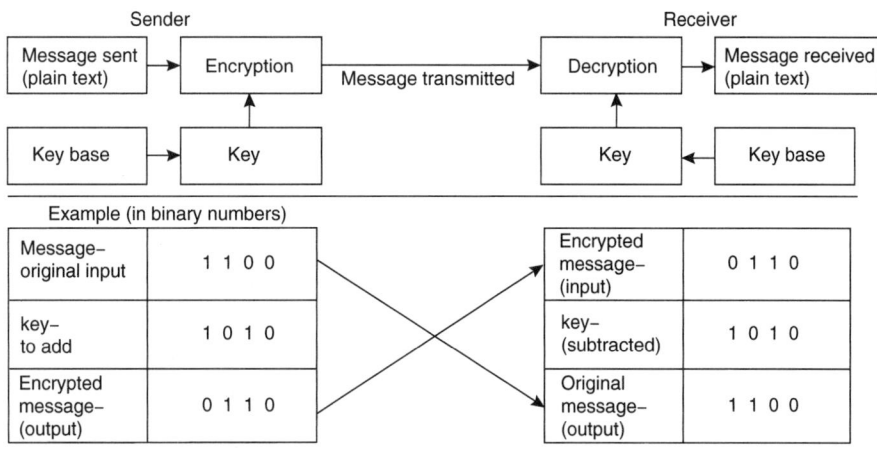

Figure 12.4 *Encrypting and decrypting data in teleprocessing*

teleprocessing, much research has been done on cryptographic systems. But experts disagree about how secure even the most complex codes are. Some claim that persons bent on accessing encrypted data will be able to break all codes, using the power, speed and advanced technology of modern computers.

Cryptography can be implemented through hardware or through software. The hardware implementation can keep up with high speeds even with gigabits of high speeds. But hardware implementations are more expensive in initial costs as well as in space occupied and running costs of power consumption. In contrast, software implementations may not always keep up with high speeds but are more versatile with solutions such as encryption. Encryption can be mathematically complex and yet it can be decrypted given enough time and motivation. One solution to this problem is to embed a cipher in a document identifying both the owner and the user. The user is the buyer who can purchase a cryptographic key to decode the document. If a person purchased one-time rights, the document would turn into gibberish when put on a bulletin board or transmitted via e-mail. This protection against unauthorized transmission is attractive since encryption algorithms protecting privacy of documents can be protected by codes such as the PGP (Pretty Good Privacy). PGP was written by a programmer in Colorado who has made it available on the Internet and some non-Internet bulletin boards for no charge. This has angered the US Government and frightened its agencies of law enforcement,

defense and intelligence. They support the 'key-escrow encryption' which allows the government to hold keys to any encrypted communication to be used for occasions of national security and the fight against money laundering, terrorism and computer crime. The US government wants this code to be part of all encryption software that is exported. US exporters argue that foreigners will not buy such a code and will get encryption codes elsewhere, thereby losing the US a lot of export. Also, it has been argued that this may be a violation of freedom of speech and expression. And so the controversy rages.

Security for advanced technology

Thus far, we have discussed the security considerations that are traditional to IT. However, computer technology is well known for its many entries (and exits) of innovation. Some of these are listed in Table 12.2. Space limitations do not allow us to examine all or even some of them in any detail except perhaps one: image processing. This will illustrate the types of threats to security posed by one advanced technology. These threats include:

- Unauthorized copying and downloading of images on terminals, PCs and workstations;
- Unauthorized release of images by users (including end-users);
- Integrity of images and unauthorized modifications made to them;
- Authenticity of images stored in documents.

Table 12.2   *Technologies affecting security*

Teleprocessing
LANs/MANS/WANS/Internet
   Superhighway
   Global networks
   PCN (Personal Computer Networks)
   Wire-less communications
   Fax connections
Image processing
Smart cards
NNets (Neural Networks)
Laptop computers
Palmtop computers
Pen-based computers
PDA (Personal Digital Assistant)
Personal communicators
Pocket pals
Teleconferencing
Video-conferencing

Since image processing requires large memory and fast computing power, it uses minis and mainframes which are accessible through a network and telecommunications. This opens a wide set of threats. Also, the large computers necessary for image processing contribute to the temptation of using these powerful resources to break the security codes of other systems.

Each of the technologies listed in Table 12.2 present a unique set of benefits, exposure to risk, and potential losses, as well as a cost of implementation of a response to the threat. Management must continuously evaluate the consequences of failure to protect its assets of information and the cost of updating its protection against technological changes. To achieve this, management must maintain knowledge of the technological changes and its potential impact on security; anticipate threats and vulnerabilities; and develop protective defense measures necessary to combat the threat. The anticipation aspect is important since the protective measures must be instituted in the design stage of development and sometimes even earlier in the planning stage and user specification stage of development.

Continued violations of computer and telecommunications security should be anticipated especially from the computer virus, the danger increasing with telecommunications and networking.

Computer viruses

A **computer virus** is software that, when entered into a computer system can cause it to stop or interrupt operations. It could also corrupt data, destroy data/knowledge-bases, and cause errors or disrupt operations. The virus does not just appear or grow, it is inserted (knowingly and deliberately) into the system. It is definitely unauthorized.

A virus is a class of programs. Another sort of computer vermin is the **worm** which is a program that 'worms' its way through a system, altering small bits of code or data when it gets access to it. A worm may also be a virus if it reproduces itself and infects other programs. In contrast, there is the **logic bomb**, which is a computer program that is set to 'explode' when a specified condition is met.

The computer virus has the ability to propagate into other programs. However, the computer virus program must be run in order to reproduce or to do damage. For this to happen, a computer virus must be introduced into the system. One way a computer virus can enter the system is in the form of a **Trojan horse**, which is a computer program that seems to do one thing but also does something else. But, the virus has a new and malicious twist to it. The virus can act instantly or lay dormant in the system till it is triggered by a specified date or an event such as the processing of five programs, or the logging on as a specific user, or whatever.

How does a virus operate? Zajac (1990: p. 26) describes the Lehigh virus (named after the Lehigh University in the US). This virus consisted of seven lines of code in Pascal and was placed in a DOS command file. It operated as follows:

> ... when a user typed a DOS command, the virus would check to see if there was a non-infected ... file on the system. If so, it infected it and incremented a counter that kept track of how many other disks it had infected. The virus would then execute the user's command. All this unbeknown to the user ... when the infection counter hit four, it would totally erase the hard disk.

The computer virus has many objectives and can have many consequences. Some cases illustrating its variety are listed below.

The Pakistani virus infected untold number of PCs as it travelled around the world creating havoc and fear among PC users. The authors of this virus had the gall (or courtesy!) to announce its existence with the message: 'Beware of this VIRUS. Contact us for vaccination.' The message was followed by a 1986 copyright date and two names (Amjad and Basit) and an address in Pakistan. The virus is also known by its generic type: the BRAIN virus, named after the volume label of an infected disk which reads '(c)BRAIN'. This naming is perhaps because the authors worked for 'Brain Computer Services'.

The BRAIN virus surfaced in the US at the University of Delaware in 1987 followed a month later by the Lehigh virus.

The Cornell virus was a passive virus with the intent of collecting names and passwords. It infected thousands of mainframe computers throughout the world to enable them to be used later as deemed desirable.

In 1987, a virus appeared in a computer network in California and interfered with the scan control on video monitors and caused one to explode.

The Jerusalem virus followed the Delaware virus by two months and its first strain appeared in the Hebrew University in Israel.

In the early 1990s, the 'Bulgarian factory' replaced Israel as the source of viruses and over 100 viruses from Bulgaria have infected the Eastern European countries.

A magazine publisher in Germany distributed over 100 000 disks to its subscribers. Unknown to the publisher, each disk was infected with a STONED II computer virus.

A student in California was given a disk with a free program. She used the disk (which was infected and unknown to her) to do her homework assignments. She took the disk to her university and loaded it on the university network to continue with her assignments. Inadvertently, she infected all the students using the network and caused the system to operate incorrectly.

An employee in a large firm used a program (not known as being infected) available externally on a public network and downloaded it to her PC. She then used this program on the firm's network thereby infecting the network and erasing many corporate files.

A 'cruise' virus is similar to the Cornell virus. It is a sophisticated passive virus that could infect disks that are distributed openly. When loaded onto a system it collects information like names and passwords and can be accessed by the intruder and used for authorized access to the system. 'Once inside the system, the intruder unleashes the virus, which lurks there until an authorized individual decrypts material or enters the access sequence; the virus then attacks the material.' (Dehaven, 1993: p. 140).

The 'stealth' virus is named after the stealth bomber that attacked Iraq in the Gulf War. It cruises stealthily inside a system for a long time before it strikes. It is known to be designed to elude most of the anti-virus systems. Its versions include the 4096, the V800 and the V2000.

One may ask: How many viruses are around? The answer can be found in various studies. One study in 1991 estimated that there are over 900 known viruses. Over 60% of the 600 000 PCs studied had been hit by a virus. Of the sites infected, 38% were confronted with corrupted files; 41% complained of unsolicited screen messages, interference or lockup; and 62% reported a loss in productivity (Sanford, 1993: p. 67).

How can an end-user protect a computer system from these infectious and dangerous viruses? One answer lies in knowing the possible motivations and the sources of the viruses. The main motivations are greed for money, revenge (against an employer or a firm), and the 'intellectual challenge' to outsmart others.

There are two main sources of computer viruses: (1) an employee (insider) and (2) openly distributed programs such as those distributed by magazine publishers and software vendors. The employee is motivated by greed or by revenge. The intruder that uses the open distribution channels is mainly motivated by the intellectual challenge of breaking a system. This type of intruder is a professional who does not get the public media attention but accomplishes a sinister mission none the less. The reward for this type of intruder is a rise in ego.

The main counter-strategy for (1) is to:

- control access to programs and data/knowledge and susceptible media
- invest in people who are the greatest threat and also the greatest asset to security for they can often stop or at least discourage intruders
- monitor and control access to all 'vital' computer programs as well as data/knowledge even when provided by insiders especially those that may be disgruntled or unhappy with the organization

- control all access during the conversion phase of new systems. This is when the system has been satisfactorily tested and every one relaxes, a perfect time for the intruder to sneaking a virus into the systems. A strategy to prevent this is for a copy of the tested system to be locked-up and used periodically to check for any unauthorized insertions.

The important controls for (2) are:

- do not use unknown software
- mainframes and even PCs should have a 'quarantine box' to sample check new software
- centralize software purchasing and purchase only from an approved list of vendors
- do not use freely distributed disks unless they are 'reliable' and tested in a 'quarantined' program.
- control access to networks. Do not allow persons or 'workstations' access without the 'need' to such access. When access is allowed it is logged. This logging may not always identify the intruder but it may dissuade the intruder to regularly change the common systems passwords
- educate and instruct employees of the danger of viruses and their epidemics keep track of where your disks/tapes and programs (including updates) have come from and where they have been

Controls for both (1) and (2) are:

- latest and frequent back-ups to recover, sometimes an extra generation deep
- anti-viral computer programs, scanners and filters (programs that check for 'signatures' of known viruses and alert the user to a possible danger)
- keep abreast with the virus technology — one journal on the subject is *The Virus Bulletin* published in the UK.

A combination of filters (also called screens) and gateway(s) act like a wall against viruses or other unauthorized traffic. They are called **firewalls**, and prevent unauthorized traffic (as defined by network policies) from inside to outside or the other way around.

The main danger of all these strategies is that they may lull the potential victims into a sense of being protected. The anti-viral strategies we have are against 'known' viruses only. Corporate managers and end-users must recognize that the intruder, especially the 'intellectually motivated' intruder, may be challenged by the control mechanism into finding a new strain and a new twist to an old threat or devise a new threat to beat the systems that we have not yet heard of or even thought of.

There is also a cost to all this control and strategies against viruses. There is a possible loss of morale when employees are not fully trusted. There is also a loss in efficiency. Each layer and level of security has an overhead cost and loss in productivity and performance. In addition to calculating these costs, one must estimate the probability of attack and the value of the loss entailed if the attack is successful. This analysis is necessary before a security system is designed and implemented to combat viruses.

To guard against viruses and other threats to security, a firm needs policies that will help make the system secure, the next subject of our discussion.

Policies for security

To implement and enforce any set of security measures there is a need for policies. The objectives of these policies could be:

- specify security measures and specify and assign roles, responsibilities and accountability to those who own and those who manage the messages being transmitted
- specify who should have what resources and for what purpose
- devolve responsibility (whenever possible) to the organizational point where controls are implemented
- ensure interoperability
- ensure basis for security review and audit
- base security access levels on risk involved so that there is no or little unreasonable imposition
- ensure sharing of network in a responsible and controlled manner
- protect resources from misuse, unauthorized use, malicious actions and carelessness.

To carry out these objectives for security, there may be a need for a set of policies which can be carried out modularly in increasing progression of complexity, or instituted all at once as a comprehensive security plan. These policies (Symonds 1994: p. 479—80) can be:

Network security policy

This needs to be an overall policy stating such principles as: the responsibility for network security lies with the individual managers of the end system or subsystem except when central control is needed for the benefit of all the users. Domains for security control and management are assigned with specification of necessary authority and accountability.

Connection rules

These specify the operational implementation of the broad overall network security policy. These rules include:

- rules for construction of access media whether this be badges, cards or passwords
- set up and manage accounts for users
- check for viruses
- check for backup and integrity of telecommunications system
- set up logging procedures
- set up monitoring of successful and unsuccessful attempts at accessing system.
- set up displays for warnings of possible misuse or abuse of system.

Access agreements

These are 'agreements' hopefully mutually agreed upon but if necessarily imposed that specify:

- definition of origin, destination, dispatch and delivery of messages
- definition of authorized origin
- definition of what constitutes 'authorized' traffic
- definition of responsibility of custodian and ownership of messages
- definition of the 'rights' to access by a remote user
- acceptance of compliance of standards for telecommunication accepted by the organization
- responsibilities as to the protection and security of messages.

Administration of authorization

Authorization can be the consequence of a formal computation of values of authorization variables as shown in Figure 12.5. Here the objective attributes are such considerations as the importance and sensitivity of the authorization involved; the subjective variables include the role and record of the person seeking authorization; and the access requested will involve the type of access desired, such as WRITE, READ, MODIFY or DELETE. The algorithm will then

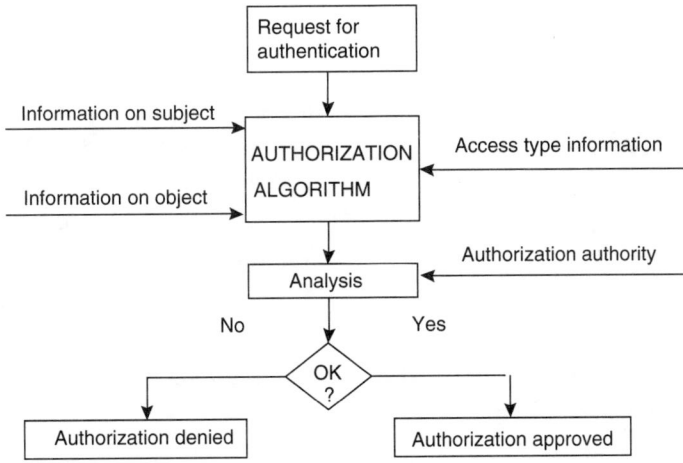

Figure 12.5 *Process of authorization*

133

determine whether a request for authorization should be granted or rejected. An algorithm may also be used to determine the level and type of authorization. This formalization is desirable in order to move the decision-making from a subjective and hence appealable decision to one that has the perception of being objective. If the decision is strictly subjective, then not only can time and energy be consumed in appealing the decision, but a lot of unnecessary bad feeling can also be generated by large egos and sensitive people.

This authorization process may seem like a lot of bureaucracy for a simple operational matter, but experience has shown that more of a Security Manager's time is spent in personal matters than one would expect. Personal emotions are often evoked on matters that involve authorization and security levels for access.

The administration of network security policies can be carried out in one of five ways:

1. centralized, where one person or a group of persons is authorized to grant and revoke authorization;
2. cooperative, where more than one person makes the decision;
3. hierarchical, where a higher level manager delegates the authority to a lower level person who then exercises all the authority;
4. ownership, where the owner of the object created assigns the authority for its access;
5. decentralized, where the authorization is assigned by the owner.

One decision not determined directly by policies but equally important is the determination of how much security is needed. This is our next and last topic for this chapter.

How much security?

Security is costly. In addition to the expense of equipment and personnel to safeguard computing resources, other costs must be considered, such as employee dissatisfaction and loss of morale when security precautions delay or impede operations. In deciding how much security is needed, management should analyse **risk**. How exposed and vulnerable are the systems to physical damage, delayed processing, fraud, disclosure of data or physical threats? What threat scenarios are possible?

As illustrated in Figure 12.6 systems and user characteristics should be assessed when evaluating risk. Opportunities for systems invasion, motives of a possible invader, and resources that might be allocated to invasion should be considered. The resources available to deter or counter a security breach should also be appraised. The amount of security that should be given to systems should be based, in part, on evaluation of expected losses should the systems be breached. One way to calculate expected losses from intrusion is by application of the formula:

$$\text{Expected loss} = L \times P_A \times P_B,$$

where

$$L = \text{potential loss},$$

$$P_A = \text{probability of attack},$$

$$P_B = \text{probability of success},$$

An insurance company or computer vendor can help management determine the value of L. Probability values are more difficult to obtain. Rather than attempting to assign a specific value (0.045 or even 0.05 may be of spurious accuracy), relative risk (high, medium or low) should first be determined and a numerical value assigned to each of these relative probabilities (for example, 0.8, 0.5 and 0.2, respectively). The risk costs can now be calculated according to the formula. For example:

Exposure	L	$\times$ P_A	$\times$ P_B	= Expected loss
1	£500 000	1.0	0.2	£100 000
2	200 000	0.6	0.5	60 000
3	50 000	0.2	0.8	8 000
Total expected loss				£168 000

Loss is determined for each exposure; the sum of the expected losses is the total risk value to the system. If P_A and P_B are probabilities for the year, expected loss is £168 000 per year.

The application of this formula will help management determine whether the addition of security measures is worth the cost and where the greatest reduction of expected losses could occur by improving security.

The figures derived from the formula are approximations at best. We simply do not have the data to calculate reliable probabilities, because the computer industry is too new to have a long historical record of invasions on which to base probability assessments. Furthermore, firms are reluctant to publicize how their security has

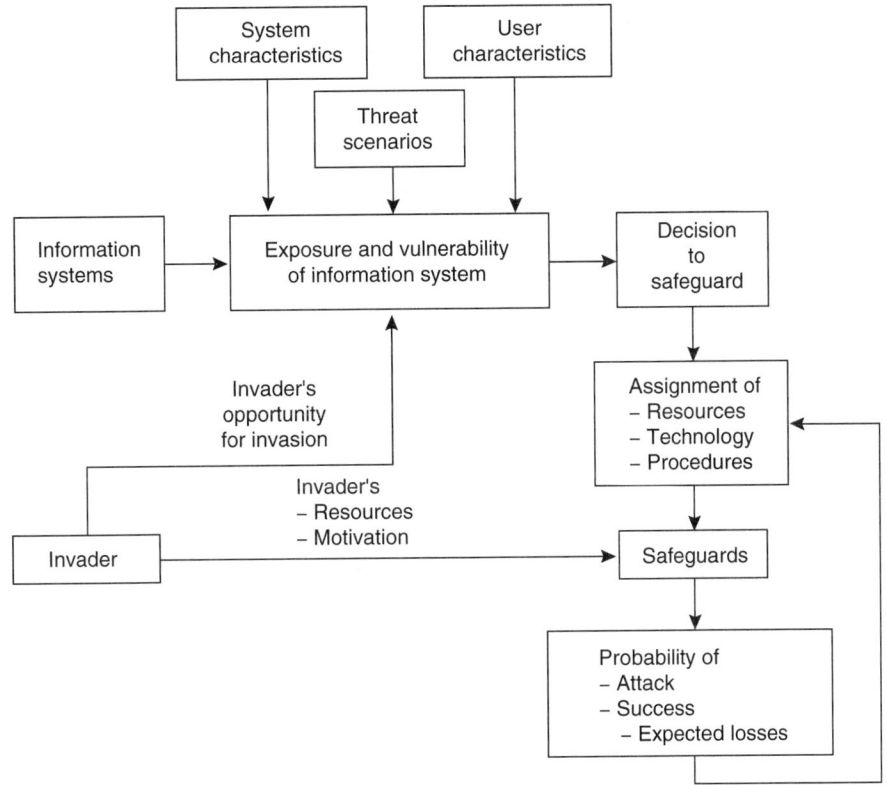

Figure 12.6   *Factors in assessing expected losses from systems intrusion*

been breached lest their credibility suffer, so news of security invasions is seldom broadcast. This means data on security infractions are incomplete. More serious, persons who design security measures are not always aware of the tricks and techniques used by perpetrators of crime to break systems security and so cannot plan countermeasures.

Summary and conclusions

There seems to be universal agreement among knowledgeable commentators that computer crime figures are destined to rise unless the computer industry and organizations that use computers and telecommunications pay greater attention to security issues and devote more resources to the protection of information systems. All known protective mechanisms in telecommunications can be broken, given enough time, resources and ingenuity. Perhaps the major

objective of security systems should be to make intrusion too expensive (in equipment, cost and risk) and too time consuming (in planning effort and time needed to actually break safeguards) to make attempted violations worthwhile.

Management's dilemma is not whether security is needed but how much. Computer crime is increasing at an alarming rate. This can be attributed, in part, to the temptation arising from the large sums of money being transferred by electronic fund transfer and to the fact that more criminals are becoming knowledgeable about computer technology and are equipped with powerful computers to help them plan and execute their crimes. There are also individuals who are challenged simply to 'beat the system'.

Risk analysis, and the assessment of expected losses and gains from security protection, is one method of helping management determine which security strategies are mot cost effective and which policies are to be revised to stay relevant despite the changing computing environment.

Advances in computer technologies (like wireless communication, image processing, tele- and video-conferencing, smart cards, neural nets and intelligent systems) pose new and unique problems of security in teleprocessing. Teleprocessing security is often part of security in corporate IT. IT management and corporate management must take steps of appointing a security officer, planning for security needs, identifying assets to be secured, preparing risk analysis, instituting policies and procedures for security, and then enforcing and monitoring these safeguards.

In the 1970s and even early 1980s, there was great concern over privacy. In the late 1980s and the early 1990s, concern for privacy has given way to concern for viruses. This is a problem that became public in 1984 and since then has become both troublesome and complex. Despite the growing number of scanners, immunizers and memory-resident activity monitors that can defend against some viruses, the parasites seem to grow faster than they can be identified and information systems are compromised. Many viruses may be in their 'incubation' stage and unknown to us. Also, the viruses are getting better and more ingenious making them difficult to identify let alone arrest. Furthermore, the problem has now become a global threat with potential insertions of a virus

that cruises through a globally accessed network. We need informational laws and judicial systems with sufficient punitive penalties to dissuade the potential intruder. And we need better hardware, and software, and procedures to beat the intruder if he or she is not sufficiently dissuaded.

Threats to teleprocessing as well as possible countermeasures in technology and in organizational policies are summarized in Figure 12.7.

Costs and benefits of security management are often intangible. For example, the costs of access can be measured, but the cost of refusing access is intangible. A rational decision on access may well be that access should be based on a 'need-to-know' basis. Often, however, the 'want-to-know' is more than the need-to-know and refusing such access can cause bad feelings, especially when the 'want to know' by a supervisor is higher than the 'need-to-know' of one supervised. The security manager must use diplomacy and much handholding to diffuse such situations if the need-to-know/want-to-know discipline is to be maintained. Thus security management is not just a problem of computer technology but an exercise in human resource management.

Corporate managers and network managers must be increasingly conscious and aware of the security of information systems for they may be

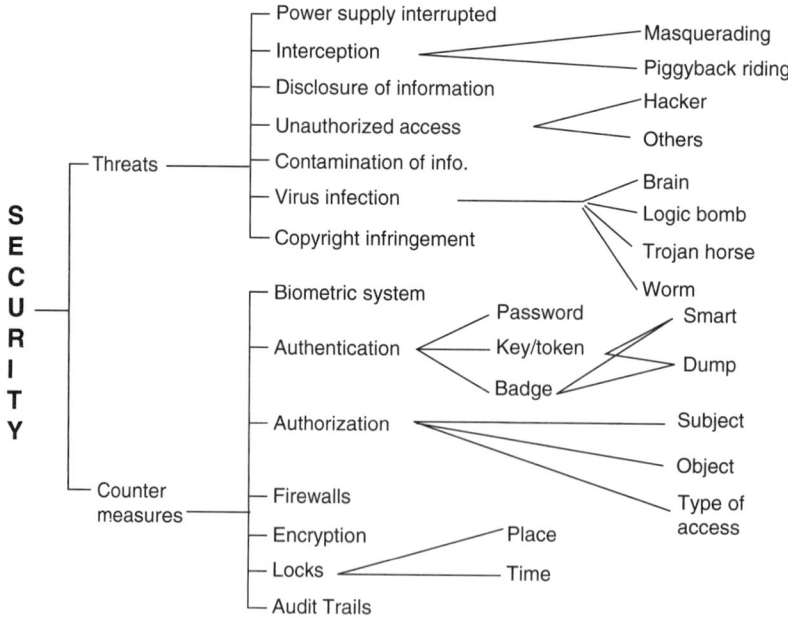

Figure 12.7 *Summary of threats and countermeasures*

held legally responsible, not just for 'prudent' protection of the company's information assets, but also for the 'prudent use of the information available to the company in order to protect its customers and employees'. (Fried, 1994: p. 63).

Case 12.1: Examples of hacking

AT&T security in the UK spotted a impostor, later known as 'Berferd', on its lines in 1991 and decided to follow him. In the next four months Berferd assaulted numerous organizations on the Internet including 300 in just one night. The AT&T lawyers decided to halt the monitoring fearing that they may be accused of harbouring hackers. At the time Berferd was in the Netherlands where hacking is legal. The Dutch authorities did nothing about it till Berferd attacked their machines.

In 1993, three friends in Milwaukee, between the ages of 15 and 22 belonged to a gang 401 (named after the telephone area code for Milwaukee in the US) penetrated three database (a bank, Sloan Kettering Cancer Center and the Los Alamos National Laboratory in New Mexico. They were caught and admitted that the only motivation was to 'have fun' and to demonstrate their technical wizardry.

Case 12.2: Examples of malicious damage

1. A 'worm' has destroyed many a computer memory and database. This use of a worm is the malign mutant of the useful worm invented by John Shoch of the Xerox Corporation in California. Shoch created a worm to wriggle through large networks looking for idle computers and harnessing their power to help solve the problem of unused resources. The malign mutant now burrows holes through computer memory, leaving huge information gaps.
2. Loss of storage can also result from a 'Trojan horse' also known to security specialists as 'logic bombs'. This happened to Dick Streeter when his screen went blank as he was transferring a free program from a computer bulletin board into his machine. Then the following message appeared: 'Got You'. Nearly 900 accounting, word

processing and game files that were stored in Street's machine were erased.
3. A 'trap-door' collects passwords as they log on, giving the hacker an updated file of access codes. The technique was used to gain unauthorized access to hospital records at Manhattan's Memorial Sloan-Kettering Cancer Center.12.4
4. A French programmer, after being fired, left a logic bomb as a farewell salute in the record-keeping software that he had been working on when fired. The bomb exploded two years later on New Year's day, wiping out all the records stored on tape.
5. A logic bomb was placed in the Los Angeles Department of Water and Power. It froze the utility's internal files at a preassigned time, bringing work to a standstill.
6. An Oregon youth in the US used his terminal to gain access to the computer of the Department of Motor Vehicles, then put the system into irreversible disarray just to illustrate its vulnerability and to prove to himself that he could 'beat' the system.

Case 12.3: German hacker invades US defence files

Suspicion that someone was wrong at the Lawrence Berkeley Laboratory was aroused when Clifford Stoll, manger of a multiuser computer system at the laboratory, noticed a 75 cent accounting discrepancy. Eighteen months of detective work followed in which Stoll cooperated with law enforcement officers to track down a hacker who used 75 cents of unauthorized time. The hacker was subsequently arrested by German authorities under suspicion of espionage.

Stoll was able to monitor the hacker's activities and observe that he methodically invaded files in some three dozen US military complexes to sift out information on defence topics. But the hacker's identity remained a mystery until Stoll's girlfriend suggested setting up a trap, a fictitious file on the Strategic Defense Initiative (also known as Star Wars) was inserted by Stoll in the Lab's computer. The hacker, whose interest was aroused when he spotted the file, stayed on-line long enough to be traced.

German authorities believed that they cracked a major ring that has been selling sensitive military, nuclear and space research information to the Soviets.

Source: Karen Fritzgerald (1989). The quest for intruder-proof computer systems. *IEEE Spectrum*, **24**(8), 22—26. See also the delightful book by Clifford Stoll (1989): *The Cuckoo's Nest: Tracking a Spy through the Maze of Computer Espionage.* Doubleday.

Case 12.4: Buying the silence of computer criminals

The computer industry research unit in the UK reports that the practice of offering amnesties to people who break into computers and steal funds is widespread. Rather than prosecute, the corporation keep silent on the crimes if part of the money is returned and the swindler reveals how the fraud was carried out. Employers fear that business might be lost if customers learn that their computer security is flawed.

In one such case, a programmer who legally diverted £8 million to a Swiss Bank account gave back £7 million for a non-disclosure agreement protecting him from prosecution. According to a member of Scotland Yard's fraud squad, employers who make such agreements may end up in court themselves, prosecuted for preventing the course of justice.

Source: Lindsay Nicolle and Tony Collins. The Computer Fraud Conspiracy of Silence. *The Independent*, 19 June 1989, p. 18.

Case 12.5: The computer 'bad boy' nabbed by the FBI

Ken Mitnick was long known for burrowing his way in the most secret silicon nerve centres of telephone companies and corporate computer centres. He even invaded the North American Defense Command computer and DEC stealing $4 million worth of software. In 1989 Mitnick was caught, convicted and put into in a low security jail. The judge ordered Mitnick to participate in a treatment program for compulsive disorders. Then Mitnick was on parole and escaped. He was back violating sensitive computer systems. His first mistake was to invade and steal software from Shimomura, a computational physicist, on of all days Christmas

day. Mitnick's second mistake was to taunt Shimomura by mocking voice-mail messages. This angered Shimomura (30) who then cooperated with the FBI in a search for Mitnick (31). Also cooperating with the FBI was the Well network of 11 000 users. Mitnick had violated Well's system in January 1995. Then, on 16 February 1995, Mitnick entered a system 5000 km away in California and wiped out all the accounting records of one of Well's subscribers. It was later learned that Mitnick had made a typing error and accidentally destroyed the accounting records. But Well's management did not know that and they decided that they could not survive any more of Mitnick and had to cut him off (and thereby warn him) or risk their entire business. They tried to contact the FBI but the FBI was on its way to arrest Mitnick and had shut off their cellular phones for fear of alerting Mitnick. Soon thereafter, at 1.30 a.m., the search ended in Mitnick's flat where he was arrested.

Mitnick faces thousands of dollars in fines and decades in prison, and without parole. The FBI is pushing for a harsh sentence to deter future computer criminals.

Source: *US News & World Report*, **118**(8), Feb. 27, 1995, pp. 66—67, and *International Herald Tribune*, Feb. 18—19, 1995, p. 3.

Case 12.6: Miscellaneous cases using telecommunications

VIRUS OFFENDER CAUGHT AND PUNISHED

Robert Morris is the son of a well respected computer expert in the US. Robert inserted a virus in a computer network that impacted negatively on over 6000 users. He was caught and tried. He was sentenced to three years' probation, 400 hours of community service and a fine of $10 000.

Source: Ralph M. Stair (1992). *Principles of Information Systems.* Boyd and Fraser, p. 628.

BUYING JEWELS IN RUSSIA

A consultant for American bank used his knowledge of password codes and transferred money to Switzerland to buy jewels in Russia. His mistake was that he bragged about it to friends and was caught. He was on bail and made another unauthorized transfer and was caught again. The jewels were sold and the bank was the first to make money out of a fraud committed on it.

BREAKING A SECURITY CODE

Ronald Riverset, Adi Shamir and Leonard Adelman are authors and managers of the RSA public key encryption. In 1977 they offered a $100 reward to anyone who could break their RSA-129 code (named after the 129 digits (429 bits) long code). In 1993, Arjen Lenstra, a scientist at the Bellcore Research Institute took up the gauntlet and in May 1994 won the bet and the award.

The breaking of the code has not exposed telecommunications to every hacker or computer criminal because the encryption code has been enhanced by using 512 to 1024 bits. It is estimated that a 1024 bit RSA key would require 3×10^{11} MIPS-years to crack.

Source: *Byte*, Sept. 15, 1995, p. 154.

Supplement 12.1: Popular viruses

It is estimated that there are some 600 viruses cruising networks. The six most often encountered during a one-month study done in the UK and reported for their percentage occurrences are:

Forms	19%
Parity Boot	16%
NYB	9%
AntiEXE	7%
Sampo	5%
Jack Ripper	5%

Sources: *Virus Bulletin*, Abingdon, Oxfordshire, UK; and *Computerworld*, Oct. 9, 1995, p. 49.

Bibliography

Bates, R.J. (1995). Security across the LAN. *Security Management*, **39**(1), 47–50.

Bellovin, S.M. and Cheswick, W.R. (1994). Network firewalls. *IEEE Communications Magazine*, **32**(9), 50–52.

Bird, J. (1994). Hunting down the hackers. *Management Today*, July, 64–66.

Cadler, E. (1991). Security strategies for the 1990's: security as an enabling technology. *Computer Security Journal*, **VII**(2), 19–25.

Cobb, S. (1995). Internet firewalls. *Byte*, **20**(10), 179–180.

Dehaven, J. (1993). Stealth virus attacks. *Byte*, **18**(6), 137–142.

Fried, L. (1994). Information security and new technology. *Information Systems Management*, **12**, 57–63.

Ganesan, R. and Sandhu, R. (1994). Introduction (to special section on securing cyberspace). *Communications of the ACM*, **37**(11), 28–31.

Gassman, H.P. (1991). Computer networks, privacy protection, and security. *International Journal of Computer Application in Technology*, **4**(4), 203–206.

Hafner, K. and Markoff, J. (1991). *Cyberpunk: Outlaws and Hackers of the Computer Frontier*. Simon & Schuster.

Landwehr, C.E., Bull, A.R., McDermott, J.P. and Choi, W.S. (1994). A taxonomy of computer program security flaws. *ACM Computing Surveys*, **26**(3), 211–254.

Molva, R. Samfft, D. and Tsudik, G. (1994). Authentication of mobile users. *IEEE Network*, **8**(2), 26–35.

Murray, W.H. and Farrell, P. (1993). Toward a model of security for a network of computers. *Computer Security Journal*, **IX**(1), 1–12.

Nash, J.C. and Nash, M.M. (1992). Matching risk to cost in computer file back-up strategies. *The Canadian Journal of Information Sciences*, **17**(2), 1–15.

O'Mahoney, D. (1994). Security considerations in a network management environment. *IEEE Network*, May/June, 12–17.

Parsons, T. and Hsu, A. (1997). Find the right firewall. **2**(2), 61–74.

Peukett, H. (1991). Enhancing the security of network systems. *Siemens Review*, H & D Special, Spring 19–22.

Salamone, S. (1993). Internetwork security: unsafe at any node? *Data Communications*, September, 61–66.

Sandhu, R.S. and Samarai, P. (1994). Access control: principles and practice. *IEEE Communications Magazine*, **32**(9), 40–48.

Sanford, C.C. (1993). Computer viruses: symptoms, remedies, and preventive measures. *Journal of Computer Information Systems*, **XXXIII**(3), 67–72.

Sherizan, S. (1992). The globalization of computer crime and information security. *Computer Security Journal*, **VII**(2), 13–20.

Svigals, J. (1994). Smart cards – a security assessment. *Computers and Security*, **13**(2), 107–114.

Symonds, I.M. (1994). Security in distributed and client/server systems – a management view. *Computers and Security*, **13**(6), 473–480.

Wallich, P. (1994). Wire pirates. *Scientific American*, **27**(3), 90–101.

Zajac, B.P.Jr. (1990). Computer viruses: can they be prevented. *Computers and Security*, **19**(1), 25–31.

13

NETWORK MANAGEMENT

Network bottlenecks could be choking your company's power to operate efficiently, and so to grow, to export, to survive. Managing the network becomes a strategic issue for the whole company ... As we begin connecting departments together, the network becomes a mission critical resource.

Chris Bidmead

The old world was characterized by the need to manage things. The new world is characterized by the need to mange complexity.

Stafford Beer

Introduction

In the 1980s, we witnessed the strong emergence of PCs (personal computers) as stand-alone computers on desktops. With trends towards decentralization and deregulation of the telecommunications industry in the US, innovation increased and prices dropped, especially the prices of PCs. As PCs became more robust and user friendly, the number of PCs on desktops of most corporations increased. However, many corporations could no longer afford the desired peripherals, databases and computing power that they needed. They had to pool and share resources. Many of these resources were dispersed and had to be connected. The solution to the connectivity problem was networks: LANs for local area networks, MANs for metropolitan areas and WANs for wide area networks. In this context, a network is a set of nodes connected by links and communication facilities that have both physical and logical components. You need to manage and control networks where zipping 'across your network are thousands of packets of information. Information that's vital to your organization. Losing even some of that data can cost your company millions of dollars.' (Derfler, 1993: p. 277).

The increase in the number and use of PCs led to the complexity of networks. There were bottlenecks in message flow, concurrent access, defective devices, devices that flooded networks with junk signals and slowed systems performance, loose connections, overloaded components, and incorrectly connected components resulting in

crashes of networks. To restore order to this chaos, there was need for an organization of management of networks that achieved most if not all of the following objectives:

- Increase productivity of end-users;
- Facilitate cooperative work between end-users;
- Function smoothly, continuously, efficiently and effectively without loss of integrity and security of system;
- Be robust against errors and misuse;
- Detect fraud activity but verify and account for selected legitimate activities;
- Monitor system to identify potential fault conditions and 'fix' faults in real-time with minimum loss of performance;
- Provide statistics (on system and its components) necessary for planning and control of system;
- Perform maintenance when needed and preventive maintenance to avoid or at least reduce breakdowns of system;
- Plan hardware/software configurations for growing needs of applications and information;
- Control devices remotely if necessary as in cases of failures;
- Implement and enforce security zones around sensitive resources;
- Be able to effectively integrate within the corporate information system.

Achieving some of the above objectives is the function of management of networks and also called network management. It is the subject of this chapter. In it, we shall examine the functions of network management identified by ISO (International Standards Organization) as:

1. Fault/problem management;
2. Performance management;
3. Security management;
4. Configuration management;
5. Accounting management.

In addition, we will examine the software necessary for network management, the managing of the human element in networks management, the development of networks and the acquisition of resources needed. We will examine how our networks can be safe and running. We conclude with an overview of the process of network management and look at future trends in networks and network management.

Management of networks

Fault/problem management

Perhaps the most important and certainly the most urgent function of network management is to detect and fix problems on a network. These problems could be a loss of performance, impaired transmission or a systems crash. They could immobilize an organization and the problems must be quickly detected and corrected.

Fortunately, we have many aids and tools to help identify and trace a problem and to provide statistics for maintenance and planning. These tools include pollers, monitors and analysers. A **traffic poller** sends echo packets to a specific device to check if there is any fault in the transmission line. A **traffic monitor** is concerned with actively checking for faulty devices. In contrast to the poller, a traffic monitor passively 'listens in' while displaying histograms of traffic patterns. Then there is the analyser. One type of analyser is the packet analyser which identifies the device that is clogging the network. Like the traffic monitor, the analyser is passive, but provides more information about the packets. The analyser for a network, sometimes called the LAN analyser, provides more information and for the entire network. It can search for duplicate network addresses, isolate failing nodes, identify the probable cause of failures, and collect clues on how the network can be improved.

There is a wide range of services offered by a **network analyser**. On the low end, we have the **software analyser** that will be adequate for a low density of traffic (around 50% of capacity), trouble shooting between two network nodes, one or two protocols, and with limited functionality. But for more complex systems, one needs the hardware–software platform handling high density traffic which may be around 140 different protocols blazing across the network at any one time. Such analysers can perform remote monitoring and control, using a large number of predefined filters. A filter allows you to sift through traffic for a selected set of parameters such as addresses, frame types and even devices by the manufacturer.

Analysers can be specialized like the **protocol analyser** which can trigger a specific action when a preset threshold is exceeded. Some analysers are portable ones and are dispatched to a trouble spot. This may not be adequate in a continuous process production line especially with a high value added product in which case the analyser is dedicated to a function or process and resides in a probe attached to the LAN, providing frequent or continuous information to the central network management. If the analysers use a distributed client–server architecture, its tasks (of monitoring, filtering, analysing, offering graphic representations of data and alerting errors to the operator) can be distributed between the embedded analyser and the manager software at the centre.

In addition to the tools discussed above, there are other facilities for fault management such as the **trouble ticket**. It records the time, date, place, operator, alarm or action taken, with each problem that occurred. Also recorded is the equipment involved and its vendor. Such information is not only useful in tracing the cause of the problem, but in helping maintenance and in trying to ensure that the problem does not occur again.

In addition to the trouble ticket and analysing tools, there are other pieces of information necessary for fault management. These include aggregated and disaggregated statistics on errors counts; traffic densities; network performance; as well as notices, alerts and alarms generated. This information should be available in easy to digest format in addition to graphical map representation of the network to locate problem areas.

The network administration staff in testing the system and reactivating it, may need control over the network, its links, and devices so that they can be initiated, closed down or restarted remotely from the central console. They should also be able to divert traffic from failing lines and devices without the end-user being inconvenienced or even knowing about it.

The problem of detecting faults can become complex because many network systems (an organization may have more than one) not only have many different devices, but many of these devices may be from different vendors. Each hardware device is associated with specific network software and protocols, as well as specialized techniques to interpret the alarms and reports generated before other appropriate diagnostic and tests are performed. However, through the use of loop-backs and tests, some faults can be isolated all the way down to a particular segment of a network, modem or other device on the network. Also, invalid configurations and front-panel tampering on various types of equipment can be detected, often remotely from the central network management console.

Performance management

The management of performance involves measuring, monitoring and metering of variables relevant to operations, maintenance and the planning of a network. The variables include network (and its components') availability (which is the mirror image of downtime of the system); response time (the elapse time from query to output); utilization of various devices and resources: both hardware (like CPU, disks and other memory devices, bridges, gates, routers, communication cards, buffers, repeaters, modems, multiplexers, switches, clients, and servers); and software (like NOS, software utilities and application programs), as well as traffic density by segments of the network. Such statistics when properly analysed can identify bottlenecks (current and potential), other potential problems, and areas that need expansion or contraction, and also help predict future trends for planning and budgeting of future systems.

Metering and monitoring is not just 'Big Brother' snooping but provides data and statistics necessary for operations. For example, statistics on the use of licensed software can be very helpful in deciding how many copies of each licensed software to acquire rather than support clients with dedicated copies that may seldom or never be used.

Monitoring can be selective and activated by a set thresholds. For example, the monitoring of a file system may start when it is 80% full, or for a printer when the print queue is longer than five or ten minutes' wait. Likewise, licensed software can be monitored when, for example, 85% of its privilege has been exceeded.

Security management

We have discussed security management in another chapter but that was largely from the view point of planning and design. During operations, security management may require enhancements to existing features and the addition of new features. Hence there is an overlap in our discussion but that overlap could serve as a review.

Many of the problems of security (and their solutions) are common to IT, as are the cases of controlling access to a database, an applications program, or even a device and peripheral. The principles of security management were discussed in the previous chapter. There are, however, special problems of security that arise in telecommunications and networking because of the exposure of messages during transmission. Messages are vulnerable to wiretapping, the electromagnetic interception of messages on communication lines. There may also be eavesdropping, passive listening, or active wiretapping involving alteration of data, such as piggybacking (the selective interception, modulation, or substitution of messages). Another type of infiltration is reading between the lines. An illicit user taps the computer when a bona fide end-user is connected to the system, but while 'thinking' leaves the computer idle and unattended. This is a tempting occasion for penetration of the system, which along with other unauthorized uses of computer time, can be quite costly to the organization.

Security management involves the definition of the jurisdiction of network zones, which are available to everyone or only to users in selected zones. This capability, for example, prevents the users in one department from using an optical scanner and printer in another department.

Another approach to security management in networks is to build a **'firewall'** referred to briefly in the previous chapter. The term is taken from fire-fighting, where it refers to the technique of preventing a fire from going beyond a line of defence. We do that in business when we need to restrict access to certain buildings. One approach would be to demand an identification and logging of all those visiting the building. A more secure approach would be to page the person being visited who would then escort the visitor in the building at all times. A still more secure approach would be to not permit the visitor to enter, but to leave a message (or package) at the front desk. In networking, all these approaches to building firewalls are used depending on the preference of the network manager. Access is controlled especially at the connections between networks such as a proprietary network and a public network.

Security management is also responsible for backing up the system in the event of a failure or a disaster on the network and to do so quickly and without loss of valuable data/knowledge.

Another problem is with viruses, mentioned briefly earlier. This is not an issue new to IT and arises even with stand-alone systems that do not use networks. However, if the virus gets on a network the contamination can be great because the propagation can be fast and spread widely. A lot of network software is able to scan for known viruses. But, the population of 'unknown' viruses is increasing (six were added each day in 1991) and it seems that a network system is never completely secure. Network management needs to be increasingly and continuously vigilant.

Empirical data shows that a virus often infects a corporate database through a corporate end-user swapping floppies, especially an end-user who has a computer at home and swaps floppies from home to the office. Other common viruses on networks are the boot sector virus, the file-infecting virus and the memory resident virus. Fortunately, there are numerous virus detecting communications software products with different types of scanners (including signature scanners), memory resident activity monitors and immunizers. In addition, there are many protective procedures that should be adopted against viruses. These include: not putting executable files on the server in directories where end-users can change them, restricting dial-in access, and not leaving the computer on and unattended all night or during the lunch break.

Another responsibility of security management is to record all information that may be useful in the future. This means that careful logs must be kept of all failures, intrusions, and unauthorized access, and qualifying each with identifiers that will trace each problem and help avoid them, or at least minimize risks of security infringement in the future.

In security management, it is important not to overspecify or underspecify. If security is lax, then there are violations by intruders and there may be some who are waiting for the opportunity. If security is too tight, then it is not only unnecessarily expensive to implement, but also has a high psychological cost of alienating the end-user who may then either bypass the system, ignore the systems procedures (such as being careless with the passwords), or just not use the systems as much or not use them at all.

Configuration management

All network systems need to be initialized and then reconfigured. Reconfiguring may mean the addition, withdrawal or modification of an existing configuration. In practice, to configure a device means assigning it a zone number and network number for routing purposes. These numbers are then used in collecting statistics on the operations of the devices.

Reconfiguring a system is also necessary when one needs to reroute networking traffic, either through different links and connections or through different permutations or combinations of equipment devices.

Another function of configuration management could be the reconfiguration of the application programs and their updates. In the early days of computing, a network manger went around each office cubicle with a computer, installing new software packages or their updates. Fortunately, there is now network software that does the distribution automatically across all legitimate workstations and clients. This saves a lot of time and energy. Also, the programs loaded onto each client node automatically draws updates from the server.

Reconfiguration can also include the rerouting or bypassing of traffic from overloaded or failed (or failing) links or congested devices to others that have slack. When the quality improves, the original configuration is restored. This automatic rerouting is very important for 'mission critical' applications where unnecessary delays and outages cannot be tolerated.

Routers needed for reconfiguration are protocol-specific, which means that more than one router is often required. New technology, however, is delivering multiprotocol routers which are capable of routing several protocols simultaneously.

Accounting management

This is not accounting in the financial and auditing sense but an accounting of network assets, both hardware and software. The accounting of each asset should be by physical location as well as by ownership so that each asset can be located quickly when needed (as in the case of a system crash) and for purposes of planning and operating the system.

Accounting of software is also important when the software is under licence. Such systems will keep track of use under the licence so that violations are avoided or at least kept to a minimum. This tracing of licences is important for the purpose of paying for the licence as well as for ensuring that no legal requirements are violated. It is therefore sometimes necessary to enforce the limitations of use and is often done by warnings when thresholds are repeatedly violated.

Accounting of assets is also important for financial accounting where the use of resources may be charged back to the user for payment toward maintenance and upgrading.

Software for network management

We have discussed the many functions of network management. But how are they implemented? In the early days, much was done manually. Nowadays, almost all such work is done by network software. Selected features of network management software are displayed in Table 13.1. Some of these features are available in stand-alone packages. Some features are 'bundled' in other packages. And some suite packages have most of the features.

These features match most of the features in the list of functions discussed above (though not in the same order) as being part of the functions of network management. However, the functions needed by any one network will be unique. It is therefore up to the network management personnel to select one or more software packages or to acquire a suite of integrated programs which may be more expensive but more comprehensive.

Such a choice is not unique to network management. Every user of a PC has to face the decision of buying the best software packages for each function like word processing, spreadsheets, database management and even networking, or, alternatively, buying an integrated packaged suite that has all the desired features plus much more. The suite choice is expensive and not optimal for each function, but it does

Table 13.1 *Selected features of network management software*

Diagnostic analysers
Server monitoring and reporting
   CPU utilization
   memory utilization
   log-on and security
   provides configuration information
Network traffic monitoring
Application metering
   enforces blocking when licence is exceeded
   queues users when licence is exceeded
Application distribution
Notification and alerting
Virus protection
   server
   client
Software distribution
Hardware and software inventory
   allows setup of licence limits
Reports
   predefined
   user-defined
   filtered
Automation and scheduling of tasks
   client automation
   server automation
   remote control of client PCs
Software support
Integration
User interface
Database
Free technical support

avoid some of the problems of making the different packages compatible with the operating system available and with each other. Furthermore, the format for each package is consistent and one does not have to learn a different set of commands and icons for each package. This is the equivalent problem faced by network managers: acquire a suite that does all or most of what you want and pay more, but do not have the hassle of packages being incompatible and having to learn more instructions and formats.

User management

Besides hardware and software, there is another important element in every network system: the end-user. Some of them are professionals with considerable knowledge of computers and telecommunications. There are, however, many who have little or no knowledge of computers or of telecommunications. They have to be trained and educated. For some, use of networks can be a cultural shock. Take for instance an office with a large volume of interdepartmental mail that was traditionally delivered manually by a messenger. Now this is to be replaced by e-mail. The advantages do not appear instantly, nor are they obvious, but the immediate problem is having to learn about telecommunications and networks and having to change the lifestyle at the work place.

End-users need to be trained not only in accessing and navigating networks, but also on the policies and procedures of handling data files; on trouble shooting and recovering from a crashed system; on the nature of network maintenance and network security; on protocols (rules and procedures that must be followed in transmission); and on the functions and limitations of routers (facility that selects and provides a path for a message).

End-users should also be trained on computer programs for network management. For example, there is a profiler program that collects information provided by the end-user. If not completed correctly by the end-user, then programs and updates due to the end-user will not be forthcoming.

The timing of the training is also important. For example, training on the applications specific to a LAN should be done before the LAN is installed or else the 'shock' resulting from an unfamiliarity with the new application can be demoralizing, causing a drop in productivity.

Introducing personnel to networking involves some of the same problems of introducing lay persons to computers. The change has to be managed with great care, some psychology and a lot of patience. It would help if the network system, especially the interfaces, were end-user friendly. What does this mean? An example would be a system with a fast response time, tolerance of common human errors, and one that keeps the end-user fully informed. For example, if there is a system crash, breakdown or slow-down in the network, the system should acknowledge the problem and estimate the time it would take to come up again. Many end-users are tolerant and understanding of problems provided they know what is happening and what to expect.

Development of networks

Thus far we have assumed the existence of a network which must be reconfigured, made secure, accounted for and maintained. But, how about an organization without a network? Or, how about one that has a network but wants another LAN, or considers the existing LANs to be inadequate and needing to be replaced? Then we need a project which must be developed. This requires a development methodology. The methodology appropriate for a complex project like a network is the SDLC, Systems Development Life Cycle. The activities of the SDLC for a network are similar to other projects in IT: feasibility study, user's requirements specification, design, implementation and the system made ready for operations. During operations, there are evaluations. If a modification is required and it is minor, then the system is maintained. If the modification required is deemed to be major, then the system is redeveloped. These activities and their iteration and recycling is shown in Figure 13.1. There are some differences in content with a typical IT project and these differences will now be examined.

The first decision to be made is what type of network is needed. The choice is between LAN, proprietary network, ISDN, public packet switching network and a public switched telephone network. There are also choices within these types of network, like the type of LAN that is best for the needs at hand. Also, the origination of the network (should it be centralized or be a client—server system?) should be considered.

Another important consideration in the design of a network is to decide whether the entire system is to be supported by one vendor or by many vendors and whether to make it an open system. The advantage of having one vendor is mostly in the short term since this will avoid the hassle when something goes wrong and one vendor blames another. With only one vendor the system is well integrated and easy to install, train on, maintain and manage. However, not being tied to one vendor has the great advantage of being able to select the best (or cheapest) component and 'plug and play'. This is the advantage of the open architecture (where architecture is the style of construction of structure including layers in the construction) and open systems, where there is interoperability between components available from different vendors (preferably internationally). In contrast, a closed architecture

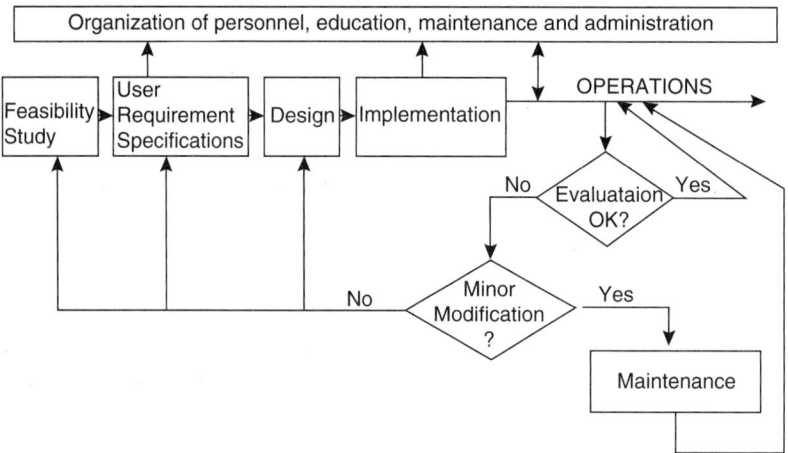

Figure 13.1 *Development of a network*

is proprietary technology developed by firms in the telecommunications and computing industry like SNA by IBM, DecNet by Digital Corporation, XNA by Xerox and those developed by governments like the DDN in the US).

The important design decision of selecting an architecture often depended on where you were and what equipment you had. If you are in the US and have IBM equipment, the chances were that you would have selected SNA. If you had a UNIX operating system, then you would have selected TCP/IP. However, if you were in Europe, you would have preferred OSI. The OSI is developed by the ISO (International Standards Organization) and was strongly supported by many countries in Europe. Europeans also supported standards developed by CCITT (Consultative Committee on International Telegraphy and Telephone) and ECMA (European Computer Manufacturers Association). But all over the world, including the US, there was strong pressure to adopt international models such as the OSI model. The OSI model would add to flexibility, reduce response time, reduce costs and add to the availability of standard network products. Network managers had to decide what products to add so that these would cause the least disruption when a standard was ultimately adopted.

Another important set of decisions concerns software. At this design stage, we are not concerned with software required for management of networks which was discussed earlier, but software required for operating a network. Such software is also concerned with protocols such as SNMP and CMIP. The **SNMP** is the Simple

Network Management Protocol which defines systems/network management standards for primarily the TCP/IP-based networks. The **TCP/IP** (Transactional Control Program/Internet Protocol) is a system designed for the US Department of Defense and is a commonly used communications standard. In competition, though, is the **CMIP** (Common Management Information Protocol), currently the only internationally ratified network management protocol standard for the OSI. The CMIP was intended to support business at different levels (local, metropolitan and wide area networks, i.e. the LAN, MAN and the WAN). The CMIP, however, has not yet caught on. It is likely, though. that in the late 1990s, the CMIP and the SNMP protocols will converge. Meanwhile, there are several network management platforms, such as OpenView, that are based on standards including SNMP and CMIP. OverView is actually a family of over 130 solutions from over 80 vendors.

Other important design decisions include the selection of bridges and gateways, security features, back-up and recovery procedures, transmission mode, links to carriers, switches, and even surge protection.

The implementation of a network will most likely involve acquiring ('buy' decision) some network software and the rest is developed in-house (the 'make' decision). Most of the hardware needed will be acquired and only seldom made in-house. In all cases, there are well tried and true procedures of resource acquisition in IT that are applicable to network development.

During operations, there are evaluations. Here the decision needs to be made as to whether a modification needed is a minor one for maintenance, or a major one for redevelopment. This is an important and difficult decision. Such maintenance decisions are not unfamiliar to IT personnel, but the terms of defining a minor or major modification are different. Consider, for example, three requests for modification of a network:

1. My LAN connection worked fine for a year and now is down. Can you please help?
2. I would like a LAN connection for my assistant who now is located in Room 39 of Building 131.
3. The current network is inadequate for my needs. I think that I need a token ring network for my department. How soon can I have it?

Of the above three requests, the first is clearly a problem with maintenance and can most likely be covered by the maintenance budget and existing maintenance personnel. The second problem is minor if Building 131 is wired for a LAN and is connected to the corporate LAN. Otherwise, it is a major problem of modification and may well require redevelopment. This request needs further investigation and is placed under advisement. The third request requires a major modification entailing new project development and new funding.

The problem facing network management is to distinguish as clearly and quickly as possible which is a minor and which is a major modification. What network management needs, then, is a set of guidelines for policy and procedures relating to maintenance of networks so that the maintenance process is clearly stated and known to all end-users.

Resources for network management

From the description of the nature and functions of network management it becomes clear that it is an important responsibility. It is also a difficult one, especially if there are hundreds if not thousands of nodes connected to the network and when the operations, and indeed the survival of the enterprise, may depend on the network(s) for the enterprise. The resources needed are hardware, software and personnel. These are discussed below.

Hardware and software

Hardware and software are needed to provide information necessary for network management. What is the information that is needed? Much of it is needed for operations at the console on an on-line real-time mode. The console operator should be able to see the network displayed (preferably in colour) at any level of aggregation, identifying the current and potential bottlenecks which may be flashing. The status of each crucial physical and logical component can be traced (each physical and logical unit may well be identified by its parameters such as name, state, physical location, etc.). All monitoring information and even statistics should be available that are either menu driven or command driven. The console operator should be able to get answers to questions like:

- What are the network loads on different sectors and at different times?
- What types of errors are occurring and where?
- What types of conflicts (like concurrency) are occurring and where?
- What channels are loaded or near loaded with what load factors?
- What are the waiting-line queues and to what extent are they exceeding or approaching the set thresholds?
- Where is security weak or possibly being violated?
- What printer and print driver is location 119 using? (Are they compatible?)
- What is the printer and print-queue status of B562?

Answers to some of the above questions and more may come without a query. That depends on the network management systems program being run. Some systems may even have a simulation program that would give answers necessary for smoothing the load and increasing the efficiency of the system. What-if questions asked may include:

What if I added a workstation (or ports) at point 119? How would this affect the service time and length of queue?
What if I added a router on segment 5?
How will the performance be affected by the addition of specific hardware or software?
What is the maximum distance travelled by end-users if I were to add a workstation at point 146? (The answer to this question will require

a database of floor plans, site plans, wiring closets, and equipment inventory by location.)

The answers to some of the simulation questions may be long term solutions that are important to planning, but such capability is often useful in problem-solving and decision-making for a network.

Some consoles have the ability to send and receive messages with the receiving being prioritized. Some of these incoming messages are recorded and appear as reports for later analysis. Some systems have alarm filtering. Some perform functions of monitoring and reconfiguration automatically. Some systems can even predict network faults based on historical data. Unfortunately, however, there is no system that is fully integrated that provides an end-to-end management, partly because different components of the system are manufactured by a proliferation of vendors. As a result, a network management centre for a large and complex system would have multiple consoles each manned by a trained operator and performing one or more functions.

Many of these problems of network management will come close to resolution as we approach a truly open system (some systems use a buffer layer to approach an open standard architecture). The open standard will not only provide flexibility in operations, but will decrease delays, improve response times and reduce costs.

One desired feature of a network system on the wish list of many network managers is that network management systems programs be intelligent (by using AI techniques for making inferences). We already have some intelligent components like the intelligent router, which could route a message between Paris and Frankfurt through New York. What we do not have is an intelligent hub and intelligent network management systems.

Personnel

To perform the functions needed in network management and to manage its resources, there is a need for staff headed by a **network manager**, also called a **network administrator**. The staff can be organized either centrally or in a distributed fashion. The principles involved in the selection of a central or distributed organizational configuration is no different from those relevant for a distributed IT organization. The personnel involved are of course different and must be grouped for functional efficiency and effectiveness. The organization will also depend on the size and complexity of the network organization, the organization culture of the enterprise, and the personalities involved both in IT and in corporate management. In the case of a small organization, there may be just one person (carrying a screwdriver) responsible for a network; in complex networks the number of personnel involved may be in the tens.

The functions of network management have been mentioned as being reconfiguration, monitoring and maintenance. These functions could result in additions and subtractions of nodes on the network. This can be a difficult political decision in cases of a zero-sum game. i.e. additions must be balanced by subtractions. Which one goes and which one stays? A somewhat similar problem arises in assigning software and security levels. This may not be a zero-sum game, but withdrawing or not assigning high security levels of new software can cause ill feelings. Security levels can be a status symbol, and if a higher level of security (or new software) is given to a subordinate on a need basis, but not to the supervisor, serious problems of unhappiness and loss of face can result.

Such decisions, along with the other decisions needed for planning, developing and operating a network, fall on the network administrator who, in addition to being a technician, must be comfortable with software (especially operating systems software), have a working knowledge of hardware, be knowledgeable of accounting and budgeting practices, and also be a politician.

To perform the functions of network management, the staff have to be organized. We discussed organization in an earlier chapter but the emphasis then was on starting a teleprocessing department. However, as the volume of processing increases and as it becomes more diverse and complex, there is need for adapting the organization to changing conditions. One such configuration, shown in Figure 13.2, is for a network of a medium sized company. In it, each person may represent one function or multiple functions. For example, the librarian or the security officer may be part time jobs assigned to a person most inclined and trained for the task or the assignments made in order to 'smoothe' and distribute the load. Also, one person may do other jobs as the need arises. For example, a software or hardware engineer assigned to maintenance

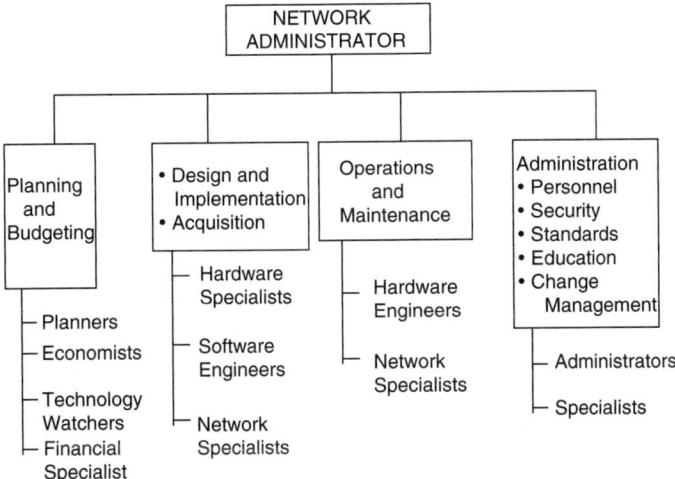

Figure 13.2   *Organization for network planning*

may well help out in the installation and implementation of a system and may even contribute to a feasibility study. This rotation of work is good for both the personnel involved (providing variety of work), but also for the organization (providing back-up and knowledge necessary for integration).

Network management must not only be concerned with the operations of networks, but also with the planning of networks. This may include facilities planning. For example, if the organization is to build a new building, should it be wired for networks? If networking is needed after the building is built, then wiring will be much more expensive. Likewise, putting cable conduits underground in order to connect buildings for networking is much cheaper and easier if planned before the buildings are built. Sometimes, such planning is not possible, but when possible, networking should be planned because such planning ahead is not only cheaper but much less disruptive for the organization.

Training is another important function of network administration. Typically, this means the training of the end-user and education in telecommunications and networking for corporate management and other corporate personnel. This is important, yet the principles involved apply to other IT functions such as those discussed earlier.

Network personnel, including the network manager and network planners, must be cross-trained on the central network console. Also important is the training and upgrading of network management staff, including the network

administrator. Their field is ever changing with changes that are technological with organizational and social implications. Network management must keep well abreast of these developments and new networking products as they occur and in some cases before they occur. Some of the changes in telecommunications and network technology will now be discussed.

Network management personnel have a virtual job that changes all the time, not only because the enterprise environment changes, but also because their technology changes. This may be said of many an IT professional, but more so for telecommunications personnel because their technology is ever changing and changing very fast. Telecommunications is one of the sectors of computer technology that is changing the fastest and this can have profound implications, not just for the enterprise, but for the nation and indeed all of us.

Networking is not just networking of telecommunications and computer equipment, it is the networking of people. It not only brings people together through e-mail and bulletin boards, but it can facilitate group work and impact on decision-making and problem-solving.

Successful networking will require changes, changes required by corporate management and end-users as well as those required by maintenance (preventive or otherwise). Also, there is a lot of swapping of data, for example, from the client to the server and vice versa. In all cases, this should be transparent to the end-user. They often get nervous with obvious changes and worry how they will be affected. The satisfied end-users are often ones that are not required to change what they do not initiate and want. Also, they do not always want to know about the bits and bytes of the system. When they need to learn about the system, they should be given what they need and when they need it, not too much before being needed and certainly not too late.

Training and education should be reactive, but also proactive. Education and training, as well as communication between the end-users and the technical personnel, are important strategies to reduce the stress of systems change. The relationship between the provider of communication services and customers of services must be reactive as well as proactive. There is often tension between the provider and customer because the customer often feels that the provider is not giving all that is possible and something is being held back. The customer and end-user may also feel that the provider can potentially reach further into their domain than they would like. The provider has more control over the customer's equipment than the customer would prefer. The customer may feel that their equipment is not restored fast enough and long after the systems performance is degraded.

There is sometimes a tension between end-users and corporate management on one side and network personnel on the other side. Why? Partly because telecommunications and networking personnel have great power since

possession of information leads to power. These personnel are what Boguslaw calls the 'computer elite'. They have complete access to the corporate and enterprise information that enters the network. This situation arises with other IT personnel especially database personnel. But, telecommunications personnel have great power of another kind. They can often start and stop any client and any server. They can 'listen' into any message that comes in or goes out. These personnel may not use their power and commit computer crime but they still have access to enterprise and corporate information that they may misuse to affect decision-making and problem-solving by corporate management. One way to overcome this problem is to build trust among the end-users and corporate management so that such misuse could never happen. Some management find it prudent to bond all telecommunications personnel.

Selecting the best equipment configuration for networking is another important responsibility of network management. A state-of-the-art configuration might consist of a multimedia open system with clients and servers using PCs, workstations and smart terminals, as well as cellular and wire-less devices, connected to LANs, MANs and WANs. The other extreme could be the stand-alone system around a mainframe with a telephone and PBX connection. Alternatively, there may be in-between configurations based on

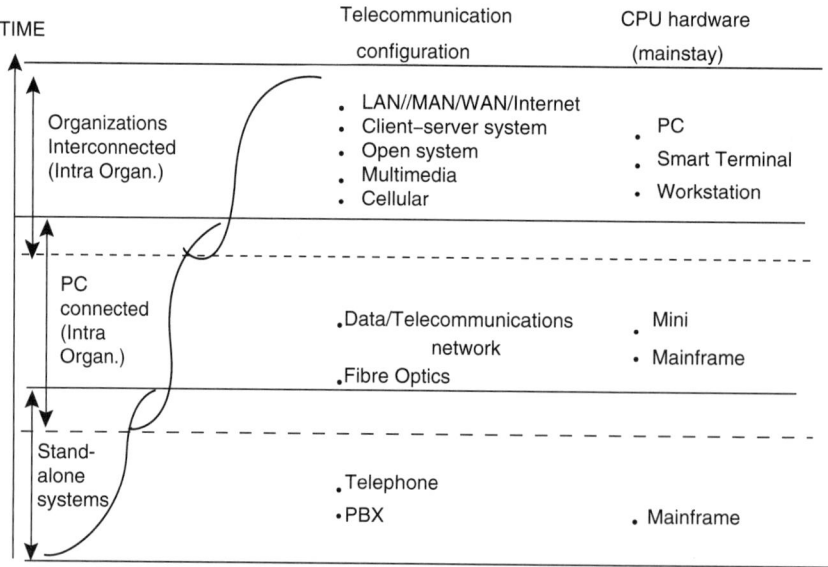

Figure 13.3 *Evolution of networks*

Table 13.2 *Summary of functions of network management.*

Fault/problem management

Troubleshoot, diagnose and predict potential
  faults
  outages
Identify, classify, analyse and report faults/outages
Initiate selected automatic 'fixes' and restorals

Performance management

Measure/meter/monitor
Network availability
Response times
Downtime
Hardware utilization
Software utilization
Traffic
Predict trends for planning, budgeting and
  maintenance

Security management

Control of access to network
Define network security zones
Detect and protect against viruses
Erect and maintain 'firewalls'
Backup and restore
Report unauthorized use and misuse

Configuration management

Initialize network entities
Reconfigure
  systems (add/subtract/modify)
  rerouting of traffic through links and devices

Account management

Manage assets of inventory
  hardware
  software (application programs)
Track and enforce licences

User management

Management of change
Training
Make system end-user friendly

mainframes and minis using fibre optics and a data telecommunications network. In many large organizations, there is not one but more than one configuration, each serving different units of the organization. Each set of configuration has its own growth curve depending not only on the network personnel but also the end-users. By the time the top part of one curve is reached, another more advanced configuration curve is found that is overlapping and reaches higher levels of performance and service. Knowing on which curve you are and when to switch to another growth curve or even leap-frog a curve is an important decision that network mangers must make.

Figure 13.3 shows the evolution in network configuration.

Summary and conclusions

A summary of functions to be performed by network management are listed in Table 13.2.

Case 13.1: Networking in the British parliament

Networking has made a wide range of information services instantly accessible to members of the British Parliament. It has a FDDI (Fibre Distributed Data Interface) backbone and spans three buildings including the Palace of Westminister. The standard services include e-mail, word processing, bulletin boards, and access to a CD-ROM library containing archives of national newspapers and Hansard. The system already has over 400 users but is designed to serve up to 4000 end-users.

The plan for the ultimate network is still being implemented. The going is slow because cabling through the twelve inch thick walls of the venerable building is a slow process. Also, any cabling should allow for the future. However, some of the future services planned do not face technological problems but purely political and social ones. For example, take the implementation of videoing all the happenings and debates in parliament into the offices of the Members of Parliament. This is technologically possible. But is it politically desirable? If implemented, would Members of Parliament stay in their offices rather than go to the floor? Here is a good example where network management is not a just technological matter but a political one also.

Source: George Black. All-party networking, *Which Computer?*.

Case 13.2: Analyser at Honda auto plant

The Honda automobile production plant in the US needs analysers for maintaining its large network. Since it has a high value added product it cannot afford delays in its monitoring efforts and in detecting failures in its token ring network.

Management wanted probes connected to its critical LANs all the time, 'so we can get immediate alerts of problems and already have the diagnostic equipment in place.' The site has a LANvista system that comprises a master console with four remote token ring 'slave' probes.

Jeffers, at Honda adds: 'Using the master console, we can toggle a view of any LAN, or all the LANs, to any network activity, or perform diagnoses ... Because probes are continually monitoring, we may already have data on problems — which is important if something has crashed to the point where we can't recreate the event'.

Because of the client—server architecture used, network management 'can make changes to the settings on the remote units, such as changing thresholds and trap requests, from the master console, can monitor and control multiple screens from a single console.'

Source: Daniel P. Dern (1992). 'Troubleshooting remote LANs'. *Datamation*, **38**(4), 53—56.

Supplement 13.1: Prices of LAN management software in 1995

Problem management (including the providing of monthly reports)	$17K
Remote monitoring	$15K
Administration (manages user IDs/passwords and resources)	$ 5K
Backup and recovery	$ 5k
Performance tuning (tracks baseline performance, analyses trends and recommends improvements)	$ 5K

Source: *Computerworld*, Nov. 27, 1995, p.54.

Bibliography

Antaya, D. and Heile, R. (1995). Digital access devices: criteria for evaluating management options. *Telecommunications*, **29**(6), 51—52.

Boehm, W. And Ullmann, G. (1991). Network management. *International Journal of Computer Applications in Technology*, **4**(1), 27—34.

Briscoe, P. (1993). ATM: will it live up to user expectation. *Telecommunications*, **27**(6), 25—30.

Broadhead, S. (1992). Network management. *Which Computer?*, **15**(10), 11—125.

Chapin, A.L. (1994). 'The state of the Internet. *Telecommunications*, **28**(1), 13—18.

Corrigan, P. (1997). Fundamentals of network design. *LAN*, **12**(1), 93—103.

Derfler, F.J. Jr. (1993). An eye into the LAN. *PC Magazine*, **12**(1), 277—300.

Henderson, L.B. and Pervier, C.S. (1992). Managing network stations. *IEEE Spectrum*, **29**(4), 55—58.

Kosiur, D. (1991). Managing networks. *Macworld*, February, 152—159.

LeFavi, F.A. (1995). Network quality assurance: a checklist. *Telecommunications*, **29**(7), 57—60.

Marx, G.T. (1994). New telecommunication technologies require new manners. *Telecommunications Policy*, **18**(7), 538—551.

Muller, N.J. (1992). Integrated network management. *Information Systems Management*, **9**(4), 8—15.

Nachenberg, C. (1997). Computer virus-antivirus coevolution. *Communications of the ACM*, **40**(1), 46—51.

Rigney, S. (1995). LAN tamers. *PC Magazine*, **14**(20), 237—267.

Schneier, B. (1994). *Applied Cryptography*. Wiley.

Steinke, S. (1997). Tutorial on SNMP. *LAN*, **12**(2), 21—22.

Strehlo, C. (1988). A 10 point prescription for LAN management. *Personal Computing*, **12**(7), 109—118.

14

RESOURCES FOR TELEPROCESSING

This is a Man's Age. The machines are simply tools which man has devised to help him do a better job.
Thomas J. Watson (of IBM fame)

I don't think the mind was made to do logical operations all day long.   *Patrica*

Introduction

We have discussed some resources needed for teleprocessing such as the media for transmission, the devices for connectivity, network management resources, and telecommunications personnel. We also recognized that the traffic is rapidly changing from data to text and now to multimedia. As the complexity of the traffic increases and as the volume of traffic increases we need more capacity for the processing of all this traffic. At the client end, we have PCs and workstations. At the server end, PCs and even minis were no longer adequate and we needed mainframes and file servers. But even this processing capacity may not be adequate and soon we will need parallel processing and perhaps even neural computing. It is these computing processors that we will examine in this chapter.

There are two good reasons for discussing parallel processing in a book on telecommunications. One is that parallel processing will be the faster way to perform a large volume of computations simultaneously; and two, because parallel processing is very appropriate for pattern recognition which is the procedure for voice and image processing and other applications of AI (Artificial Intelligence). Such processing along with data and text is the multimedia processing that will be important in the future mix of teleprocessing traffic.

Some processors will come with their own operating systems for telecommunications. But in addition we need software at the client-end, referred to a comms software; software in the connectivity devices; and software that facilitates distributed processing of applications programs, referred to as middleware. Such software we

will examine in this chapter. However, we shall not discuss applications software because that is independent of telecommunications and beyond the scope of this book.

We start with parallel processing.

Parallel processing

One solution to the demand by AI systems for raw computational power is to move to **parallel architectures**, i.e. hardware that can use more than one processing unit on a given problem *at the same time* (in other words, processors working in parallel). So, if the AI system has to search many options in order to find a 'good enough' one, then perhaps it can search more than one option at once, i.e. the search can be executed in parallel. This is the sort of promise that parallel processing hardware offers for AI. But, at a finer level of detail, there are several rather different styles of parallel architectures, and there are a number of very different ways to exploit parallelism in AI programming.

To begin with, there are both coarse- and fine-grain parallelism to be found in AI problems. At the so-called coarse-grain level a problem is composed of a small number of significant subproblems that can be processed in parallel. Thus, at the top level of the management decision-making task, there will be a number of quite distinct alternative possible decisions that could be made. A coarse-grained parallel implementation of such a problem may be able to use a small number of processors all working in parallel, each one processing a different alternative decision. At some point each independent processor must then report its findings as to the worth of the specific decision it processed (this might be in terms of how well it solves the original problem, its cost and side-effects, etc.) to a central processor, which will then choose the best decision to make.

Notice that there is always a need for some processor in the system to assume exclusive control from time to time. In the case of our simple example, this centralization of control was only necessary at the end of the task (and presumably at the beginning in order to decide how and what alternative decisions to explore initially). But, in general, there is a continuing need for this sort of processor intercommunication throughout the process of complex problem-solving. In addition, it is often hard to predict, in advance, exactly when and what will need to be communicated between the processors working in parallel.

Hence the difficulty in designing systems to exploit such coarse-grained parallelism. It is relatively easy to construct computer hardware composed of a number of processors that are both capable of working independently in parallel and of intercommunication. But it is very difficult to structure programs so that they can effectively exploit the coarse-grained parallelism inherent in a given problem. The construction of accurate and efficient simple sequential programs (i.e. conventional programs) is in itself a very demanding task. To construct parallel programs appears to be much more difficult. So, in this class of parallel system's work, the hardware is well in advance of ability to use it effectively.

The answer is, of course, for the computer system itself to detect and exploit the parallelism inherent in a problem as and when it occurs. Then the programmer's task would reduce to the conventional one of providing a correct sequential algorithm (with perhaps some indications of parallel possibilities), and the computer system would, according to its sophistication, parallelize the algorithm for more efficient computation. It is quite conceivable that an operating system could exercise this sort of judgement on a program that it is running, and, although there are a number of system-development projects working towards this sort of goal, such systems are still largely a future event.

Coarse-grained parallelism, with its relatively small number of parallel processors, is sometimes known as **mere parallelism** to contrast with **massive parallelism**, which, as the name suggests, involves large numbers of parallel processes (also

known as **fine-grained parallelism** — a term that emphasizes the elemental nature of the parallel processes). These are shown in Figure 14.1. Computer hardware is available for directly supporting this fine-grained parallelism. These machines (such as the Connection Machine; see Waltz, 1987) offer the programmer a large number of quite primitive processors, as opposed to the previous category, which made a small number of powerful processors available for the programmer's use.

A first point to clarify concerns the possibility of using such massive parallelism when we have just explained the great difficulties involved with the effective use of only a few parallel processors. In fact, these massive parallel machines are easier to use than the previous ones we have discussed. And this is because the primitive processors either do not need to communicate with each other while executing, or because the necessary intercommunication is simple and predictable in advance.

When does this type of parallelism arise? It actually occurs in a number of somewhat different ways in AI problems. It commonly occurs in vision systems and also in many different types of so-called **neural-network models** — the **new connectionism**.

You should know that computer processing of visual input amounts to the processing of thousands of simple data items known as **pixels**. Furthermore, the early stages of this visual processing — e.g. the noise removal and edge finding — often involves a very simple computation that has to be repeated on each pixel in the image. And finally, the operation to be performed on each pixel typically only involves the few nearest-neighbour pixels. Each quite simple operation can then be performed independently of most of the other pixels in the visual image, and so we have an application of massive, but simple, parallelism.

Such parallel hardware is so important for efficient image processing (and pattern recognition in general) that there is a whole range of different hardware products designed for just these sorts of applications. So we can confidently predict the coming importance of this type of parallel computer in business applications of the future.

The new AI subfield of connectionism is another potentially major application of massively parallel computers. At the moment, this is largely an unfulfilled potential, because, although there are many connectionist models

One step at a time:
traditional sequential
architecture

Several processors
operating simultaneously
and in parallel
(mere parallelism;
coarse-grained parallelism)

A network of processors
interwoven in complex and
flexible ways in massively
parallel systems
(fine-grained parallelism)

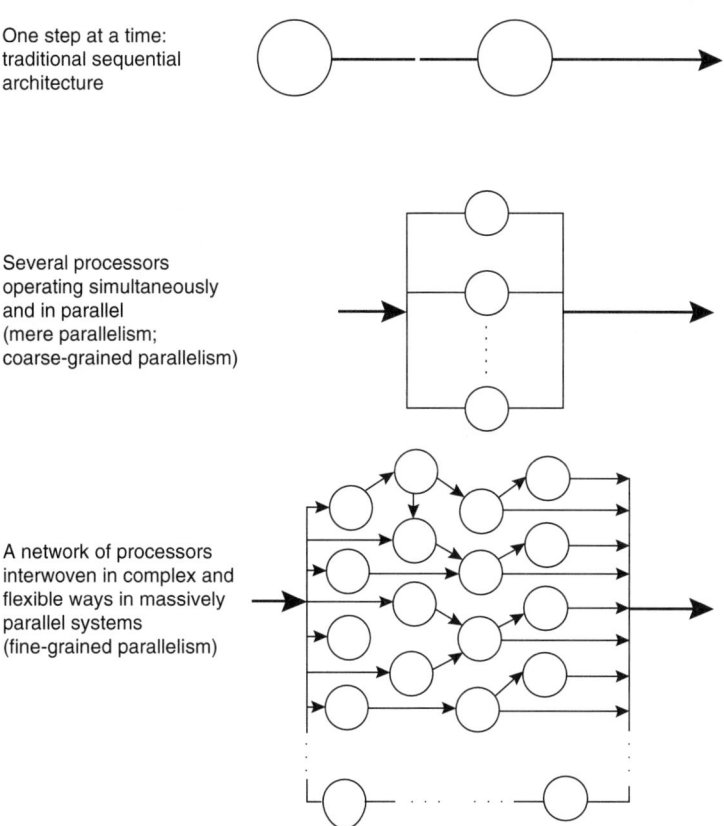

Figure 14.1   *Sequential, parallel and massively parallel processing*

being built, most of them are being implemented on conventional, sequential computers — we expect this to change. There are many, very different, schemes that are being explored under the banner of connectionism or neural computing.

The unifying theme in this AI subfield is that the main computational effort is distributed over a large number of primitive processors, networked together and operating in parallel. The analogy with brain structure and function, with its network of neurons operating in parallel and communicating via on/off pulses, is easy to see but not very productive. Currently, the analogy is only superficial; there are no deep similarities between human brains and so-called neural computers or connectionist models.

Nevertheless, neural-network models seem to offer problem-solving potential that is, in some significant ways, superior to that obtainable from more conventional implementations of AI problems. The neural computers themselves have proved to be powerful pattern-recognition devices.

Software for telecommunications

There are many types of software (besides applications software) needed for telecommunications and networks. These include: operating systems for networks, called the network operating systems (**NOS**); software on devices especially smart and intelligent devices; software that facilitates linking to say a LAN (MAN or WAN) called telecommunications or **comms software**; and then there is **middleware** that that is the enabling layer of application program interface software that is in the 'middle' of the network and the application. We shall discuss each below.

Network operating systems

Just as an operating system for a PC enables the user to access peripherals and manipulate data off a disk drive, likewise a network operating system enables a PC connected to a network to use peripherals. It controls access and use of a network and ensuring correct and hopefully efficient use of these resources.

Comms software

We have implied the use of comms software many times in earlier discussions. For example, in the discussion of modems we asserted that modems have the capability of error checking and compression. Well how does a modem have these capabilities? Through comms software. The comms software enables the linking of a computer to a LAN to perform on-line computing remotely. It has a combination of features that may include the following:

- Allow for a variety of transmission codes. For example, a PC would use the ASCII code, that is, a seven-bit code. In contrast, a mainframe may use a six-bit EBCIDIC code. The comms program must recognize differences if any and make the necessary conversion for compatibility. Another code used in a comms program is the control code, like a parity bit. This guarantees proper transmission and arrival.
- A low level of communication may be emulation as in say a terminal emulation where a terminal will bypass the local computer's resources (like CPU, memory) and rely on a remote host computer for all the processing tasks and the terminal then behaves like a computer to the end-user.
- A second level of communication will be having the ability to send and receive data. Sending a file from a local computer (or client) to a remote host computer is (or server) uploading. Receiving a file that has been transmitted by a host is downloading.
- X-ON/X-OFF support which allows flow control for accurate uploads and downloads.
- Answering an incoming phone call at an operatorless computer system, also called autoanswering.
- Dialling a phone number without assistance, also known as autodial.
- Recall and perform a logon sequences of instructions for a remote computer system.
- Compress data to reduce time and charges for transmission.

- Encrypt and decrypt data for security purposes.
- Check file names on a disk without leaving the comms program.
- A wide range of communication speeds.
- Set user-set defaults specifying speed, parity, etc. and not have to reset every time.
- Changing settings while still connected to another system.
- Break signal, which allows the sending of a signal instead of a control-key and letter combination to interrupt a mainframe computer.

A combination of features will perform different functions. Some of the many functions include:

- Dial-up mail;
- Remote file access;
- Network access;
- Host links;
- E-mail;
- Connection to a BBS (Bulletin Board System);
- Connection to the Internet;
- Internet e-mail;
- On-line services

We shall discuss the last five applications in later chapters when we discuss telecommunications applications. There are, however, other applications that are telecommunications-independent. These include spreadsheets, word processing, and data management. The comms program may need to be integrated with these applications in addition to telecommunications applications. In fact, modern computers include software with integrated applications or at least integrated comms programs and modems. The author bought a PC in 1993 that had a comms program with its inbuilt modem (including a fax modem).

Comms programs do meet most network problems. However, if 'you have a specialized communications need, there's probably a comms package designed to meet it. But the best bet for both the personal and the professional user is still a general-purpose package.' (Derfler, 1995: p. 210).

Software for linking devices

Most software for linking devices performs a specific function. However, some devices, like say a router, need to perform the function of routing messages optimally if possible. There are some operations research models that compute optimal solutions. Also, devices could be smart not in the sense that it has a computer in it and can compute as a stand-alone device, but in the sense that it has intelligence in the sense of AI. Thus it needs to be able to make inferences given decision rules. This makes the device quite valuable because it no longer is 'dumb' in the sense that it must do exactly what it is programmed to do but in addition it is able to adapt to changing conditions.

Middleware software

Middleware is the software enabling layer that supports multiple communications platforms and protocols. In many computing enterprises, data is scattered all across various incompatible networks, computers, operating systems, network architectures and protocols. Middleware is an application program that has **APIs** (application program interfaces) that enable working with separate programs that interoperate even when they are running on different LANs, MANs or WANs with varying protocol stacks. Middleware has file-transfer capabilities and can hook into e-mail software for development of mail-enabled applications as well as support object-oriented programming techniques.

One can look at middleware in the context of the seven-layer ISO network architecture as shown in Figure 14.2. Here the network traffic can be seen to split into different sessions, each corresponding to a separate applications program. The middleware is the interface between the sessions layer and the application programs. The APIs enhance the sessions layer functionality without reducing its functional complexities but makes it easier for the programmer at the applications end.

Middleware functions will vary with the vendor. The essential services offered will include distributed computing with security, balancing of load, location transparency, error recovery and transaction management.

There are many types of middleware:

Database middleware, which gives the programmer a single API that can be used to access different databases.

Remote Procedure Calls (RPC), which enable the access by calling desired procedures located in different places on the network. When an RPC is invoked, a program executes a procedure call very much like executing a subroutine in our standard 3GL procedure language like Fortran or COBOL. The call is received by the remote server and the results are returned to the sender. RPCs are not very popular in a client–server environment because procedures are not the best way that processes communicate in advanced operating systems software.

Object Request Brokers (ORB), which enables access to services offered by any object anywhere on the network. The objects could be programs or they could be resources.

Structured Query Language (SQL) Orientation, which means that the middleware will support the high level languages SQL found

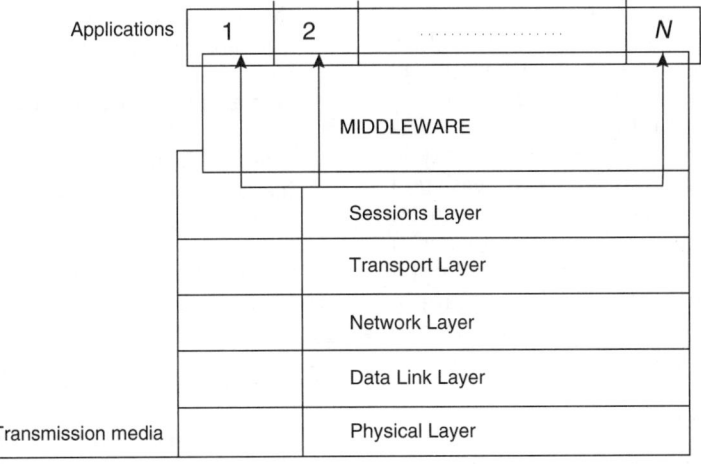

Figure 14.2 *Middleware in the layered architecture*

in many DBMSs (Database Management Systems) and other dialogue languages systems.
Message middleware, which enables programs to communicate via messages which is important in distributed computing environments. A message middleware could establish a communications sublayer that could support other middleware such as the RPCs, ORBs or other software that needs to work in more than one place simultaneously.

For message middleware, a programmer who wants to communicate between programs such as send data from program A to program B will send a command like SEND, 'Data point', 'Data destination', i.e. specify the sending and receiving data point after the command 'SEND'. After this call is made, the message-based middleware uses literally hundreds or thousands of lines of code to set up a session with the remote destination node, initialize the necessary protocols and ensure any recovery handling that may be necessary. This is a great savings to the applications programmers and makes the process so much easier and user friendly.

Message middleware is inherently asynchronous and ensures that a process is never blocked and processes the messages concurrently. In contrast, the database middleware and the RPCs are synchronous and may stay idle until it gets a clearance and permission to transfer. Thus the messaging middleware is more efficient than the database middleware and the RPCs.

Message middleware comes in two types: processes messaging, which requires both the sending and receiving processes to be available at the same time, or message queuing, which does not require simultaneous processes and can stack them in a queue. The message middleware can vary with queuing characteristics: transactional queuing, that is, survive failures; persistent message queuing, which survive some failures and do not lose the message; and non-message persistent queuing, which do not survive across any failures.

Message queuing implies a queue manager, a software program that is responsible for supplying the message queuing services used by applications including the ability to store messages in a queue for later delivery.

Summary on middleware

Summarizing middleware, we can say that middleware offers a high-level interface to

Table 14.1 *Summary of Middleware characteristics*

APPLICATION ENABLING SERVICES
Transaction workflow
E-mail
EDI (Electronic Data Interchange)

DISTRIBUTED SYSTEMS SERVICES
DISTRIBUTED SERVICES
Location
Time
Security

COMMUNICATION SERVICES
Database access
RPC (Remote Procedure Call)
OBR (Object Broker Request)
SQL (Structured Query Language)
Message middleware
Process-to-process messaging
Message queuing
Transactional message queuing
Persistent message queuing
Non-persistent message queuing

cross-platform file servers and provides interoperability to total enterprise applications and network services (even something simple like printing). Also:

... the basic chore of middleware is to move data ... the APIs enhance sessions layer functionality while reducing its complexity ... the goal is to find the API that has just the right level of abstraction and functionality for the job at hand ... When middleware is at its best, the effect is that of a distributed operating system not just as simple interprocess communications API ... No single vendor of message-delivery APIs support all platforms, protocols, and programming languages, but between them they address all conceivable mini, micro, and mainframe over just about every protocol. (King, 1992: pp. 63, 64, 66).

Different characteristics of types of messaging is summarized in Table 14.1.

Summary and conclusions

As teleprocessing grows in volume and complexity, processors will need to be powerful and versatile. Currently, mainframes are used as servers of teleprocessing services as well as for applications that need teleprocessing. With the teleprocessing of large volumes of real-time data and multimedia traffic like video-on-demand, the processing needs will increase. Even supercomputers may not be adequate. Parallel processing and the emergent option in parallelism – neural computing – may become desirable. We may also see 'programming currently available hardware in the form of a distributed-memory multiple instructions stream multiple data stream computer ... a multicomputer'. (Norman and Tharnisch, 1993: p. 263).

Along with processors for telecommunications, we need hardware network connectivity hardware like bridges, gateways and routers, but we also need protocols and a network architecture. All these resources have been discussed in previous chapters. What had not been discussed thus far was software. Some software like operating systems software and applications software and software needed at the connectivity devices are not discussed and considered beyond the scope of this book. What is discussed is middleware which is the software between the network and the application. This software with its APIs are especially necessary for distributed computing where there are a diverse set of platforms, programming languages and protocols involved.

Modern teleprocessing has a wide variety of options for each of the resources needed. In hardware, one has the whole spectrum of scalability from microcomputer PCs to supercomputers and some parallel processing. In architecture and protocols, there are choices in the US besides the OSI model and its protocols which has many supporters in Europe. In connectivity, there are topologies, access methods for LANs, MANs and WANS. And for software, there is the choice of integrated packages with comms software, multiple operating systems and many programming languages, together with middleware software and its many types. This array of choices is good for the consumer and end-user but poses a difficult decision-making process for the network manager.

Whatever the choice made by the network manger, we now have APIs that are universally available to all applications for both client–server and peer-to-peer interactions. The distributed environment with its many interfaces and services as well as development tools and

Table 14.2 *Summary of Services and Languages in a Distributed Environment*

SERVICES
Transactional management
Transactional queuing
Messaging queuing
Messaging middleware
Protocol stacks
Error checking
Resource recovery
Security

DEVELOPMENT TOOLS
3GL (Third-Generation Procedural Languages)
4GL (Fourth-Generation High Level Dialogue Languages)
SQL (Structured Query Language)
CASE (Computer-Aided Systems Evaluations), integrated or otherwise
OO (Object-Oriented Systems)

programming languages available to the application programmer in a distributed environment is summarized in Table 14.2.

Case 14.1: Replacement of mainframes at EEI

On 2 March 1995 the IBM mainframe 4381 was replaced along with its 7.5 gigabytes of disk storage running VM. It cost $250 000 per year and had reached its cost-effective life. It had a negative salvage value and thousands of dollars had to be paid to haul the equipment away. The controllers (IBM 3174/3274) and telecommunications hardware which was older than the 4381 proved to be worth more and collected a dew hundred dollars as salvage value.

The computer room was much quieter with Tricord servers running NetWare and various other PCs running e-mail and the communications programs. The servers now had greater storage capacity and the system ran on a token ring topology reflecting the IBM heritage and the utility industry that EEI (Edison Electrical Institute) represented.

EEI took the existing Statistical Analysis System (SAS) and its COBOL applications running on VM and rewrote them in the programming languages Magic (from Magic Software Enterprise Ltd.) and Btrieve (Btrieve Technologies Inc.) on the LAN. The new 'LAN-based applications allowed EEI to improve service to its members, avoid duplicate mailings, and increase the quality of its databases'.

Source: *InfoWorld*, **17**(16), April 17, 1995, p. 64.

Supplement 14.1: Top world telecommunications equipment manufacturers in 1994

Vendor ranking	Revenues (US $ billions)
Alcatel	15.94
AT&T	14.28
Motorola	13.41
Siemens	12.78
Ericson	9.65
NEC	9.08
Northern Telecom	8.87
Fujitsu	4.92
Nokla	3.68
Bosch	3.23

Source: *International Herald Tribune*, Oct. 4, 1995, p. 14.

Bibliography

Bernstein, P.A. (1996). Middleware: a model for distributed systems services. *Communications of the ACM*, **39**(2), 86−88.

DeBoever, L.R. and Max. (1992). Middleware's next step: enterprise-wide applications. *Data Communications*, **21**(9), 157−164.

Derfler, F.J Jr. (1995). Do we still need comm software? *PC Magazine*, **14**(5), 201−204.

Dolgicer, M. (1994). Messaging middleware: the next generation. *Data Communications*, **23**(7), 77−83.

Dragen, R.V. (1997). Java tools get real. *PC Magazine*, **12**(1), 181−214.

King, S.S. (1992). Middleware. *Data Communications*, **21**(3), 69−76.

Linthicum, D.S. (1995). Serving up Apps. *PC Magazine*, **14**(18), 205−236.

Norman, M.G. and Thanisch, P. (1933). Models of machines and computation for mapping in multicomputers. *ACM Computing Surveys*, **25**(3), 263−302.

Rao, B.R. (1995). Making the most of middleware. *Data Communications*, **24**(12), 89−99.

Rose, R. (1997). Personnel training. *LAN* **121**(2), 65−71.

Schreiber, R. (1995). Middleware demystified. *Datamation*, 41−45.

Yoon, Y. and Peterson, L.L. (1992). Artificial neural networks: an emerging new technique. *Database*, **23**(1), 55−58.

Zahedi, F. (1993). *Intelligent Information Systems for Business: Expert Systems with Neural Networks*. Wadsworth.

15

NATIONAL INFORMATION INFRASTRUCTURE

One of the greatest contributions technological development can make is in systematically, carefully, intentionally building a national infrastructure for information exchange.

George H. Heilmeister

Many of the technologies and players needed to conduct the information infrastructure are already in place.

Andy Reinhardt, 1994

Introduction

Every country and even many cities have an infrastructure, be it for roads, electricity, water or sewage. It requires an investment that no one person or firm can afford but it is built and supported by the government and its services are offered to its users free or for a price. Every so often, every few decades, advances in technology require an infrastructure that is very expensive but which has the potential of having a profound impact on society. One such infrastructure was the building of the interstate freeway system in the US under the Highway Act of 1956. It was 41 000 miles long and cost over $50 billion by the time it was completed in the 1970s. The infrastructure highway enabled one to travel from most points to other points in the US without stopping for a single traffic light or pay a toll. If you got tired on the way, you could stop for a rest and a picnic on rest areas located strategically on most freeways. And if you wanted to rest for the night, you would get off the freeway on one of the many exits where hotels and motels and restaurants have arisen to tempt the weary traveller. This freeway system has created new businesses and eliminated others. In fact, it has made towns out of ghost towns along old highways and created new towns around the new freeways. It has encouraged travelling for vacations and business and in many ways has changed the life and landscape of the country. And some three decades later, the effects of the interstate highway are still to be seen. And now comes the possibility of another infrastructure, that for information, which may have an equal if not greater impact on our lives. On this information highway you can cruise for business, pleasure, adventure, entertainment (including films and movies) and knowledge. It may not create (or destroy) towns and cities but they will change if people take to teleworking and reduce the need for high rise buildings and parking lots. You may no longer need to go to library buildings but can download books and articles that you want. Health-care, education and shopping could all be electronic and a very different experience.

In short, life will be different with an information infrastructure. There may well be a sea-change. The impact will be profound but unlike the building of the interstate highway, the autobahn in Germany and the motorway in the UK, the information highway may not be so well planned or orderly. It may well be like the free-for-all competitive struggle of construction of railroads in the 19th century with highly competitive companies (and international alliances) striving to get advantages and be the best and first to succeed.

There are questions that arise such as: Where is the information highway headed? Will it be a 'free' way or a toll highway? Will it use the analogue telephone or the digital system or both? Will the use of the infrastructure be controlled and regulated by the government? What will be the role of the government in relation to the businesses and industries involved? Who will be the major users of the infrastructure.? What applications will be transported? Will there be universal access or universal service or both? Will privacy be secured and property rights protected? What will the system cost? Who will pay for it? When and how will it be implemented?

In this chapter we will attempt to answer the above questions. We will start by providing a scenario of what an NII (National Information Infrastructure) may look like. We will then examine the advantages, limitaions and obstacles for its implementation. We will rely on the opinions of experts and chief executives in the field of telecommunications, networking and computing (Pelton, 1994). But first, we take an overview of what is happening in constructing NIIs around the world. We then look in greater detail at plans in the US.

NIIs around the world

We do not yet have an NII anywhere in the world, but there are many countries that are laying a foundation for NIIs. One is Japan where NTT, Japan's largest common carrier, plans a $410 Action Next Generation Communications infrastructure with fibre backbone and ISDN by 2015. Its current fibre backbone already serves intercity and intracity communications for professional and business applications. But Japan does not have a large cable market, nor does it have the software or applications for the consumer and home market. It is being said that what Japan will have in 2015 is a very large and beautiful lake but no fish to inhabit it nor any boats or recreation facilities to take advantage of its capacity.

Neighbouring countries like Hong Kong and Singapore are upgrading their telecommunications capabilities. Singapore has declared the national objective of being a wired island.

In France, there is Bete (Broadband Exchange over Trans-Eurpoean Links) with research facilities connected in France and Switzerland. There is also Brehan, a complete communication system for teleconferencing, video transmission, LAN interconnection and circuit emulation. The best known project is Minitel. It started as a computerized telephone book in Paris. Your author recalls wanting a phone connection in Paris and was asked to wait for three years. Instead he inherited the phone that was in the apartment but also inherited a phone number that belonged to the occupant some twenty years ago. The phone system was archaic. France then took the bold step to leap-frog existing technology and go for the next generation. It calculated (not allowing for inflation) that the cost of a terminal would be less than a Parisian telephone book and so gave every one a terminal with the ability to access the telephone directory. Thus Minitel now has the same access to homes as did the telephone. Now Minitel offers other services including video text and is an excellent base for an NII.

Germany has installed fibre in about 80 large cities and this is the basis of VBN (Vernittelandes

---

Table 15.1 *NII related projects in Europe*

DECT (Digital European Cordless Communication) are standards being developed for the cordless telephone, cordless switchboards and office networks.
ESPIRIT (European Strategic Planning for Research in Information Technology) is concerned among other things with compression techniques for interactive media.
GEN (Global European Network) is expected to be absorbed in METRAN (Managed European Transmission Network) which supports transmission across Europe at rates up to 155 Mbps.
GSM (Global Systems for Mobile Communications) ... Twelve years and about 6000 pages later, the standard will be DCS 1800 for 1.8 Ghz band. A third generation European standard planned for 2002.
IMPACT (Information Market Policy Action) is an effort to establish services in the areas of interactive multimedia and geographic information.
OPN (Open Network Provision) is intended to make liberalization possible by eliminating the technical obstacles in the way of the deadline for the privatization of telecoms in the year 2000.
RACE (Research and Development in Advanced Communication in Europe) is focused on integrated broadband communications and image/data communications.
VSATS (Very Small Aperture Terminals) is concerned with devices that appear on roof tops and behind garage doors have been agreed upon and now in the testing stage. A committee is now studying devices for satellite news-gathering systems.
TERA (Trans-European Trunked Radio) incorporates the best in trunk technology for standards for voice and data communication in optimized packet data services and for private mobile radio use.

Breitbandnetz) with broadband connections for video-conferencing via satellite with connections to international video-conferencing networks. BERKOM (Berlin Kommunikation) is a pilot project that has applications for business as well as for publishing and medicine.

In the UK, one will find a very special case when compared to many PT&T countries in Europe: it has a liberalized telecommunications industry since 1981 allowing the telephone and TV on the same network and making fibre optics profitable for private companies in the UK, with cable companies making half their money from telephony. One nation-wide project is Energis, owned by 12 electric companies that piggybacks the fibre optic network on the power grid. There are other test beds including one by BT in Ipswich. BT is also replacing all its copper wire with high capacity fibre.

There are many projects with joint participation of many countries in Europe. These are listed in Table 15.1. In addition to participation with countries in Europe, there are many projects with participation from Europe and the US including PEAN (Pan American ATM Network) with AT&T as the American participating with 18 other countries in a high speed network for data, text, voice, video and image transmission. And then there are projects in the US itself between US companies and by US companies themselves.

NII in the US

The US has been on the path of an NII for the last thirty years or so. Not by design but by accident: an unintended consequence of the cold war. With the cold war between the US and the USSR heating up, the American Department of Defense wanted a communication system that would not be paralysed in the event of unfriendly action and at the same time a network systems that will enable its researchers in academia and research institutions to be able to communicate between themselves. And so was born the ARPANET which was the test bed for many of the topologies and switching methods that we use today. And then there were developments in architecture and protocols giving us SNA and TCP/IP. Alongside network products were being developed, which included hubs, routers, switches, LAN connectivity products and networking software (*Data Communications*, 1994: 9–208). But perhaps the biggest milestone in the path towards an NII is the Court Judgement in 1982 which broke up the monopolistic telecommunications industry in the US, allowing it to go into computing and at the same time allowing computer companies to go into telecommunications. This liberalization (like that in the UK) freed the creative juices of the industry. We also saw a number of networks being implemented, some public and some private but all from private enterprise. Some of the private networks are international, large and very effective like those implemented by DEC, Texas Instruments and IBM. Then there were networks built by individuals in the computing profession (with many contributions like CERN from Europe) that gave us the Internet, a successor to the ARPANET. The Internet in the mid-1990s became the *de facto* national and even an international network. We shall discuss the Internet in Chapter 20 and the global network in the next chapter. For the remaining part of this chapter we will look at plans for an NII in the US, examine its need, its limitations and the obstacles in its path.

Who are the players in this business of the information highway infrastructure? One may be tempted to say 'the government' since the government is often responsible for infrastructures, even in the US, as in the case of road highways. But not so in the case of the information highway in the US, at least not when it comes to implementation of the infrastructure. Even so, industry would rather have the government participate actively because the government has the power of regulating the industry. The government responded in the mid-1990s when it announced goals for connecting all schools, medical facilities, libraries, community centres, businesses and homes in something like the scenario depicted in Figure 15.1. Also, the government wanted the system to be safe from criminals and abuse and it wanted the right to tap the system for what it may consider security reasons. There was also another condition: the access to computing had to be 'fair and equitable' and essential services were to be available universally.

Governments must choose points of intervention carefully. It must bring creativity to the problems of supporting universal service. It may have to settle for universal access (i.e. access for the poor, disadvantaged, minorities and the economically deprived) to the NII.

Does universal services and universal access mean that the poor and the disadvantaged had

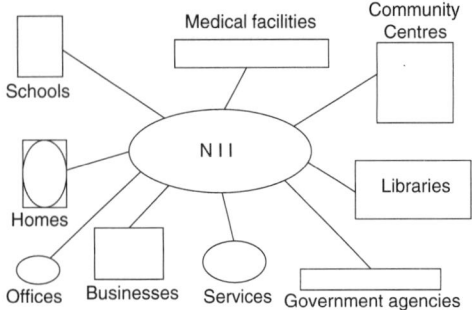

Figure 15.1 *A scenario for an NII*

to be subsidized? And if so, will the government do that or must the industry bear that burden? This is not explicit but there is support for the NII at high levels. The US in the early 1990s had a president and vice-president who did know the difference between a potato chip and a computer chip. They each even had an e-mail address.

And there was support in the legislature too. In the early 1990s, the leader of the House of Representatives in a TV address to the nation had a strand of fibre optics in one hand and compared it to the copper bulky cable that now connects much of the capital and called for modernization of telecommunications. So the vibrations are good but they are in a cloud of uncertainty that makes the industry cautious and nervous.

And who is involved in industry? There are of course the hardware manufacturers that gain in producing the many computers, microprocessors, interfaces and interconnecting products. Even the multibillion dollar software companies are interested in developing software that will give access to the NII and the global network with a click of the mouse and, with a few more clicks, access to an encyclopedia or video-on-demand.

The software industry in the US should not be discounted for it accounts for twice what is spent on hardware, which by one estimate will be around $700—800 billion (Pelton, 1994: p. 27). But the biggest players are the carriers of the gigabits of information that will roar down the information highway. They are the carriers of telephone, TV and cable. Each has billions invested in their technology and are pushing it hard for national adoption. The war is ON.

Watching the rise of NIIs are industries that stand to gain from an infrastructure, including those of advertising, entertainment, banking, publishing, game producers and software houses. All industries look to an NII for increasing

productivity, creating new products and services, improving competitiveness, and increasing trade.

Since the stakes and risks are high, test beds are being used to determine consumer preferences. Some of the pilot studies testing these preferences are listed in Table 15.2. An example of the type of question being tested is the consumer demand for any film at any time. This would require a vast library available electronically at all times, which could be very expensive. Or would consumers be happy with a limited set of films say the top ten ranked at the time? This would be relatively easy to provide, for each film could be run at ten minute

Table 15.2 *Pilot studies in the US*

FEDERALLY FUNDED PROJECTS

AURORA	Has 2.4 Gbps channels using packet transfer mode.
BLANCA	Test bed links for FDDI LAN with SONET-based ATM switches.
CASA	Uses fibre optic lines at 633 Mbps for remote supercomputer use.
MAGIC	Employs SONNET links and ATM to create a gigabit WAN.
NECTAR	Uses ATM to link gigabit LANs and WANs to one another and to supercomputers.
VISTANET	Concentrates on medical imaging with gigabit networks.

COMMERCIALLY FUNDED COLLABORATORY ON INFORMATION INFRASTRUCTURE

	Attempts to find solutions to practical problems of end-user interface and network navigation
NIXT	Prototyping data highway concepts using the Internet, FDDI, frame relay and ATM.
SMART VALLEY	To promote data superhighway through private and public partnership.
XIWT	To work out the kinks in bringing gigabit technology to the business desktop and to the home.

intervals and all ten films shown on ten different channels, so that there would never be more than ten minutes to wait. And then, of course, a large number of permutations are possible. Selecting the right option in the correct mix is a serious problem, but a managerial one not a technological one.

This is not to say that there are no serious technological problems. There are many and the most basic is the selection between telephone, cable and TV as the medium for information transfer. The telephone has the advantage of being the oldest with an entrée into many million homes.

But the telephone has a heavy bag of disadvantages. One is that there is a heavy investment in the analogue equipment when the world is going digital. Also telephone systems are based on copper wire, a $60 billion investment, when fibre has a much greater capacity (for the same weight and volume) and is easier to maintain. Copper cables can be converted to fibre but at a high conversion cost.

TV has many advantages, one being its entry into most homes in the country by being the most popular home appliance for a very long time. And it promises the choice of 500 channels. But TV has a serious problem for telecommunications: the image on the screen. The quality of the image is not so good on the screen for data and text as it is for pictures and it is text that will occupy a large part of the traffic in the future especially with the recent exponential growth of e-mail.

Cable has a much greater capacity than the telephone, but it is mostly a one way street, what is scheduled (being determined by management of the cable company) with little if any interaction by the consumer. You cannot interact by asking questions about a product and placing an order. Cable firms are in general financially weaker and are heavily indebted. But they have the great advantage that with the addition of a box they can upgrade their system to pump torrents of digital data in addition to its images. Video-on-demand (one in every five people in the US rent a film video once a month) and interactive video games are also possible but all these enhancements will cost more, around $250—350 per home.

Neither cable nor TV has the nifty switching facilities or the communication skills and experience of the technical personnel of the telephone companies. It would seem that cable and TV have many characteristics that are complementary to the telephone and that they should cooperate and produce a better joint product and better services.

Instead it seems likely in the highly competitive world of the US that these industries will compete in each other's markets. This may well expand the market but this may still be somewhat of a zero-sum game where one's gain is the other's loss. The winner in the final analysis may not win on technological grounds but on how they finance expensive projects and how they package their technology with a content that the end-user wants to consume. The technology that the end-user may well want is at least one channel that is ubiquitous, multimedia, interactive and end-user friendly. The experience for the user should not be difficult or frustrating; in fact, it should be an adventure and even some fun. But what mix of technology and content will the consumer prefer? 'What is truly impossible to foretell is how much ordinary people will pay for the new offerings ... But at present, no firm quite knows what people are willing to pay for, nor how to deliver it to them at an attractive price.' (*Economist*, Feb. 25, 1995, p. 63).).

A comparison of telephone, cable and TV for important technological characteristics is shown in Table 15.3. These carriers are testing their packages of technology and content (like news, weather maps, shopping, etc.) in different test beds across the nation. The decision on which package to offer will not be made mostly on technological grounds but on consumer response. If you give the consumer a choice, you will get a response in the market-place. Testing this response are numerous pilot studies, a selection of them listed in Table 15.4. Some of these projects have since been abandoned and others may be on the way to being abandoned. So there is a high risk because the

Table 15.3 *Comparison of telephone, cable and TV*

	Cable	Telephone	TV
Affordability	Fairly good	Good	Fairly good
Availability	Good	Very good	Excellent
Bandwidth	Good	Very poor	Good
Acting	OK	Excellent	OK
Capacity	Very high	Not high	High
Content	Questionable	Depends on user	Regulated somewhat
Ease of use	Very good	Excellent	Very good
Security	Not so good	Poor	Not so good
Reliability	Not so good	High	High
Popularity	Not much	High	High
Interactivity	Not possible	High	High

Table 15.4   *Sample trials for interactive services on TV*

Year	Location	Company	Targeted end-users in 1995
1994	Florida	Time Warner	4,000
1994	Omaha	US West	40,000
1994	New Jersey	Bell Atlantic	7,000
1994	New York	Nyex	800
1994	New Jersey	Bell Atlantic	7,000
1995	California	Viacom	Not available
1995	Washington	TCI	2,000

funding involved for each player is not just in the billions but tens of billions of dollars. And so the carriers are looking for partners in this adventure of high risk and high profits. Also, management is concerned about control so that they may protect their existence and survival. One example is the publishing industry. They recognize that the telecommunications technology can soon transfer a 20 volume encyclopedia in a matter of seconds. So why would anyone own a set of encyclopedias or even buy books or go to the marbled library building when they can scan any book, decide if it is worth reading, and then download it from a national digitized library, all the while sitting comfortably at home? The same logic applies to films. Why go to a cinema when one can see a film-on-demand? And so the moguls of the film industry in Hollywood are concerned. They have the content but not always the control over the distribution of the content of the highway. Thus there has arisen a set of incredibly complex potential alliances of computer manufacturers, software houses, carriers, publishers and Hollywood studios. Often one firm is with another in one alliance and competing against the same firm in another

alliance. What is at stake is the confluence of the telecommunications industry, the computer industry, the publishing industry and the film industry. Each player has to decide that mix of content and distribution that they want to control, all this within a cloud of uncertainty of governmental intervention and consumer response.

One cannot push the government lest they make too many regulations, but one can test the consumer response. The main options are compared in Table 15.5. They are being tested in different pilot studies, some of which are listed in Table 15.2. In these projects, what is being tested is not just technology and content but also the managerial skills and viability of alliances with a mix of partners. There have been failures of alliances, which may partly be due to the stock prices and foreign exchange rates dropping, but may also be due to the difficulties of managing complexity of international multi-industrial alliances. This may have been the reason for the loss of $3 billion by Sony in its purchase of the Hollywood giant Columbia Pictures, and the $7 billion sale of the film company MCA by the Japanese computer giant Matsuhita. At the time of the purchase in 1990, Matsuhita called the MCA purchase a pillar in the multimedia strategy for providing entertainment software that could use electronic hardware.

Significant mergers and alliances are not always with content as witnessed by the BT (British Telecommunications) which is making strategic acquisitions and alliances in the telecommunications and computer industry not just in Europe but in the US. The search for the right alliance and the proper mix of services offered with adequate control was on and most likely will go on right through the 1990s.

Table 15.5   *Alternative strategies and their characteristics for NII*

	Cable	Telephone	Internet
Topology:	Unswitched	Star or circuit switched. Good switching capability	Packet switched, router
Protocols:	Proprietary analogue	ATM, ISDN	TCP/IP
Backbone:	Analogue, fibre, satellite	Fibre optic	NSFnet
Orientation:	Entertainment	Communication (verbal)	Communication (written)
Key Users:	Homes	Everybody	Business, government and individuals
Relationship:	One-to-many	One-to-one	Many-to-many
Interactivity:	None	Some	Some

Issues for NII

There are many issue involved, both technolog-ical and organizational. One technological issue concerns the choice of a network architecture and set of protocols. In the US there is strong compe-tition between SNA and TCP/IP. The former has the clout of its designer IBM, but the latter has the support of the operating system UNIX and the reality is that it is adopted by the ubiqui-tous Internet. In Europe they do not have such a difficult choice. Being the home of international organizations for standards it has the OSI by ISO. Europe also has an adoption of the B-ISDN.

There is also the problem of bandwidth man-agement and determining what bandwidth will be required and where. Who will want interac-tive computing and who will not? Who will use information for consumption and who will want it for communication? Or both, and then what would be the mix?

Another technological decision relates to having an ATM that can allocate bandwidth on demand, handle interactive multimedia in large quantities at high speeds, and assign priorities to data even at the cell level and all without any loss of content.

As for organizational issues, there is the prob-lem of determining who controls the content of what is transmitted. Surely those who own the films will want its control but how about the control of crime in films? How about pornog-raphy? And how about the content of what an individual user can send and receive on the NII? Can someone put a computer program on the net that he or she has written even if designed to break a government code? Each of these issues is contentious and is being contested in the courts. Many of these issues are addressed in the US Telecommunications Act of 1996 along with the provision of a V chip to provide parents with con-trol on what their children can and cannot see. But these control issues of the 1996 Act are being hotly contested in the courts.

Industry wants government to be intervention-ist but only in selected areas like the hooking up of medical specialists with rural clinics; the connec-tion of schools in rural areas to educational insti-tutions; and the connection of poor communities to community centres and national libraries.

Surely the government has some role to play but a balance has to be found between control of information and entertainment and its freedom of expression. How much protection should the couch potato be given if he (or she) wished to see mud wrestling all day long? Or gamble half the day away?

Whatever the decisions, whether they be tech-nological, organizational, or political, they should result in the NII being open, easy to use, afford-able even for the poor, multipurpose, seamless in accessibility and protective of privacy, as well as having a rich and balanced information content. Governments and industries should move towards common standards that are not too rigid, so that technology can grow and innovate without one seg-ment getting an advantage over another.

We conclude this chapter with a positive note from a CEO in the communications industry: 'We are poised on the verge of an information infrastructure that I believe will serve us well in the next century. Its development is inevitable.' (Heilmier, 1993: p. 34).

Summary and conclusions

We started this chapter by comparing the interstate highway in the US with the informa-tion highway infrastructure. It may be no coin-cidence that the information infrastructure is called a highway because there is a desire for it to be as successful as the interstate highway was and still is. There are of course commonal-ties, but there are also differences. The interstate highway was ordained and paid for by the govern-ment. It had origins and destinations in a static system that is bounded by geography. It had a grand design. In contrast, the information high-way is not ordained by law or the government but may well evolve through private enterprise and populous support. The infrastructure for the information highway has origins but no destina-tion and no assurance of where it is going or what traffic mix it will carry. It has no specific design except that there must be a seamless dynamic web of networks that will be equipped with intel-ligent ATMs that will direct (switch) the multi-media traffic roaring down at gigabits per second to where it needs to go without any loss of con-tent integrity, privacy or security. For local con-nectivity we may have local telecommunication networks in addition to some private networks, LANs, as well as wire-less and wired telephones.

The NII that will emerge may well not be a monolithic system but many highways all inter-connected seamlessly together, but with gigabits per second speeds and even tera bits per second capacity. Also, the system will most likely be open with universal or near universal protocols and access. Ideally, we may even hope for the integration of 6GCS (sixth-generation computer system) with communication carriers, suppliers, as well as with the electronics industry to offer content that is rich and varied.

An NII may lead the country to greater productivity and even greater competitiveness in business and industry. It may create more jobs and may improve our health services, enhance our knowledge, and give us greater access to entertainment. But if this mix of services is used mostly for playing games or entertaining the couch potato, then the information infrastructure would have failed. It would also fail if it did not meet the real-world needs of its end-users and citizenry, or if the system was too expensive or too complex to use. The infrastructure would also be a failure if it was inequitable and left out those who could not access the infrastructure, be they children, business people, civil servants or blue collar workers. It is here that government can play a part and subsidize for services that give access to schools, community centres, and hospitals that would otherwise be inaccessible. The government may also subsidize pilot demonstration projects that are too risky for private enterprise, thereby providing important feedback on end-user needs and acceptance thresholds. This would help raise the levels of interoperability and end-user friendliness of interfaces, and prepare the public for the changes in technology that may well affect the way they work and play. Furthermore, the government should not withdraw its support for free enterprise in computing or support monopolies or the separation of the telecommunications and computer industries as happened in earlier days in the US.

Standards will be crucial, not just to an NII nationally, but to it internationally, enabling it to be connected globally. This is the subject of our next chapter.

Case 15.1: Alliances and mergers between carriers

British Telecommunications PLC plans an alliance with VIAG Intercom AG in Germany aiming to compete with Deutsch Telecom. The joint venture will offer domestic and international private virtual networks for voice and data traffic with international connections to be handled by

Concert which is a result of a joint venture with MFS Communications in the US. VIAG Intercom will use its 4000 kilometres of fibre optics owned by a regional power company that VIAG owns.

MFS Communications in the US has won a limited licence to operate a fibre network in Paris, the carrier will free users from having to purchase leased lines from France Telecom.

Telecom Finland Ltd has signed an international agreement with MFS to operate an ATM backbone extending to various European cities including St Petersburg, Russia.

MFS is operating a similar network (as with Telecom Finland) in Frankfurt, Germany, to link with an international backbone to networks in London, Stockholm and the US.

MFS has won a international carrier licence to own an end-to-end subsea fibre between the US and Sweden.

Source: *Data Communications*, Feb. 1995, p. 18, and July 1995, p. 48.

Case 15.2: Share of the European VAN market

VAN, Value-Added-Network, includes e-mail, EDI, reservation transactions, videotex, card authorization, and enhanced fax. In Europe, these VAN services are dominated by PT&Ts as shown in their share of the world market:

France Telecom	20%
DBP Telecom	13%
Telefonica	10%
Reuters	7%
BT	4%
IBM	4%
Swift	3%
Unisource	3%
Stet	3%
Geis	3%
Others	31%

Source: Ovum Ltd (London), printed in *Data Communications*, Aug. 1993, p. 18.

Case 15.3: Telecommunications law in the US

In 1995, the telecommunications industry in the US had revenues of over $700 billion, around 6% of the entire US economy. The growth came despite it being burdened under the Federal Communication Act of 1934. At that time there were only 3 TV-cable companies and one telephone monopoly. In the 1980s, there was some deregulation with the dismantling of the one monopolistic AT&T into seven large local telephone companies. Then came 400 long-distance telephone companies (dropping rates by 65%); numerous TV and cable operators; 30 communications utilities; the increase in our daily lives of computing, information processing and satellite services; a proliferation of PCs; use of digitized voice-mail and the merging of data, text and voice transmissions; and the popularity of e-mail and the Internet. Such changes in environment required a change of the 1934 Act, which resulted in the Telecommunications Act signed in February 1996. The 1996 Act is a comprehensive rewrite of regulations regarding the communications and computer industries, including the deregulation of long-distance and local telephone companies, cable operators and computer manufacturers, allowing each to compete on each other's otherwise protected turf. The 1996 Act allows for cable rates to be gradually deregulated by 1999 (allowing the high rates in the short run to provide capital for integration and expansion), and also extends the limits on the number of TV stations allowed in a single company from 25 to 35.

Between 1934 and 1995, the FCC (Federal Communications Commission) supervised the monopolistic carriers. With the 1996 Act, the FCC will encourage creativity and a free market economy for communications and computer industries whilst insisting that all the competitors will be interconnected and will share certain resources and facilities. One problem is that long-distance companies do not have the wires into homes and must rely on local providers. This reliance represents 45% of total long-distance revenues. Economies can be achieved by integrating long-distance and short-distance transmission resulting in efficiencies and cost reductions that could (theoretically) be passed on to the consumer.

The contrasting philosophical approach of the 1996 Act compared to the 1934 Act is that in 1934 the government tried to predict all the possible future problems and formulate regulations for each of them; the 1996 Act recognizes that the telecommunications and related industries are too volatile and shifting to enable prediction of the future and so the industries are allowed to develop and innovate unfettered with detailed regulations to be legislated as and when the need may arise. The Act has already resulted in a flurry of acquisitions, joint ventures, alliances, partnerships and mergers in attempts to 'bundle' services of the residential phone with long-distance calling, TV, cable, fax, video, wireless and cellular communications, paging, on-line information services, Internet services, entertainment like video-on-demand, and interactive services. Conceivably, most if not all services will be available with 'one-stop-shopping' through one carrier or utility in one bundled package and charged for in one monthly bill.[1]

The 1996 Act will not formally institute an NII or even a formal national infrastructure in the US. However, one stated objectives of public policy is to have computer access to all homes, libraries and health-care facilities, especially in the poor and rural areas, by the end of this century. The FCC has been empowered to encourage computer access to all schools, libraries and health-care facilities, especially those in rural and isolated areas, through inducements and incentives to the appropriate industries.

It is expected that the 1996 Act will encourage investors, reduce the inhibitions of entrepreneurs threatened by regulations, offer open markets resulting in reduced rates, and spur innovation that will bring new products and services to the market-place, thus giving consumers greater communication choices and rate options. The Act will trigger an explosion in learning by people of all ages, at all income levels, and in all areas (urban and rural). It is also expected that soon medical doctors and other professionals will be able to offer multimedia state-of-the-art services and consulting advice to all persons who need it irrespective of where they are. For the telecommunications industry, the Act allows integration, mergers and expansion. For the consumer, they will have integrated and end-user friendly services at a lower price (it is predicted that phone prices will drop by 20% in three years).

There are two additional features of the 1996 Act that were extensions to the 1934 Act. These relate to the V-chip and the display of indecent material on the Internet.

The V-chip provision requires that all TV equipment manufacturers install a microcircuit (in all its equipment with 13 inches (or more) of their diagonal screen) costing about $1 each in order to enable parents to 'block' material that they consider unsuitable for children. To help parents make the blocking decisions, broadcast programme Rating Boards are expected to assign a 'grade' classifying each of their programmes based on violence and 'indecent' content.

The indecent provision makes it criminal to 'knowingly' publish material that is 'indecent for minors' on public networks like the Internet. Violators of this provision face a fine of $25 000 for individuals and $500 000 for companies in addition to a possible jail sentence of up to two years. The term 'indecent' is open to interpretation and it is feared that it could well include anything on, say, abortion, rape, sex, nudity, obscenity, and even crime, violence and terrorism. The term 'indecent' is to be defined by the FCC in the 'public interest'. This will not rule out challenges in the courts. The day the bill was signed into law as the 1996 Act, the American Civil Liberties Union and 19 other organizations filed a suit challenging the law as being unconstitutional by violating the First Amendment, which allows freedom of expression. Also, 49 national organizations (including one in New Zealand) organized the 'Coalition to Stop Net Censorship'. The Coalition argues that the law will chill free speech such that public discussion would be diluted to the level of that which is acceptable only to children.[2] Other organizations also protested, including some 150 international Web sites by turning their pages black in mourning for 48 hours.

Judge Greene, known as the telecommunications Czar in the US for having made many important rulings concerning communications, including the 1984 decree to break-up AT&T, is concerned that the 1996 Act may not 'prevent domination' by a few large corporations over 'what is rapidly becoming the central factor of American life'.[3]

While the resolution of the legal battle over the 1996 Act is still uncertain, there is less uncertainty about the economic consequences: the restructuring of the communications industry and the likely dominance of the US in the global telecommunications market.

Sources: 1. *The Washington Post*, Feb. 2, 1996, p. A15.
2. *Gopher* at://gopher.panix.com:70/0/vtw/exon /faq.
3. *Wall Street Journal*, Feb. 12, 1996, p. B2.

Supplement 15.1: Milestones towards the development of an NII

1966   ARPANET, birth of networks.
1972   SNA, the first network architecture.
1976   X.25, the first public networking service.
1978   ETHERNET, the first LAN standard.
1980   Internet, the dawn of networking for the masses.
1982   Modified final judgement and the break-up and deregulation of AT&T.
1985   Deregulation of the telecommunications industry in the UK.
1993   British Telecom buys 20% of MCI and the rush to merge and acquire telecommunication assets starts globally.
2015   Planned completion of 'Next Generation Communications' infrastructure in Japan.
2015   Planned date for a nation-wide fibre infrastructure in the US.

Bibliography

Benhamou, E. (1994) NII development: where do we go from here. *Telecommunications*, **28** (1), 23–24.

Heilmier, G.H. (1993) Strategic technology for the next ten years. *IEEE Communications Magazine*, **31** (2), 30–37.

Jackson, D.S., McCarroo, T. Ressner, J. and Woodbury, R. (1995). Battle for remote control. *Time*, Spring, 69–72.

Kay, K.R. (1994). The NII: more than just a data superhighway. *Telecommunications*, **28** (1), 47–48.

Lippis, N. (1994). The new public network. *Data Communications*, **23** (17) 60–64.

Mercer, R.A. (1996). Overview of enterprise network developments. *IEEE Communications*, **34** (1), 30–37

Pelton, J.N. (1994). CIO survey on the national information infrastructure. *Telecommunications*, **28** (12), 27–32.

Reinhardt, A. (1994). Building the data highway. *Byte*, **19** (3), 46–74.

Taylor, M. (1996). Creating a European information infrastructure. *Telecommunications*, **30** (1), 27–32.

Telecommunications in Europe: creating new links. *International Herald Tribune*, 14 October, 1993, 13–24.

16

GLOBAL NETWORKS

The world has become a market-place for information... Computerized data recognizes no border check-points, customs duty or immigration officers.

Wayne Masden

Introduction

In earlier discussions, we mentioned or implied the many problems like bandwidth capacity limitations, control of 'content', as well as privacy and security considerations. All these problems are magnified in global networks. There are also problems of global telecommunications relating to developing countries. In addition, there are problems unique to globalization of telecommunications like international standards, international protection of intellectual property, transborder flow of information, and the brain drain resulting from global outsourcing (contracting of computing services) abroad. These are some of the topics that we shall examine in this chapter.

An NII is a subset of a global network, and since we do not have any NIIs it would be logical to conclude that we do not have a global network. And yet, we do, at least a *de facto* global network: the Internet. We will examine the Internet in great detail in another chapter and we will see that it was not the consequence of any grandiose careful design or any design at all. Also, it has no formal organization for its maintenance and control. It certainly was not expected to carry the traffic for businesses, though you can do some business on the Internet and can order flowers in one country to be delivered in another. However, you would be wise not to make any large financial transaction on the Internet, as it is not yet safe or secure for monetary transactions. We have e-mail, but we do not have universally accepted **e-money** (electronic money). If we had, we could cruise the souks of Istanbul and the bazaars of Bombay. We could drop-in the Victoria and Albert Museum in London

or the Louvre in Paris for an interactive look while still sitting in our chairs in Australia or New Zealand. We could get sick in one country and have lab results sent to another country for an international expert's opinion. We could order a rare book, a best seller or a professional book available anywhere in the world and read it in the comfort of our homes. What would that do to global levels of education and development? What would that do to our paradigm of higher education? How will telecommunications affect the development of developing countries? Will it improve communications and bring the world together or will there be a gap between the information rich and the information poor? How will that affect our life-style and standard of living? We shall examine these implications and the applications of telecommunications in four chapters later in this book. In this chapter we will be concerned only with applications that have an important global dimension. We will make no value judgements; we are more interested in the technological feasibility and prerequisites of these applications. For example, the solution of the international protection of intellectual property rights is a prerequisite to having free access to books in a digital library.

It is these subjects that we will address in this chapter. We examine one important consequence of global networks, that of global outsourcing, and then discuss some outstanding issues in global networks, namely transborder flow of data, the protection of intellectual property and the viability of a secure means of making transactions on the global network. First, however, we will take an overview look at global networks.

Global networks

A **global network** connects PCs, workstations and data centres around the world through fibre optic, satellite and microwave links. There were in mid-1990s, over 70 countries with full TCP/IP Internet connectivity, and about 150 with at least e-mail services through IP or via more limited forms of connectivity. Many of these countries in the 1990s are seeing the convergence of the industries of telecommunications, information and entertainment, along with the rise of multimedia traffic across their national borders.

Some applications are mission critical, where timely data is crucial, as with the Adameus reservation centre in Germany, but the 'killer' applications are e-mail and messaging. Global systems for information systems processing is not yet viable partly because of the lack of standards in important areas, because of the absence of consistent end-user interfaces, because there is little well trained personnel (administrative and end-users) to absorb the technology, because of the narrow spectrum of services offered globally, and because there is no global predictable legal and regulatory framework for fair competition and incentives for investment in telecommunications and networking. This list implies that there are prerequisites for a successful global network. More specifically, the prerequisites include: technological infrastructure, solution of international issues, government support, control of content, and international standards. These will now be discussed below.

Technological infrastructure

What is needed for a telecommunications infrastructure is a reliable technological infrastructure that includes a uninterrupted power supply, a trained cadre of technical personnel, and experience in computing. A physical telecommunications systems is necessary so that it can be coupled with at least grassroots connectivity like the Fidonet, which has simple protocol for storing and forwarding messages and a strong error checking mechanism for ensuring correct transmissions.

The technological infrastructure includes equipment for telecommunications. Their existence varies greatly in different countries with varying priorities. In Russia, there are $40 billion in projects to lay fibre optics. In Thailand, the emphasis is on cellular phones. In Vietnam the declared desire is to go digital. In Hungary, the drive is towards an advanced cellular digital communication system. And in China, it is estimated that they need over $50 billion including the costs for an increase in its telephones by some 80 million. For some countries, backwardness may be an asset. When a country goes from little or no infrastructure to the latest technology, it will leap-frog entire stages of development. This happened with the telephone industry in France in the 1970s and the 1980s.

International issues

An organization with a global network must understand and follow the many different regulatory standards concerning transmission of data across national borders. It must also observe working hours that allow for the different time zones and be familiar with the different languages used.

Government support

The success factor for any information systems application is the support of top corporate management. For global systems it is necessary to have the support of the government and its PT&Ts where much of the power resides. They may have to build and maintain a national backbone and help provide gateways for other nations. The attitude of the governments relating to open systems and privatization of the telecommunications industry is important. Though there are some 40 countries that have some privatization and liberalization, there are only two, the UK and the US, that have a liberalized telecommunications industry. The deregulation of AT&T (in the US) in the 1980s saw the rise of MCI and Sprint and the plummeting of rates and increase in services in the US.

Not all countries recognize that privatization is important or even relevant for them. Many countries attach low priority to privatization and claim that attending to the more basic needs like food and hunger, clean water and shelter, poverty and debt payments, are more important that the distribution of data and information. Poor countries with high unemployment assign a very low priority to labour saving devices and computing and telecommunications fall into that category.

Some countries think that free distribution of information can be more detrimental than beneficial to their environment, especially the free movement of pornographic or violent material. Other countries acknowledge the need for free information flow for their industrialization and development and may even help them to leap-frog stages of development.

Despite the contradictory advantages and disadvantages of regulation and privatization, many countries have found a balance in favour of at least privatization and universal access despite its PT&Ts having the monopoly of 100% of its voice service market. A good example of this is Singapore that has made many moves towards making the Internet more accessible to its citizens despite its strong concerns about the importation of pornography. Other examples in Asia are Indonesia and Malaysia, with Australia and New Zealand coming close to being in the zone of 'no-regulation'. Europe is not too far behind despite the slowness in the adoption of the European Commissions proposals to pull down barriers to privatization and liberalization of their telecoms.

The monopoly of telecom companies often give them power over pricing, and so prices cited by information providers may vary up to tenfold across countries. In Pakistan and the Philippines prices for telecommunications are raised in order to attract foreign exchange.

Control of content

Many countries question the value of the content of much that may roar down the global networks. To them, the advantage of 2000 Newsgroups or 500 channels is irrelevant when they have a very high signal-to-noise ratio anyway and bandwidth is very expensive. They argue that telecommunications are important to high tech personnel but of marginal value to countries with low tech people where 'small is beautiful'. They also argue that telecommunications will produce advertisements for products of developed countries which not all countries can afford or need. They do not necessarily want to be influenced by the life-styles and cultural values of the sending country. This is an old argument that has been around the UNESCO for years and runs into the philosophical argument of censorship and freedom of speech and expression and the right of a government to determine what is best for its population. 'Many countries believe that

data pertaining to economic potential and structure is a national resource and that their governments should be able to exercise control over its collection, use and distribution.' (Woody *et al.*, 1991: p. 36).

Standards

Acceptance of international standards is important for interoperability of telecommunication devices and hence interconnectivity. The acceptance, however, can come only after international standards are agreed to and this is where the bottleneck often appears. Getting different countries to agree to international standards is often not a technological but a political issue which requires the harmonization of conflicting national policies and local regulations, a process that is slow and frustrating at best.

Consequences of global networks

The satisfaction of prerequisites for global networks can result in many applications. These are discussed in four chapters in the next part of this book. The most common application globally is e-mail and file transfer. These are very end-user friendly. Putting a 'home-page' on the Internet requires the use of a simple language called HTML which defines the look of the page, so that it can be read by anyone on the Internet irrespective of their hardware and OS configuration. However, some applications cannot become common until some outstanding issues concerning global networks are resolved. For example, we cannot have the advantages of digitized library (despite what this could mean to education and development) without resolving the problem of protection of intellectual property. Likewise we cannot see businesses using global networks very much unless data and funds can be transferred securely and safely across national borders. These issues will be examined later in this chapter.

Important consequences of global networks are: the impact of telecommunications on developing countries, global outsourcing, transborder flow and the protection of intellectual property. We start our discussion with the consequences of telecommunications on developing countries.

Telecommunications and developing countries

Telecommunications can help developing countries (and developed countries) through increased trade including that with multinational companies. However, for some developing countries this may result in a negative balance of payments because the imports may tend to be larger than the exports resulting from telecommunications. However, if the imports (of raw materials and capital goods) were judiciously selected then they could improve productivity and gross national product in the long run. Developing countries could also benefit from businesses and industries now possible because of telecommunications. One example is financial services. Telecommunications also facilitates all business dealings and their conclusion through EFT, electronic fund transfer.

Another very important use of telecommunications is for faster and more reliable communications through e-mail and file transfer using global networks like the Internet. This is partly between administrators and managers in developing countries who once studied in developed countries and can still benefit greatly by maintaining communications with their contacts, teachers and professional mentors in developed countries. Traditionally, the contact was through mail, telephone and even diplomatic pouch. But at best this is slow (even on the phone where getting an international connection is not fast or easy even after allowing for changes in time) and not always reliable.

Also, there are many consultants in developed countries who worked in developing countries and have much knowledge and even dedication to these countries but all this is lost because to a lack of communication. Now these consultants can keep in touch with their contacts in developing countries by giving advice and even downloading materials and computer programs on the Internet as and when needed. Now communication with experts, consultants and professional mentors can be fast and even interactive or at least on-line on the personal and professional levels. These contacts can refer the query to others when necessary and provide information of the latest state-of-the-art technology.

> The process of entering the global community is as much one of entering a new culture as it is one of learning about and exploiting the new set of technologies ... future generations

of information services and products that live on the network, including intelligent user agents and knowbots, are not likely to find prenet analogues as easily. To understand the electronic network culture and be most effective in exploiting it, one must live in and explore it. (Sadowsy, 1993: p. 46).

To achieve electronic communication, the developing country must have a basic telecommunications infrastructure. This infrastructure is partly technology which will include availability of adequate and reliable bandwidth, reliable telecommunication media, connectivity between critical network links, a fairly uninterruptible power supply, supply of spare parts and access to maintenance services.

This infrastructure is capital intensive and may not be affordable by countries that have to consider the needs of foreign exchange for more mundane but essential imports for daily living like food and medical supplies. These basic needs have a higher priority than the desire for interactive communications, interconnectivity and better navigational tools for the Internet, even if this means access to an information-rich environment.

The need for a telecommunications structure cannot always be justified on the platform of communications. Not all developing countries want unfettered communications on the Internet for various reasons. Some do not want material that they consider pornographic and what developed countries may consider artistic. Some countries may also want to restrict information that they consider harmful to their society or may object to material that they consider subversive. For example, in 1995, Chinese dissidents in the US sent a sea of anti-governmental material on the Internet on the occasion of the anniversary of the Tiananmen Square incident. The government was caught by surprise and may well prevent that from happening again. But there may be other surprises in store. As with security, you can build a 'firewall' but there are many creative people around that want to break that wall for purposes that may be ideological or even just the fun of breaking the rules. Controlling content is also a problem in developed countries where parents want to control the material that their children see on the Internet. This is possible by software that can prevent material with obvious screen names and a history of sending certain types of material. A 1996 Telecommunications Law in the US requires a V-chip that allows control over content. But controlling content world-wide even with international agreements may be very difficult.

Discriminating information coming from MIT (Massachusetts Institute of Technology) as to whether it is technological or ideological would be difficult.

The electronic infrastructure also includes the knowledge and ability to navigate, locate and retrieve information over the world-wide global networks. Also, you need the equipment to send and receive electronic information. In some developed countries there is a PC on most office desks and homes and the typewriter is no longer in local production. In many developing countries, they do not even have typewriters on every desk. And when they do, it is very much part of their way of communicating. Changing from the typewriter to the PC for daily communications requires a sea change in attitudes and life-style.

Global outsourcing

Outsourcing is the subcontracting of services. In the early days of outsourcing, it was mainly in the preparation of data, But data collection has since been largely automated. However, the need for outsourcing has not diminished since it has now shifted to programming. In such services the developing countries have a comparative advantage. Yet all the advantages, risks and limitations of outsourcing within national borders applies to outsourcing abroad plus much more. It is the extra additional advantages and limitations that is the subject of rest of this section.

The most obvious advantage of global outsourcing is the cost advantage. The MIPS (millions of instructions per second) cost to labour cost ratio in a developing country is almost 10—50 times that of a developed countries (Lu and Farrell, 1990: p. 290). This is partly because the cost of computing equipment required (numerator of ratio) is high in developing countries and because of the high cost of transportation, import tariffs and the many middle men involved. The other reason for the high MIPS cost ratio in developing countries is that the cost of labour (denominator of ratio) is low in developing countries even for skilled computer specialists. However, these costs may drop over time partly because of the large increasing pool of computer specialists. In India alone, there are over a quarter of a million new graduates per year that

are technically trained that speak English and are 'abundantly available at very low salaries' (Apte, 1990: p. 293).

Good programmers and analysts are not the preserve of developed countries. One American hiring programmers in Austria and Eastern Europe in the 1970s was asked why he was looking outside America when Americans had the best departments of computer science and were the leading manufacturers of computer equipment. The answer was that many programmers outside America have fewer resources of computer memory and computer run time and so they have to work harder and write more efficient programs. This statement may well be true of many a developing country in the 1990s. However, many of the professionals that are involved in outsourcing of computer services are family businesses with little or no experience of large complex information systems development.

There are other advantages of global outsourcing of computer development projects by developed countries. These include having access to a pool of professionals in developing countries without having to pay generous fringe benefits or having to contend with the unions of developed countries. Also, global outsourcing offers knowledge and access to entry into the market of the vendor country.

Countering the advantages of global outsourcing are the many risks, disadvantages and limitations involved. The risks include political and social instability of the vendor country, changing national laws regarding foreign investments and national political ideologies that are sometimes hostile to collaborations with foreigners.

The disadvantages of global outsourcing are that in developing countries it requires a much greater degree of monitoring (and control) of quality, process and time of completion. There are also additional costs of telecommunications, travel and the costs of training vendor personnel to the organizational and national culture of the host country. There could be potential problems with foreign exchange, time zones, and the negotiation of legal contracts for the outsourcing. There is also the problem of finding appropriately trained information systems personnel with an adequate infrastructure of telecommunications and other required computing resources. Moreover, there may be resistance among systems professional in the developed countries, especially in times of recession when outsourcing personnel may become potential competitors.

One important obstacle in global outsourcing to developing countries is the lack of experience in many types of information systems development projects in developing countries. Akinlade of Nigeria, has the following observation:

> In a typical developing country, several application generators are available so that developing non-trivial software systems from scratch is not only unnecessary but is also uncommon ... the software engineer in a developing country may never be able to put into practice, by working on real-life, large, complex systems, the theory that he has learned ... his working environment does not motivate him to play an active part in the country's evolution of this very exciting discipline. (Akinlade 1990: pp. 71–2).

There is one important problem arising partly from global outsourcing: it causes a brain drain from the vendor country to the developed country. There is already a brain drain resulting from students of developing countries that do not return from developed countries where they go for their advanced education. The developed country does not always discourage this brain drain because in many cases they pay for the education through scholarships and jobs. To add to this, there is the brain drain resulting from global outsourcing. This, argue the developing countries, is unfair and even an act of stealing by the developing countries. The professionals involved are mostly educated in the developing countries and trained on local problems. They go to the developed countries to meet their clients in outsourcing and are then hired by their clients. For the developed countries these resources are relatively marginal, but for the developing countries they are crucial for their economic development. Some developed countries do not help the problem by liberal immigration policies. A skilled programmer or analyst can get an immigration visa to the US, and there are firms in the US that aggressively import computer programmers and analysts and facilitate their immigration, which only increases the brain-drain problem.

Given all the drawbacks and limitations of global outsourcing, it is still attractive to many developed countries to have their software constructed abroad, including networking software and even all network services. The basic players and content of global outsourcing are summarized in Figure 16.1. One can expect global outsourcing to grow vigorously in the late 1990s and early 21st century. What greatly inhibits greater global outsourcing are the constraints in transborder flow and the lack of protection of intellectual property. These topics we will now discuss.

Transborder flow

Changes in the world political scene offer opportunities for trade and communications across national borders. Large masses of people in counties like the former USSR, Eastern Europe and China have joined market

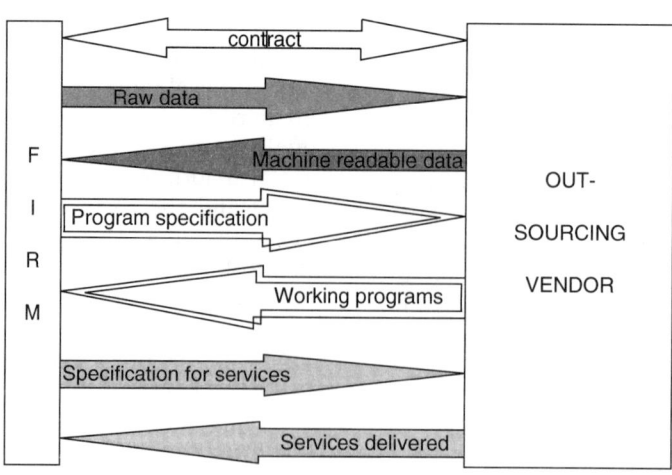

Figure 16.1 *Types of outsourcing*

economies. Other countries, like India, that were market-oriented economies, but unfriendly to Western companies in computing, are now more open and less afraid of foreign domination. The GATT agreement of 1994 has reduced trade barriers and acknowledged intellectual rights of the computing industry. If these rights are respected by the important trading countries, then this contributes towards the opening of new markets of information-based services and will lead to globalization of IT. This growth will come partly from multinational companies that will prosper and the international firm will be the norm rather than the exception. Information systems will share data across the many barriers of borders and geography whilst overcoming the risks of foreign exchange and national political stability. There are, however, still problems posed by cultural differences between managers, end-users and developers and the wage differential between different countries.

Since transborder flow is the concern of multiple countries, it has been the subject of discussions in international bodies. In 1985, the 24 countries of the OECD (developed countries) adopted a Declaration on Transborder Data Flows, in which they agreed to:

1. promote access to data and information and related services, and avoid creation of unjustified barriers to the international exchange of data and information;
2. seek transparency in regulations and policies relating to information, computer and communications services affecting transborder data flows;
3. develop common approaches to dealing with issues related to transborder data flows, and, when appropriate, to develop harmonized solution;
4. consider possible implications for other countries when dealing with issues related to transborder flows. (Gassman, 1992: p. 204)

Part of the problem of seamless transborder data flow is the concern for the security and privacy of data that flows across national borders. Sixteen out of the 24 members of the OECD countries have passed guidelines in 1988 which state eight principles (Gassman, 1992: p. 204):

1. collection limitation,
2. data quality,
3. purpose specification,
4. use limitation,
5. security safeguards,
6. openness,
7. individual participation,
8. accountability.

Telecommunications and networking could bring instant international communications across national borders though there may still be restrictions on transferred flow of data and information. International telecommunications has helped global outsourcing to the advantage of developed countries for it has made developed countries more accessible. But there is a downside to it: developed counties are attracting computer personnel from the developing countries and so there is a brain-drain in developing countries. Furthermore, developing countries may export (by design or accident) computer viruses across national frontiers. Such a virus made in Pakistan devastated billions of bytes of data all over the US. Also, telecommunications across the oceans has encouraged unauthorized access to computer systems such as the espionage for the former Soviet KGB by hackers in West Germany. The threat of espionage from the former Soviet countries may have decreased but international sabotage is still a threat. There is a story that the software bought by Iraq for the control of its air defence system contained a computer routine that was activated remotely and destroyed the Iraqi control and command system just before the Gulf War. If this story is true it may never be confirmed. But it does raise grave possibilities such as the 'bugging' of a country's software system by its enemies or even terrorists. There is also the danger that espionage will shift from the political to the economic stage. And so the need for security is a very important concern of all managers of development, not just the concern for international violations of security but even violations of a firm's database. This is not only of concern to multinational firms but all firms that have secret and confidential data and are vulnerable to unauthorized access. Firms must also be concerned with their proprietary software despite the GATT agreement of 1993–94 that protects copyrights of software, because there is some uncertainty about the observance and enforcement of the protection of intellectual rights. Every year billions of dollars' worth of software is pirated and development managers must continually try to secure their systems against such piracy.

Protection of intellectual property

Transfer of technology can be inhibited by the lack of protection for **intellectual property** which includes patents, copyrights and trade secrets. From the point of view of the management of development of information systems, copyrights are more important. A **copyright** under the Copyright Act of 1976 (in the US) classifies computer programs and data as non-dramatic literature works deserving copyright protection. **Software piracy** is the unauthorized copying of computer programs for the purpose of selling the illegal copies or for unauthorized commercial or even private use. Programs that would sell for $600 are being sold by software pirates for as little as $15 in shopping malls in Hong Kong and Seoul. Computer dealers download unauthorized copies of software to the hard disks of computers that they sell, or offer pirated software free as an incentive to buy hardware. Thus there is little incentive for customers to buy programs legally constructed by software houses in developed countries.

It is estimated that in 1993 the US software industry lost $2.1 billion in Japan and about $10–12 billion world-wide in developing countries because of software piracy (Gwynne, 1992: p. 15). The BSA (Business Software Alliance), operating in the US, estimates that 90% of all software in use in Asian countries is pirated (Weisband and Goodman, 1992: p. 87).

What makes software piracy different from other thefts is that its occurrence is not detectable or obvious as with say the theft of jewels. Also, software piracy is cheap, easy and efficient. There are ways to protect unauthorized copying but there are also ways that the pirate can get around it by opening up the programs and rewriting them.

The other problem is that software piracy is often not considered wrong or ethical. For some, it is like driving a car at 60 or even 80 in a 55 m.p.h. zone. Some think that it is a transfer of technology to developing countries that is long overdue. Whatever the motivation or justification, the temptation to pirate software is great.

The developed countries, however, object to the 'free ride' especially when the software production world-wide is expected to grow from $36.73 billion in 1989 to around $340 billion by 1996 (Schware, 1990: p. 101). The US is the source of much new and innovative software and is very concerned. It is proceeding at all levels: international, bilateral and unilateral.

At the international level, the US demanded the strengthening of the protection of intellectual property lacking under the multilateral Berne Convention, and this was an important objective of the Uruguay round of GATT in 1993. It was achieved and over a hundred countries have signed the agreement. But there are countries outside the GATT agreement and the adequate enforcement of the agreement, from the point of view of developed countries, is somewhat inadequate or even in doubt.

At the bilateral level, the US has agreements with many countries, but not all of these agreements have been enforced. For example, the agreement with South Korea took four years before the government brought their first copyright action. In China, there was no mechanism to enforce the copyright protection law agreed to in June 1992, a source of great friction between the two countries between 1994 and 1996.

If the effort at the international and bilateral levels fails, the US government can act unilaterally. Section 301 of the US Trade Act was amended to toughen penalties for countries that do not protect American intellectual property rights. Then, in 1993, the US government suspended the duty-free status for Cyprus on the grounds that Cyprus had imported almost four million blank cassettes, possibly to copy and then export both software and other intellectual property like films and music. This action against Cyprus is intended to be a strong signal that the US will no longer allow intellectual piracy which has existed with impunity for many years.

Outside the government, there are associations for software protection like the BSA, Software Publishers Association in the US; CAAST, Canadian Alliance Against Software Theft in Canada; FAST, Federation Against Software Theft in Germany; and INFAST, INdian FAST in India. In the US, trade associations like the International Intellectual Property Alliance and the BSA based in Washington have their own crusade against software piracy. For example, they have identified Ong Scow Peng based in Singapore with illegal programs worth several million dollars. The stated goal of BSA for Mr Ong is 'incarnation'. Thus far the attempt at incarnation is confined to individuals, but a national incarnation, even by a software industry that acts independently of its country, can be quite harmful to developing countries. In our world of ever-changing computer technology which emerges and becomes obsolescent very rapidly, it is important for developing countries to have good cooperative relationships with the leaders of computer technology in developed countries.

The actions by the US (governmental and private) is very bad news for developing countries for they may now have to pay for the software (and copied hardware) that they use. But in the long run, control of software piracy is good news for developing countries. It will increase investment and licensing by developed countries and will strengthen the computer industry (especially the software industry) of developing countries. It will also increase outsourcing by developed countries for they will be less afraid of their software and data/knowledge being pirated. So, in the long run, developing countries will benefit. But paraphrasing John Maynard Keynes, the software pirates may say: in the long run we are all dead and gone anyway, so why not make money while we can. And there are other ways to make money: doing business on the Internet.

Global network and business

The combination of processing text and images on the Internet makes it a good target for use by business. A business can pay a fee of around a thousand dollars a year and have a 'home-page' of advertising material that is theoretically accessed by around 25 million people connected by over 20 000 computer networks run by universities, businesses and governments. The customer can access the system, for example the ISN (Internet Shopping Network), by entering a command as simple as http://shop.internet.net. The customer can then access the ISN server and browse through the ISN catalogue which has up to 20 000 computer hardware and software products. If an order is placed, the customer gets confirmation messages via the Internet and receives the goods within two days if in the US. But, there are problems with shopping in the electronic shopping mall. Businesses face computer fraud and customers are afraid of giving their credit card numbers on the Internet. To solve some of these problems, there is a consortium of large electronic firms called CommerceNet which has $12 million in state and federal funds that is working towards making the on-line business easier and safer to use. Meanwhile, there is the netiquette that disciplines those that take advantage of the system. For example, spamming (flooding the

system with the same posting) is looked down upon as is the regular repetition of the same blatant advertisement. However, netiquette, as well as the protocols on the Internet, is changing all the time and any details stated here need to be updated to be valid anymore.

The greatest interest to business lies in the confluence of PCs, commerce and money. However, for business to benefit globally from this confluence and global telecommunications, there has to be a way to transfer funds globally. We already transfer funds by EFT (electronic fund transfer) but that is between banks. What is needed is a means of transferring money by individuals to any other party. Credit cards are such a vehicle but transferring funds by credit card globally on the Internet is asking for trouble. A secure transfer system must be able to confirm tens and hundreds of transactions a day. Software houses and credit companies are now working on ways to encode credit card numbers on-line without any danger of misuse. But this may take time before the system is thoroughly debugged and certified as easy and safe to use globally. Smart cards with stronger safeguards against misuse are being developed by Mondex, a consortium of British banks. Another approach is 'money tokens' which are prepaid smart cards. But what if you wish to spend more than what you prepaid for? What is needed is **e-money**, electronic money, that can be transferred electronically, safely, quickly and cheaply.

The concept behind e-money is to provide an electronic signature (like a watermark on some bank notes) with the recipient checking with the issuer. Each issuance is identified uniquely as is a bank note by a serial number. If copied, not much harm can be done for the copied signature is only good for one transaction and large transactions will be verified before being accepted.

Various approaches to e-money are being pursued by firms like DigiCash in Holland and CyberCash, an American firm working out of Austria. The electronic payment system may make it easier for money launderers and tax dodgers. Thus, the issues raised are not just for banks, but for national agencies trying to prevent such practices especially from going across their borders.

E-money can be called **digital money** because the code for the money and the parties involved are often digitized. Digital money does provide a medium of exchange but it does not solve the second role of money, that is to be a store of value. Digital money can be refused as legal tender. In our traditional monetary system, people deposit money in a bank and, against it, take loans, draw credit or even demand cash. For this to happen with digital money, we will need to have legal money in the real economy deposited for each unit of digital money. Actually this is what CyberCash proposes to do. They will hold real money in escrow for its equivalent in digital money. But this raises other questions. Should the equivalent of real money earn interest and if so, for what rate? Can digital money be used for collateral for a loan? Can it be exchanged for foreign currency and, if so, at what rates? Answers to these questions now rest with national central banks and national monetary policies. National governments may well hesitate to abdicate these decisions to cyberspace or cyberpunks.

Digital money has a demand because traditional money as bills can be copied by high quality copiers. Guarding paper money is expensive. ATM (automatic teller machines) are popular but not always safe from muggers. Credit cards are also popular but not secure against fraud or violations of privacy. The problem with credit cards is that they make it possible to construct a digital profile of the customer and trace a person's card transactions with 'decidedly discomforting intimacy'. The memories of George Orwell's Big Brother looking over your shoulder immediately comes to mind. Privacy is under threat. What is needed is not just a 'digital signature' which requires establishing authenticity that can be traced back for undesired purposes, but a 'blind signature' that is easy to use and has the elegance of anonymity of paying cash and yet retaining the store of value. This approach will protect some (especially the money launderer even across the borders) but will conflict with the government's desire to control such illegal transfers.

The problem is one of balancing the legitimate needs of privacy protection against violations of privacy and potential surveillance; the understandable desire for security against fraud, and safeguarding the national monetary system against fraud and misuse as in money laundering. There is also a conflict between the needs of financial institutions like banks and individuals. Individuals are concerned with privacy and security being compromised, while banks are concerned with a system that is cheap and fast to use for transactions across borders.

What some people think we need is digital money. Actually, many transactions today are conducted with digital money if you consider all the EFT transfers which are digitized in cryptographically sealed streams of bits of data moving between government clearing houses. What is not digitized satisfactorily yet is the last mile in the money transactions, that of payments in cash. In addition to e-money and digital money, there is also the possibility of the **electronic wallet** or **electronic purse**, which is a palm-held calculator-sized reader of smart cards that can be used at petrol stations, toll booths, retailers, fast food restaurants, convenience stores, school cafeterias and places where cash is now essential.

> You'll download money from the safety of your electronic cottage. You will use ... cards in telephones (including those in the home), as well as electronic wallet, disgorging them whenever you spend money, checking the cards on the spot to confirm that the merchant took only the amount that you had planned to spend. The sum will be automatically debited from your account into the merchant's. (Levy, 1994: p. 176).

Summary and conclusions

Information and telecommunications technologies play an important role in world trade, helping to define markets, design and manufacture products, sell goods, transport commodities, and exchange payments. At each stage of the trade cycle, data are recorded, stored and processed by IT, with output delivered to managers, workers and clients largely by telecommunications. To be competitive in world markets business people in all countries need access to information technologies. For this we need transborder flow of information, a secure means of monetary transactions and protection of intellectual property.

A definitive statement of the issues of global telecommunications may never fully be achieved because information and telecommunications technologies are continually changing through research and development. In addition, the number of applications is expanding. Finally, issues of global telecommunications are resolved in bilateral and international forums where world views and national interests continually shift, and where power politics complicate negotiations. IT managers must remain alert to these interests and power shifts and adapt to them.

There is a high correlation between poor telecommunications and many developing countries. The information infrastructure in developing countries is often inadequate for a robust and sustained support of networking activities. Developing countries that have a weak capital structure also have weak planning for investment in infrastructure and a poor payoff for telecommunications. Electronic network connections have a potentially *higher* payoff for developing countries because of the lack of reliable alternative delivery mechanisms in the developing world. (Goodman *et al.*, 1993: p. 43).

The 1990s will be a period of flux for international issues of IT. All countries may agree on the need for global telecommunications, but how they go about it will differ. The developing countries will want to leap-frog the growth curves without giving up any of their existing privileges. The developed countries will want to concentrate on integration and consolidation. The consolidation will include the assertion of their international rights, especially as regards the protection of intellectual property including compute software. There will be more confrontations like the threat by the US to impose a stiff tariff on selected goods worth over a $1 billion on China if China continued to refuse to satisfactorily protect American intellectual property rights including those of computer programs. There was a last minute compromise and there will be more. Both sides seem to be preparing for test of strength to see who blinks first. Hopefully, global telecommunications will not suffer from such blinking competition and will continue to grow and prosper in the future.

Case 16.1: Global outsourcing at Amadeus

Amadeus Global Travel Distribution SA (Madrid) is a computerized reservation system with its central processing done at Helder, Germany. It serves more that 104 000 PCs through its two large centres in England and the US and through 10 outsourced access networks at BT/Concert (UK); the Spanish travel industry; French Travel industry; SITA access network; Scandinavian travel industry; X.25 networks in Belgium and the UK; and the German travel industry. In addition it serves airlines of SAS in Scandinavia; Iceland Air

in Reykjavik, Iceland; Thai airlines in Bangkok, Thailand; Iberia in Madrid, Spain; Airinter in Paris, France; Air France in Valbounne, France; Lufthansa airlines in Frankfurt, Germany; and Finn Air in Helsinki, Finland.

Amadeus owns and manages most of the backbone equipment at the two centres in Atlanta (US) and London (UK). They are connected by separate 384 kbps transatlantic lines to Erding in Germany. They are configured with uninterruptible power supply and routine monitoring of traffic coordinated with the telecommunications and processing centre in Germany which is equipped with a cluster of IBM mainframe computers. These centres as well as the other networks offer services not just to the travel operators for airline tickets but also for reservations for theatres, car rentals, hotels and other travel related items for agencies that own PCs. Amadeus collects a small fee for each booking which in 1994 was around 135 million.

Evaluating the experience of running such a large outsourcing processing network the lessons to be learned were less technological and more of an organizational nature. Such lessons were dividing the outsourcing between at least two operators so that the competition will drive prices down; measuring availability from end-to-end including access lines; not including the SLA (Service Level Agreement) in the contract so that no adaptation to changing conditions can be made without a battery of lawyers going into play; and recognizing that the 'PTT infrastructure monopolies make outsourcing mission critical international backbones too risky at present. And the PTT alliances now being struck throughout Europe may make it even tougher for providers of managed network services to offer international outsourcing.' (Heywood, 1994: p. 80).

Source: Peter Heywod. Global Outsourcing: What Works, What Doesn't. *Data Communications*, Nov. 21, 1994, pp. 75–80.

Case 16.2: Telstra in Australia

Telstra is the largest integrated telecommunications carrier in the Asia-Pacific region and 'provides a comprehensive range of telecommunications solutions, including international data/voice network solutions, network management and service performance.' Telstra in

1994 was the 16th largest telecommunications company in the world with a telecom revenue of US\$ 9.8 million and 8.9 million main lines.

Telstra's MobileNet digital network is one of the largest in the world in terms of geographic coverage, including international roaming in 150 cities in 27 countries, a two-way paging and voice mail computer messaging service, and a fax and data service. Telstra also has an analogue coverage which is one of the world's most extensive, reaching 89% of Australia's population. The analogue coverage is planned by the government to be phased out by the year 2000 and taken over by its digital system.

Telstra is also part of the TINA consortium formed in 1993 with about 40 of the world's major network operators as well as switching and computer manufacturers. The consortium is attempting to develop technology that will enable information systems that make up a network to collaborate more effectively with each other by integrating databases and providing services.

Source: *International Herald Tribune*, Oct. 4, 1994 p. 16.

Case 16.3: Telecom leap-frogging in developing countries

ITU estimates that 4 billion out of the 5.7 billion people in 1994 did not have the basic telephone services. Developing countries have a great fear that they will never catch up with developed countries. One approach to leap-frogging the telecommunications gap is to use fixed wireless that provide customers service from a radio station to antennas in homes and offices which perform like regular phones. It is expected that the majority of the 800 million subscribers in 2000 will have such a service. They are already being installed in Brazil, Chile, Columbia, Finland, Germany, Ghana, Malawi, Mexico, Russia, Spain, Sri Lanka, Vietnam and Zambia.

The GSM, Global System for Mobile digital cellular system, is being installed in Eastern European countries and is planned for extension into 100 countries.

Some of the reasons for the inability of developing countries to achieve telecommunications parity with developed countries are:

- old infrastructures;
- know-how deficit;

- red tape, high tariffs and restrictive governmental policies;
- high costs: for example, a 64 kbps connection could cost around US \$8000 with equipment costing around US \$30 000;
- lack of funds: not all developing countries reinvest their profits into the network but spend it elsewhere. Consequently they cannot afford the telecommunications infrastructure that they so desperately need. Without the infrastructure they cannot integrate into the global market. And without integration they cannot pay for the infrastructure — a catch-22 situation.

In 1995, the 24 countries of the OECD (Organization of Economic Cooperation and Development) generated 85% of the world's telecommunications service revenues. The high income countries which account for 15% of the global population have 71% of the world's telephone lines. To overcome this disparity, the World Bank has estimated, will require \$55 billion (about 10% of the world's annual spending on telecommunications) every year over a six-year period to finance the necessary development in developing countries that include the former Eastern Europe bloc.

Source: *International Herald Tribune*, Oct. 5, 1995, p. 16, and May 17, 1995, p. 19.

Case 16.4: Slouching towards a global network

Guenther Moeller, Director General of EURO-BIT, the European Association of Manufacturers on Business Machines and Information Technology has a list of six principles for a GII (Global Information Infrastructure):

1. protection of intellectual property rights,
2. universal access to networks,
3. privacy and security,
4. access to research and development,
5. interoperable systems and applications,
6. new applications.

'Looking globally, we see a patchwork of incompatible communications networks marked with high costs, low quality services and very little interoperability between systems. What we need is a common worldwide infrastructure to communicate information at reduced cost.'

Source: *International Herald Tribune*, Oct. 8, 1995, p. 11.

Case 16.5: Alliance between French, German and US companies

In 1995, a consortium between French Telecom, US Sprint and Deutsche Telecom was announced. French Telecom is the telecommunications arm of the French nationalized PT&T; Sprint is the third largest American long distance carrier; and Deutsche Telecom claims to have the most advanced ISDN, the densest network and the most extensive cable network in Europe. The alliance of these three large companies is expected to produce a serious competitor in the global telecommunications market.

Supplement 16.1: World-wide software piracy in 1994

Total losses to international piracy US $8.08 billion that include (in billions of US dollars):

Japan	1.31
US	1.05
France	0.48
UK and Ireland	0.24
Others, especially China, Russia, Thailand, India and Pakistan	5.0

Large gainers from international pirated software as a percentage of their total software used:

China	8
Russia	95
Thailand	2
India and Pakistan	87

Source: *Fortune*, July 10, 1995, p. 121.

Supplement 16.2: Index of global competitiveness

Using 1991 Performance data, a telecompetitiveness index was calculated using 43 specific factors measuring critical areas of telecommunications. Performance in 10 categories were measured and converted on a scale of 1–10, where 10 is the best score. A few such categories measured and the overall index of telecompetitiveness (T-C) appear below:

	Infrastructure	Productivity	Penetration
Canada	7.1	6.3	9.5 8.2
France	7.8	3.0	3.7
Germany	3.4	4.3	5.6
Japan	6.0	6.3	4.2
Singapore	7.9	6.3	5.3
UK	4.7	4.5	4.8
USA	7.0	7.1	8.8

	Quality	R&D	Index
Canada	3.8	6.4	
France	4.7	2.5	5.9
Germany	N.A	5.2	4.7
Japan	6.7	5.8	5.9
Singapore	8.0 2.2	6.6	
UK	4.4	1.6	4.9
USA	7.9	5.1	6.2

Source: *Stentor*, Dec., 1995, p. 3

Supplement 16.3: Telecommunications media for selected countries in 1994

Country	Phone lines	PCs	Cable TV
	(units per 100 people)		
Argentina	4.1	1.7	13.2
Australia	49.6	21.7	–
Brazil	7.4	0.9	0.3
Canada	57.5	17.5	26.9
China	2.3	0.2	2.5
Czech Republic	20.9	3.6	5.7
France	54.7	14.0	2.8
Germany	48.3	14.4	18.0
Greece	47.8	2.9	–
India	1.1	0.1	1.1
Indonesia	0.3	1.3	–
Israel	39.4	9.4	13.3
Hong Kong	54.0	11.3	0.6
Japan	48.0	12.0	8.3
Korea (South)	39.7	11.8	5.8

Contd.

183

Country	Phone lines (units	PCs per 100	Cable TV people)
Malaysia	14.7	3.3	—
Mexico	9.2	2.3	2.2
Netherlands	50.9	15.6	37.5
Portugal	35.0	5.0	—
Russia	16.2	1.0	—
Singapore	47.3	15.3	—
Sweden	68.3	17.2	21.9
Switzerland	59.7	28.8	32.3
South Africa	9.5	2.2	—
Taiwan	40.0	8.1	14.1
Thailand	4.7	1.2	—
Turkey	20.1	1.1	0.4
UK	48.9	15.1	1.6
USA	66.2	29.7	23.2
Venezuela	10.9	1.3	1.0

Source: Extracted from *IEEE Spectrum*, **38**(1), Jan. 1996, p. 40.

Supplement 16.4: Telecommunications end-user service available in regions of the world

COUNTRY	Lines			
	ATM	Frame relay	ISDN	Private
Africa	None	None	None	Urban
Australia	None	None	None	Urban & Rural
Eastern Europe	None	None	None	Rural
N.America	Limited Urban	Urban & Rural	Urban & Rural	Urban & Rural
Pacific Rim	Limited Urban	Urban	Urban	Urban & Rural
Rest of Asia	None	Limited Urban	Urban	Rural
S.America	None	None	Urban	Urban
W.Europe	Urban & Rural	Urban & Rural	Urban & Rural	Limited Urban

Bibliography

Apte, U. (1990). Global outsourcing of information systems and processing services. *The Information Science*, **7**, 287–303.

Denmead, M. (1994). International carriers team up for global reach. *Data Communications*, **23**(17), 85–88.

Derollepot, F. (1995). Software piracy. *Personal Computer World*, Vol. 404–413.

Gassman, H.P. (1992) Information technology developments and implications for national policies. *International Journal of Computer Applications for National Policies*, **5**(1), 1–7.

Goodman, S.E., Press, L.I. Ruth, S.R. and Ruthowski, A.M. (1994). The global diffusion of the Internet: patterns and problems. *Communications of the ACM*, **37**(8), 27–31.

Gupta, U.G. (1992). Global networks. *Journal of Information Systems Management*, **9**(4), 44–50.

Guynes, J.L. (1990). The impact of transborder flow regulation. *Data Communications*, **7**(3), 70–73.

Gwynne, P. (1992). Stalking Asian pirates. *Technology Review*, **95**(1), 15–16.

Levy, S. (1994). E-money: that's what I want. *Wired*, **2**(12), 174–177, 213–215.

Liebermann, L. (1997). Website traffic management: Coping with success. *Internet*, **2**(1), 59–66.

Malhotra, Y. (1994). Controlling copyright infringements of intellectual property: Part 2. *Journal of Systems Management*, **45**(7), 12–17.

Sadowsky, G. (1993). Network connectivity for developing countries. *Communications of the ACM*, **36**(8), 42–47.

Strauss, P. (1993). The struggle for global networks. *Datamation*, **39**(17), 26–34.

Weisban, S.P. and Seymore, E.G. (1992). International software piracy. *Computer*, **25**(11), 87–90.

Woody, C.V. and Fleck, R.A. (1991). International telecommunications: the current environment. *Journal of Systems Managment*, **42**(6), 32–35.

Part 3

IMPLICATIONS OF NETWORKS

17

MESSAGING AND RELATED APPLICATIONS

Within three years, e-mail will become the critical application at virtually every Fortune 500 company... As e-mail goes mission critical, the downsized data center may return as the full sized mail center.

Daniel M. Gasparro, 1993

Money is really information, and as such can reside in computer storage, with payments consisting of data transfers between one machine and another.

James Martin

Introduction

Early messages were conveyed by telephone or by mail. The telephone was mostly a one-to-one relationship and, when it had to be a many-to-many relationship, we had teleconferencing and, with more advanced technology, we now have video-conferencing and multimedia conferencing. Conferencing is useful in bringing people together on-line and in real-time consultation, but to conclude transactions, especially in business, agreements have to be formalized in writing and then sent by mail. But traditional mail is slow and with computers it became possible to send mail electronically. This is known as electronic mail, or e-mail for short, as distinct from all other mail, which is referred to as 'snail mail' by e-mail enthusiasts. E-mail was unstructured and unformatted for personal and informal mail but, for business transactions, the messages were structured and often special forms were required. Such messages were sent electronically by EDI, Electronic Data Interchange. With telecommunications and networks business, transactions were conducted with the necessary messages transported swiftly between buyers and sellers wherever they might be and wherever the price was right. But for the transaction to be completed, money had to be transferred. Funds were then transferred electronically between computer systems and over distances using telecommunications. This required another set of messages to be transported

and is known as Electronic Fund Transfer (EFT). Meanwhile, there was cooperative processing where telecommunications and its networking was used by people working together and sharing contributions to the same data/knowledge-bases. These approaches to messaging had some overlap but were not mutually exclusive and so they coexisted but needed to be coordinated. The coordination could done by groups in cooperative or collaborative processing as in a client–server environment as discussed earlier. And whether collaborative or not, it was desirable for all messaging in an enterprise to be integrated. Approaches to such integration is referred to as the MHS, the Message Handling System. MHS is still evolving but some directions are becoming clear. We shall examine these approaches in this chapter along with some of the messaging approaches mentioned above, as well as teleconferencing. We will not discuss all types of messaging especially the telephone and the post office because they are not electronic. Nor do we discuss e-mail which is electronic. This is not because e-mail is unimportant. On the contrary, it is too important and a discussion of it will make this chapter unduly long. Therefore we shall defer the discussion of e-mail to Chapter 19 and discuss it in the context of its greatest use, that by knowledge workers and teleworkers. We also defer the discussion of video-conferencing to the next chapter where we discuss it in the context of distributed multimedia processing.

Teleconferencing

Teleconferencing is a method of electronic communication that permits interactive exchange between two or more persons through their clients or terminals. With electronic mail, a message is delivered instantaneously to an electronic mailbox, but there may be a delay before the intended recipient collects it. Teleconferencing eliminates this time lag since participants are waiting to respond to messages that are sent to the screens of linked workstations/desktop PCs, or between computers in conference rooms as far apart as one or more continents. A teleconference program handles the logistics of this communication and determines who has the 'floor' when several users want to 'speak' at the same time. Some systems include a gag feature to prevent a single person from monopolizing communication channels. Often the terminals of conference participants serve as electronic flip charts enabling the participants to access specially prepared conference materials from a database. Teleconferencing has been used extensively in discussing product development and product production, thereby having greatly reduced the product development life cycle.

Teleconferencing saves travel time and travel costs. However, many professionals like the travel and believe that interpersonal relationships are important when conducting business, so face-to-face communication is preferred over teleconferencing. The cost of teleconference equipment is high, another disadvantage. However, when the executive's time is too valuable to be spent in air travel and frequent business trips over great distances are necessary, the investment in teleconferencing is often a business choice of preference.

Electronic data interchange (EDI)

EDI is a direct exchange between computer-to-computer of separate organizations of standards business documents such as purchase orders, invoices, bills of lading and related business document necessary to perform specific transactions. The transactions are recorded on specific standardized forms that allow for specifying content as well as for checking of content and certain errors. The EDI thus differs from e-mail which is primarily text that may or may not be formatted and structured. Also, while e-mail often includes personal correspondence EDI is designed for business information exchange.

Process of EDI

The process of EDI as compared to traditional processing is shown in Figure 17.1. Traditionally, inquiries were made by telephone and then confirmed by mail. With early computer processing, the paper hard-copy agreements were converted into machine readable form for computer processing and then the transaction

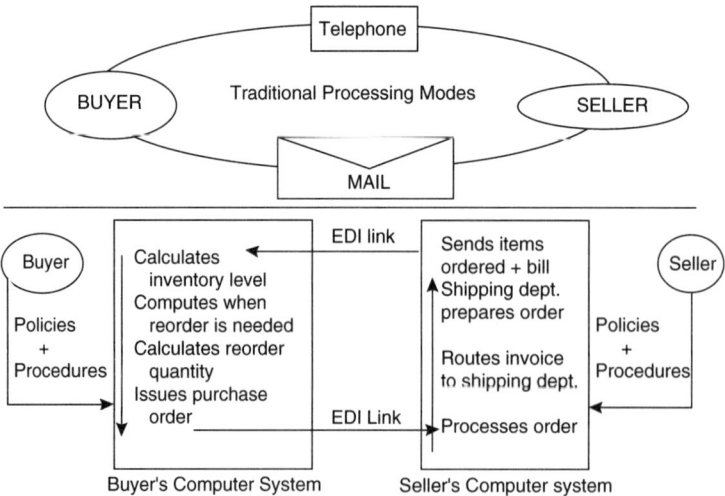

Figure 17.1 *Traditional messaging vs. EDI*

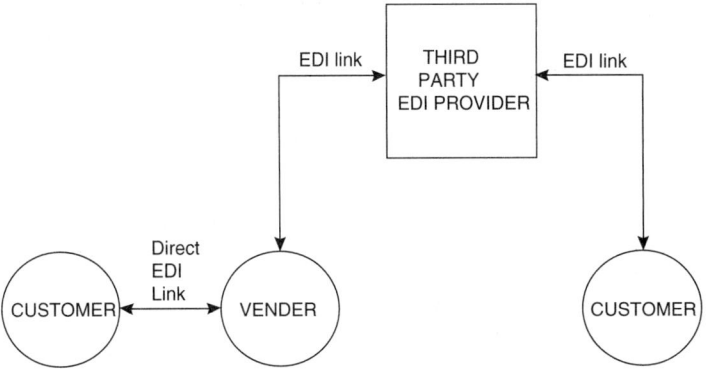

Figure 17.2 *Alternative EDI links*

was consumed. Under EDI, the process is greatly automated eliminating the conversion from hard-copy. The buyer's computer systems governed by policies and procedures of the buyer (as embodied in a computer program) determines when to buy. This determination is made by an inventory control computer program which determines the reorder point given expected demand and inventory-on-hand. The necessary quantity to be ordered is specified in a purchase order and sent to the seller of the inventory ordered. The seller's computer system acts on the order by rules embedded in its computer program. The order is processed and instructions sent to the shipping department which then ships the necessary goods to the buyer. At the same time, an accounts receivable is issued and sent to the billing department which then collects the money, may be by a cheque mailed to the seller, or the money is transferred between the banks of the buyer and seller through EFT. Thus the process is greatly automated, reducing costs of data entry (and correction of errors that do occur) and of processing as well as the time required for both data entry and processing. Automation also reduces transcription and conversion errors that would otherwise occur during the manual conversion of data from hard-copy to machine readable form. Such a system can give the seller a comparative advantage over a seller that does not have EDI and could 'lock' the customer for convenience, lower costs and reliability. The speed of processing can reduce the inventory otherwise held by the buyer and reduce inventory holding costs.

EDI between two trading partners does require that the trading partners have computers and a communication EDI link. This may not seem to be a problem at first sight because most businesses have computers. If their systems are compatible, then a direct link is possible as shown in Figure 17.2. Unfortunately, however, many computer systems are not compatible in hardware and its operating systems software. They then go through a third party provider that serves as a go-between clearing-house converting the transaction from one system to being acceptable to the other systems. This third party can also serve as a store-and-forward point as does a post office, where messages are stored, sorted and sent when desired to the destination. The third party intermediary function can cause extra cost and time and so large businesses demand that their supplier have compatible computer systems and large suppliers wanting the business will comply.

This discussion enables us to now specify the conditions necessary for a successful EDI:

The two trading partners must have compatible computer systems (hardware, software and processing procedures).

There must be an EDI transmission link, either direct or through a third-party provider.

Transaction documents and product identification must be standardized between the two trading partners for format.

Mailbox facilities must be available for storing, sorting and forwarding messages.

Standardization

We have mentioned standardization on two occasions: one, standardization of document format; and two, the standardization of EDI transmission. The former is usually settled between firms and corporations that are trading partners. This includes format for purchase order forms and packaging slips. The other type of standardization, that for EDI transmission, is more difficult because it involves not just telecommunications vendors but also many countries when international commerce is involved (Gordon, 1993). With the increase in global outsourcing and multinational firms, the exchange concerns are global. Initially, it was the exchange within a firm. In Europe, where EDI is at around one-fifth of the level of the US, there were around 5000 EDI sites in 1990. For an international EDI, there are two competing standards. One is the X23 developed by ANSI (American National Standards Institute) in the US, which covers over 80 business or transaction sets that are mostly generic with some that are industry specific (Trauth and Thomas, 1993). In contrast, there is the more recent European standard, EDIFACT (FACT is the acronym of 'For Administration, Commerce and Transport'). It is very different in approach and philosophy to the X23. EDIFACT is centralized, systematic, anticipatory, a tool of industrial policy, as well as being responsive to governmental direction and national policy. In contrast, the X23 is distributed, pragmatic, reactive, entrepreneurial and individualistic, and attempts to maximize the role of the private sector. For the X23, international standards could be only a guide whilst the EDIFACT is a single European standard adopted by all European trading partners. Whilst EDIFACT is attempting to make its standards international with or without American support, European firms doing business with Americans must follow the X23 standard. But as multinational trade in IT increases between developed and developing countries, the need for compromise may move us towards a truly universal international standard. Global standards other than for EDI are necessary not just for global outsourcing but also for transborder flow of data and the globalization of IT.

Despite the problems of standardization, EDI has great potential both nationally and globally. EDI will be used increasingly to automate transactions between computers without human intervention and the output of one application for one trading partner becomes the input for another trading partner. In one study in the eastern states of the US, it was found that over 25% costs of processing was for reentry and that over 70% of

---

Table 17.1 *Summary on EDI*

PREREQUISITES OF EDI
Both trading partners have computer systems
The two computer systems must be linked, either directly as with compatible systems or through a third-party EDI provider
Both parties must have a 'mail-box' capability
Both parties must follow standards for document format
EDI transmission

ADVANTAGES OF EDI (over traditional modes of telephone and mail)
Fast
Lower costs of data entry and processing
Certain operations are automated
Errors especially in data entry are reduced
Reliable and well established system
There are many spill-over services including those in outsourcing

DISADVANTAGES AND LIMITATIONS
Closed system
Not universally used
Real-time but can be 'slow'
Cost for software installation

---

computer output became the input for another computer system, thereby eliminating the costs of reentry of data which is an important component of the total costs of processing. This is important as trade and commerce increase and becomes more global. EDI is often associated with business transactions between buyer and seller. But there are many other transactions that are performed by EDI.

A summary of the above discussion on EDI appears in Table 17.1.

Electronic transfer of funds

In 1965, Thomas J. Watson, then chairman of IBM, made the following prediction:*

> In our lifetime we may see electronic transactions virtually eliminate the need for cash. Giant computers in banks, with massive memories, will contain individual customer accounts. To draw down or add to his balance, the customer

---

* Reprinted with permission from IBM.

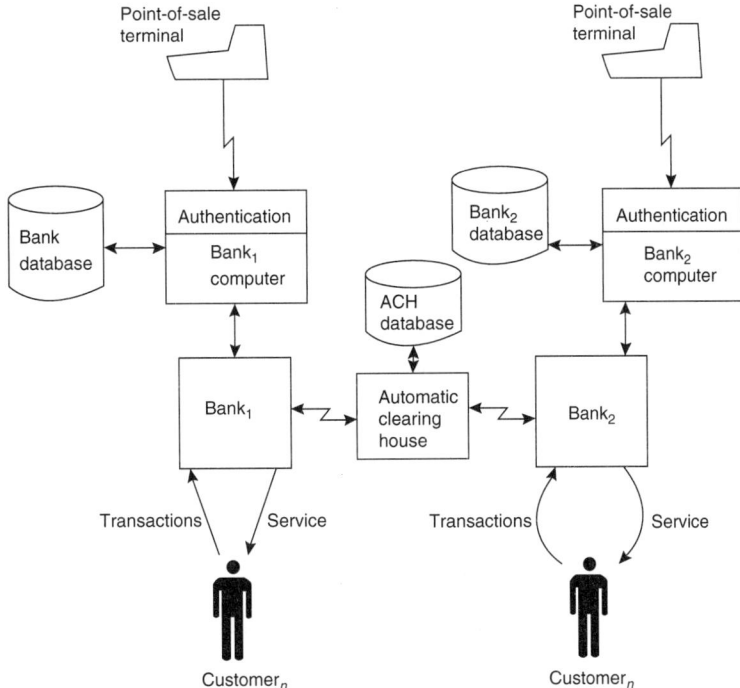

Figure 17.3   *Components of an EFT system*

in a store, office or filling station will do two things: insert an identification into the terminal located there; punch out the transaction figures on the terminal's keyboard. Instantaneously, the amount he punches out will move out of his account and enter another.

As Watson predicted, this same process, repeated thousands, hundreds of thousands, millions of times each day, now occurs. Billions 'change hands without the use of one pen, one piece of paper, one check, or one green dollar bill'. A network of terminals and memories extend across city and state lines for **electronic funds transfer** (EFT), although we have not altogether eliminated paper money and coins. Watson's prophecy has today become a reality. Billions of pounds sterling are moved daily from one set of accounts to another using computers and telecommunications without any currency exchange or paper to record and process transactions. Information on cheques (payee, amount, account number, cheque writer, depositor, institution) is converted into electronic impulses and transmitted through a telecommunications channel to the nearest **automated clearing house** (ACH), a computerized

version of the traditional cheque clearing house. (The cheque itself, that is the paper on which the information is written, is not physically moved from one location to another.) The ACH then transfers the amount to banks where the intended recipient has an account. (See Figure 17.3 for components of EFT.)

The US Treasury Department uses EFT for recurring payments, such as the transfer of funds for government employees' life insurance programs. In the private sector, many corporations deposit weekly paycheques by EFT and preauthorize account debts such as monthly interest payments and insurance payments. International systems like CHIPS (Clearing House Interbank Payments System) and SWIFT (Society for Worldwide Interbank Financial Telecommunications) have provided EFT between banks for the past twenty years.

However, paper-free banking is still far from being realized. The reasons for this are described below.

Float

When payments are made by cheque there is usually a two- to seven-day period between the time the cheque is written and the time that it is cashed. The amount of money in transit but not yet collected is called **float**. Float is, in effect, an interest-free, short-term loan to the cheque writer.

Instantaneous EFT payments eliminate float and the possibility of earning interest on these short-term loans. In fact, a company that pays bills by EFT but receives payment by conventional cheques stands to lose money. Loss of float is a principal reason why cash managers oppose EFT.

On the other hand, if all expenditures and receipts were EFT transactions, the loss of float might be balanced by early receipts. In addition, the fast processing of transfers and exact information on the status of funds might enable firms to reduce demand deposits and mobilize idle money. The conversion of liquid holdings into 'near-money' assets such as Treasury bills and commercial paper would increase profits.

Cost

Many banks do not make charges to people who write cheques provided that they maintain a given balance in their cheque accounts, whereas a service fee is charged for EFT. That is to say, bank customers are given no financial incentive to switch from cheques to direct deposits. In may be because the cheque system works well and cheque processing is low in cost that bankers appear more interested in maintaining the status quo than in promoting EFT and building a workable low-cost ACH mechanism. The transition to a chequeless society is unlikely until both banks and their customers save money with EFT.

Lack of public confidence

Most people have had personal experience of computer errors and are wary of computer reliability. The public is not yet willing to eliminate the paper backup of cheques.

In addition, the ease with which financial records can be accessed by EFT raises concerns about the security of financial data processed by EFT. When identification cards are required for EFT access, as in ATM transactions, the danger exits that stolen cards will be used to make fradulent fund transfers. Password codes can be breached by determined thieves, and systems security based on voice, hand forms and fingerprint recognition systems, more reliable

than passwords, is too costly for widespread use at present. Many people, too, object to these types of system on moral and political grounds.

The vulnerability of financial data during transmission can be protected by coding and security measures like packet switching described in Chapter 5. Fewer incidents of attempted theft are reported from EFT than from conventional cheque systems, but the potential loss per incident is much greater.

Another concern of many critics is that EFT might lead to an erosion of privacy rights. The problem arises because EFT records contain data on spending patterns and an analysis of these records can reveal personal data such as travel movements, drinking habits, debts, health status, and so on. Acceptance of EFT may have to wait until the public is assured that the privacy of individuals who use EFT is safeguarded.

Preference for cash

Many people like the feel of cash; it gives them a sense of satisfaction. Although efficient, EFT will not win public support until human needs are addressed. Perhaps user-friendly features can be added to make the use of a terminal for financial transactions outweigh the pleasure of handling cash.

EFT spin-offs

Point-of-sale terminals linked to banks and credit agencies by telecommunications is an electronic application in retailing that is technologically feasible but not yet widespread. Electronic shopping and home banking, much touted, are likewise spin-off applications of EFT that have had only limited success to date. The basic problem is lack of high transaction volume necessary to pay for these services. We next explore the promise of these systems and reasons that public reception has been poor.

Point-of-sale terminals

Electronic **point-of-sale** (POS) terminals that record sales transactions are found in many retail establishments. Most transmit sales data to a store computer which processes the data to produce accounting and inventory reports for management. Less common is the use of POS terminals to monitor charge accounts with credit limits within a given store. The sale of merchandise recorded on the POS terminal debits the customer's credit limit by the amount of the sale.

POS terminals can also be used for cheque and credit card authorizations. In this case, the terminals are part of a telecommunications network linking them with banks and credit agencies. Some POS systems immediately debit the amount of purchase from the customer's bank account and deposit the money in the store's account. This use of EFT requires a telecommunications link from the store to the customer's bank and the ability of that bank and the store's bank to handle electronic financial transactions.

Home banking

For a fee some banks provide **home banking** for people with a personal computer and modem, another EFT spin-off. Not only can bank customers access account information and pay bills through interaction with the bank's computer, but they can review bank statements, transfer funds between accounts, open new accounts, and buy or sell securities from home.

A good idea, perhaps, but home banking services have few subscribers. The problem appears to be that individuals, like their corporate counterparts, are reluctant to give up the float. Most people who own PCs lack modems: to purchase one costs from £80 to £240 — an investment that most people do not seem to think is worth the cost. Besides, home banking does not eliminate paper cheques: it merely shifts cheque writing to the bank, which writes and mails its own cheque to the payee after payment has been authorized through home banking.

Home banking is not dead, however. Many banks are shifting their electronic banking strategies to small business customers and offering financial planning software for added value. Other banks are forming consortia and joint ventures to help pay for the cost of developing and marketing home banking. The challenge to home banking suppliers is to demonstrate that the services they provide save the customer money and are more convenient to use than traditional cheque accounts.

Home shopping

With EFT, **home shopping** (also called **teleshopping**) becomes feasible. Selection of goods is through catalogues or newspaper advertisements called to the computer screen; payment is authorized on the terminal by EFT. Although the pleasure of window shopping and the ability to handle merchandise is lost, many home shopping systems allow viewers to rotate items displayed on the screen or view them in close-up.

As with home banking, home shopping has not been well received and many videotex information services that featured home shopping have failed. But entrepreneurs seem confident that once the technology improves and a broad customer base is established which will lower costs, the idea will be profitable. In the future it may become possible to key descriptions of wanted items on a home terminal and let the computer search for stores with the items for sale.

If such home shopping systems ever take hold, marketing patterns will change drastically. Merchants will have to index their goods and provide these indexes to information utilities. Shopping centres will lose customers, shop assistants will be displaced, and retail space will be replaced by warehouses that transact business electronically.

Before teleshopping becomes more widespread, however, there will probably be a surge in **electronic marketing**. Special computer programs will merge computer tapes of credit bureaux, listings in commercial and private user telephone directories, licence and association lists, mailing lists, census and other data to obtain a master customer list. Marketers will then call a large test sample from the list in order to create buyer categories, and then try to determine 'buying windows' or eligibility factors that are required for a good buyer/product match using computer software to evaluate current customer-need profiles against benefit elements of available products and services. The purpose of this marking approach would be to identify groups that might be responsive to advertising for particular products so that advertising campaigns can be directed to them.

Smart cards

An innovative use of information technology that will affect the computer, retailing, banking and credit card industries, among others, is an integrated circuit charge card, called a **smart card**.

Smart cards resemble credit cards but instead of having a magnetic strip on the back with credit information about the card holder, most contain an embedded microcomputer chip and permanent memory that does not lose its information when the power is shut off. (An alternative technology is a **laser card** which stores credit information as tiny black dots burned onto the card surface.) When a purchase is made, the sales assistant inserts the smart card into a card-reader module that is tied to a host computer which decides whether the purchase should be authorized based on account parameters, such as the card-holder's credit limit which is stored in the card's memory. (**Unified cards** add battery power, a two-line display screen and a keyboard device to a smart card so that information can be entered or read from the card without a computer terminal. Midland Bank is using a contactless card that transmits signals to the card reader so that no direct contact is needed.)

Another use of smart cards is as follows. At the beginning of each month, a certain amount of money is transferred to the card's account according to an agreement reached between the card holder and his or her bank. The first time the card is inserted in a reader each month, the amount is automatically added to the previous balance recorded in the card's memory. When paying for goods, the card will record the date and value of the transaction and a new balance will be calculated by the microchip.

An advantage of smart cards is their memory size (8, 16 or 32 kilobytes) which makes them able to store much more financial information than magnetic strip credit cards (e.g. health insurance information, telephone numbers, appointments or other records). Smart cards are used at universities to store student's records and timetables, and by the US Army to replace dog tags. In France, they are used to pay for calls from public telephones and to pay highway tolls; at Loughborough University in England, to pay for purchases in campus shops and restaurants. (In these applications the cards represent electronic money. The card is purchased for a given amount and each transaction is recorded on the card and debited until the given amount is spent and the card becomes invalid.) With their ability to store digitized fingerprints, photo and voice prints, smart cards have attracted commercial interest for application in building security. The health industry is promoting the use of the cards for medical records which will carry warnings about medications, allergies and chronic illness of the card holder plus name and telephone number of the family doctor. In case of accidents, such cards might save lives provided, of course, the card's microchip is not damaged.

Besides their capacity and versatility for information storage, smart cards are favoured for financial transactions because they authenticate transactions, saving retailers the time and cost of telephoning for credit authorization. Bad debts should shrink because the chip on the card keeps a running tab on purchases and will not approve those that exceed the holder's credit authorization.

Although a sizeable investment would be required to install the readers necessary for widespread retail use, industry experts claim that the cost would be defrayed by reduced credit card fraud (smart cards are difficult and costly to duplicate). MasterCard, which states that the big selling points for the cards are convenience, protection and flexibility, has begun testing the cards and talked of a five-year changeover from magnetic strip credit cards. Visa, more cautious, is waiting to see whether smart cards will save enough money to justify investment in them: the cost of a smart card is more than three times that of an ordinary credit card.

Smart cards are not a new concept – a key patent on the idea was granted in 1974 when the semiconductor technology was barely two years old. But the idea, like home banking and home shopping, has been slow to catch on. However, if MasterCard starts mass distribution of the cards as planned, market pressure may force other financial service companies to issue smart cards, which will in turn force businesses which rely on credit card payments to install the card readers. At that point, use of smart cards will gain momentum.

Cooperative processing

In our discussion of the client—server system in Chapter 10 we saw that it was originally designed to facilitate communications between end-user (client) and the computer processor (server). But how about communications between end-users using the same database(s) and server(s)? This configuration could improve processing horizontally between colleagues, customers, suppliers and other relevant personnel in organizations. It could also help decision-making within a firm or corporation by making the decision-making process on-line whilst sharing enterprise resources of computing as well as data/knowledge-bases. Such processing is called collaborative processing or groupwork. The processing in this environment is facilitated by software known as **groupware** and the processing known as **cooperative processing,** or **group writing,** or **group scheduling.** Such processing involves multitasking instead of the single-task processing. It utilizes client–server computing, but facilitates interaction among end-users in addition to the dissemination and routing of the necessary shared data/knowledge from groupware servers as well as application and file servers. Groupware binds the separate activities of end-users and decision-makers into an ensemble of cooperative processing. It retains the advantages of distributed processing and downsizing.

Groupware is possible because of the many electronic tools used on a LAN that may or may not use a server but facilitate horizontal communications. These tools include:

- Group calendering and scheduling, which enables the scheduling of meetings and travel, thus facilitating concentration on other productive tasks.
- Group document handling, including groupwork software utilities and development tools.

   More intensive use of e-mail messaging.

- Group meeting and teleconferencing. This enables some eye-to-eye contact and instantaneous reactions, which, unlike video-teleconferencing, can be spontaneous, taking place any time and at any place.

- Group decision-making and problem-solving support through downloading of some decision-making from the boardroom to the desktop. Cooperative processing captures the flow of ideas (and discussions in teleconferencing) and creates a continuous database and knowledge-base of all information relevant to decision-making and problem-solving.

Personal productivity programs such as e-mail, spreadsheets and word processing can still be used along with a DSS and EIS but integrated with group work seamlessly and without any special commands and procedures.

Cooperative processing will not be the best substitute for interpersonal relations of face-to-face meetings and the telephone, but it can reach a much larger audience responding to specific questions. It can overcome the barriers of time and place and offer continuous communication at one's need and convenience.

Cooperative processing has many advantages discussed and implied above, but it also has organizational implications that can be far-reaching: it can affect the structure of traditional decision-making and lead to adhocracy.

Cooperative processing compared to traditional processing is summarized in Table 17.2

Message handling systems (MHS)

There are other message handling approaches like the Bulletin Board Systems **(BBS)** and the

Table 17.2 *Cooperative processing compared with traditional versus client–server processing*

	Traditional processing	*Cooperative processing*
Tasking:	Single	Multiple tasking with human interaction
Architecture:	Open	Closed
Software:	Software on client and server are independent	Software on client and server are integrated
Applications:	Distributed	Host-based and enterprise-wide
Control:	Resides with workstation	Resides with host or server computer
Security and integrity	A problem because of the distributed control	Has better integrity and security control because it is centralized
Path	Hierarchical	Many cross-currents leading to problems of cooperation and collaboration
Time and place	Same time and same place, i.e. face-to-face meetings. Same time and different places, i.e. video-conferencing	Same time and different places. Different times and different places

Newsgroups. These are often services (along with e-mail, electronic shopping, etc.) offered by Information Service Provides and will be discussed in Chapter 19. In this concluding section of this chapter we are concerned with the integration of message handling systems that include e-mail, BBS, Newsgroups and workgroups of collaborative processing.

There are four main approaches to the integration of message handling. The most popular and successful in the mid-1990s was Software Notes by the firm that produced Lotus. This was in direct competition with the other contender in this field, IBM. Then, in 1995, IBM made a hostile and successful takeover of Lotus. Whether the creative programmers and management responsible for the success of Notes will find the large umbrella of IBM and miss their informal and 'small' environment of Lotus will not be known for years. This leaves two other main messaging vendors in the field: Microsoft with its Exchange system and Novell with its Collaborative Computing Environment. All these approaches are basically extending messaging into discussion databases, group scheduling and electronic forms within a overall and overreaching strategy for its deployment throughout the enterprise. Many vendors have acknowledged the corporate needs for downsizing and using the client—server platform for the enterprise left the mainframes behind. All the vendors incorporate multimedia processing as part of the broadband approach to information transfer. Electronic messaging is thus fast becoming an infrastructure providing much of the services that the traditional operating system provided. It is one of the technologies that manages, synthesizes and communicates information.

Despite the commonalties between the messaging vendors, there are some differences in the architecture of the systems. IBM had a middleware platform with a multiprotocol backbone for scheduling, workflow automation and other message-based applications. Lotus had a backbone for routers, an interenterprise connectivity, and a tight integration of its e-mail program and Notes. Microsoft had a back-end server and front-end client with an integrating platform for enterprise-level communication. And Novell offered cross-computability with any server through the application components provided by its groupware (Bragen, 1995).

Summary and conclusions

Message handling has seen an evolution in technology and organization. The earliest messaging systems were centralized both in the private and public sector. The public sector technology consisted of the telephone and the mail. This was the first generation of message handling that lasted till mid-1980s. Then the second generation that lasted till 1990 saw teleconferencing, e-mail, EFT and EDI. The third generation of technology saw video-conferencing, cooperative processing and the developing MHS, the integrated massage handling system. This evolution is summarized in Figure 17.4.

EDI can be characterized as an application-to-application or computer-to-computer processing using teleprocessing and networks that automated many an office and business process. These processes included price generation, product availability reporting, purchase ordering, invoice processing, shipping and receiving of goods, accounts receivables and payable, and responses to FQAs (Frequently Asked Questions).

However, EDI was appropriate for only highly structured and well defined processes and hence did not include many of the processes in a typical office and business. It could not handle problems that needed a human dialogue and is not considered a very end-user friendly system for clients and suppliers. There are also problems with standards especially for the transmission of EDI documents. Despite these difficulties, EDI is very cost-effective and has a fast response time for applications where it can be used.

EFT is concerned with the transfer of funds once the paperwork has been satisfactorily completed by EDI. One advantage of EFT is that a large number of transactions involving large sums of money can be processed quickly and efficiently. The accelerated income velocity of money under EFT has a side-effect. Corporate mangers must contend with the loss of float and the possibility that security and privacy of financial transactions may be breached. Also, there is resistance to EFT which can be traced to the fact that people like to have cash literally 'in hand'. Not all groups of potential end-users have the same reaction to EFT because they are affected by the advantages and limitations in different ways as summarized in Table 17.3.

Technologically, EFT and EDI may lead us to a cashless society. The application of information

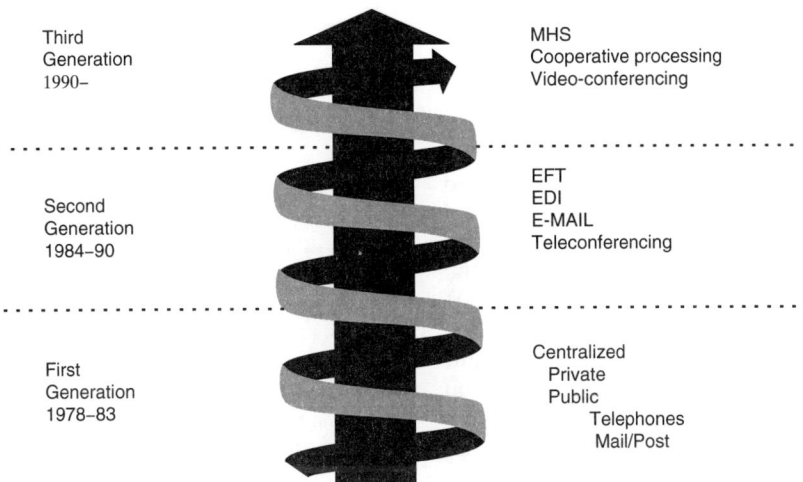

Third Generation 1990–		MHS Cooperative processing Video-conferencing
Second Generation 1984–90		EFT EDI E-MAIL Teleconferencing
First Generation 1978–83		Centralized Private Public Telephones Mail/Post

Figure 17.4   *Evolution in message handling*

Table 17.3   *For and against EFT*

	For	Against
Corporations:	Better cash management Security problems	Loss of float
Retailers:	Quick credit approval and transfer	Capital costs
Financial: institutions	Lowers transaction Fear of monopoly	Costs of ATM
Consumers:	Direct payments/ receipts Home transactions now possible	Loss of float Invasion of privacy possible

technology and telecommunications to home banking and home shopping is moving slower than many had expected. One reason is that ATMs (Automated Teller Machines), POS (Point of Sale) terminals and the consolidation of back-office automation in banks have high priority and are consuming the scarce technical resources. The cost of EFT and EDI is also a drawback. The replacement of plastic cards with smart cards may help reduce the use of cash and cheques if they can be secured, especially when the transactions are between individuals and businesses as is the desire on networks such as the Internet. We will return to this topic in Chapter 20 on the Internet.

Case 17.1: GE bases global network on teleconferencing

General Electric Corporation (GE) has launched a two phase, three-year plan to use video-teleconferencing to promote information sharing between its US operations and its international joint-venture businesses. Implemented by AT&T, France Telecom and British Telecom, the network will link 20 offices with offices in 25 other countries, providing seven-digit dialling for GE offices world-wide. The network which will be managed by GE's network-control centre in Princeton, New Jersey (US), will comprise many T-1 links and three intelligent nodes at Princeton, London and Paris with the Rembrandt series of Codes by Compression Labs Inc., as the computer system driving the video-teleconferencing. (A T-1 link is a telephone line, leased from a common carrier and can carry both voice and digital messages at 1.54 megabits per second.)

Teleconferencing will help GE customize its products for specific markets in different countries and implement new production and engineering techniques. The firm will be able to use a few dozen internal video-conferencing rooms to hold meetings with 30 to 40 customers and suppliers around the world. It hopes this technology will help the firm solve day-to-day problems as if the factory were in the same building rather than thousands of miles away.

Source: *Information Week*, June 5, 1989.

Case 17.2: Electronic data exchange (EDI) in the UK

Over 2000 companies in the UK are engaged in EDI, sometimes known as 'paperless trading'. They use networked computer-based systems to process orders, invoices, freight and forward notices, and customer declarations, in order to save manpower, reduce errors and minimize the amount of paperwork their offices handle. EDI appears to be a major facet of the proposed Police National Network. Slashing paperwork should help reduce the time that prisoners are held on remand before trial by speeding the time it takes to process and exchange information between the police, customs and Home Office.

Two factors contribute to Britain's lead over other European nations in EDI activity: the liberalization of British telecommunications and the ease with which value-added network suppliers can enter the market. (The latter coordinate EDI networks.) Nevertheless, less than 1% of all business transactions in the UK are via EDI.

Sources: *Financial Times*, July 19, 1989, p. 18 and *Computer Weekly*, Dec. 1989, p. 8.

Supplement 17.1: Costs of message handling and related processing

	1974	1994
Telex		$25+
Fax		$10–20
Electronic messaging		$1.70
Processing (MIPS)	$90	$20+
Storage (gigabytes)	$10	$3
Sending 1 megabyte of data across the US	$0.95	$0.20

Source: *Fortune*, July 1995, pp. 119, 121.

Cost of EDI in 1994 (with an Internet service provider

Start-up costs	$50
Telecom costs	$14/month
Access costs	$45/month
(based on 500–2000 documents/ month)	

Source: *Information Week*, Oct. 2, 1995, p. 70.

Bibliography

Aldermeshain, H, Ninke, W.H. and Pile, R.J. (1993). The video communications decade. *AT&T Technical Journal*, **72**(1), 2–6.

Bragen, M.A. (1995). Four paths to messaging. *PC Magazine*, **14**(8), 139–151.

Coopock, K. and Cukor, P. (1995). International video-conferencing: a user survey. *Telecommunications*, **26**(4), 33–34.

D'Angel, D.M., McNair, B. and Wilkes, J.E. (1993). Security in electronic messaging systems. *AT&T Technical Journal*, May/June, 7–20.

Ellis, C.A., Gibbs, S.J. and Rein, G.L. (1991). Groupware: some issues and experiences. *Communications of the ACM*, **4**(1), 38–58.

Emmelhainz, M.A. (1990). *Electronic data interchange: a total management Guide*. Van Nostrand Reinhold.

Friend, D. (1994). Client/server vs. cooperative processing. *Information Systems Management*, **11**(3), 7–14.

Gordon, S. (1993). Standardization of information systems technology at multinational companies. *Global Information Management*, **1**(3), 5–15.

Journal of Information Technology, **9**(2), 1994, 71–136. Special issue on 'Organizational Perspectives on Collaborative Processing'.

Labriola. (1995). Desktop videoconferencing: candid camera. *PC Magazine*, **14**(8), 221–236.

Morley, P. (1992). Global messaging: the role of public service providers. *Telecommunications*, July, 19–30.

Muiznieks, V. (1995). The Internet and EDI. *Telecommunications*, **29**(11), 45–48.

O'Keefe, R. (1995). Do the business. *The Internet Magazine. net*, Issue 10, 64–66.

Riaczak, K. (1994). Data interchange and legal security — signature surrogates. *Computers and Security*. **13**, 287–293.

Richardson, S. (1992). Videoconferencing: the bigger (and better) picture. *Data Communications*, June, 103–111.

Senn, J.A. (1992). Electronic data interchange, *Information Systems Journal*, **9**(1).

Thuston, F. (1992). Video teleconferencing: the state of the art. *Telecommunications*, January 63–64.

Trauth, E.M. and Thomas, R.S. (1993). Electronic data interchange: a new frontier for global standards policy. *Journal of Global Information Management*, **1**(4), 6–17.

Wright, T. (1992). Group dynamics. *Which Computer?* **15**(7), 38–48.

18

MULTIMEDIA WITH TELECOMMUNICATIONS

In an age when computers will be voice, touch, joysticks, keyboard, mice and other devices; television is inherently passive, a couch potato medium.

George Gilder in Microcosm, 1989

Have nothing in your house that you do not know to be useful or believe to be beautiful.

William Morris

Introduction

Multimedia is nothing very new conceptually speaking; however, there are many applications of multimedia. A list of current and potential applications appears as Figure 18.1. In some PCs, with most workstations, and interactive TV, multimedia comes as a standard capability. Then there are specialized devices with the ability to compose music and do computer-aided learning (CAL). But all these devices do not necessarily require telecommunications. Some applications do use telecommunications like telecommuting (using computers for work at home), e-mail that delivers multimedia documents, teleshopping, telenews, telebanking and financial applications, cooperative processing, and games. But multimedia in these cases have only enhanced existing applications that existed without telecommunications. Then there are applications that are not feasible without telecommunications such as video-on-demand, the digital library, video-conferencing, distance learning and telemedicine. It is these latter applications that we will examine in this chapter. But to appreciate multimedia applications we need to understand the nature of multimedia, its characteristics that are relevant to computer processing, and the resources needed for these applications. This is where we start this chapter.

Multimedia and distributed multimedia

Multimedia is an extension of the traditional media of numeric data and text. Later there were graphics added on but that did not have the resolution and quality of pictures. The addition of graphics, images and pictures, as well as audio, animation and full feature films, are what is often referred to as multimedia. The extension of multimedia processing is when it is available from a distance through networks and telecommunications. Then we have access not just to the local library of multimedia but to all the libraries strewn across the country and indeed around the world.

Multimedia not only adds to knowledge but also to the friendliness of the computer interface. It is so much easier to communicate with all the senses of sight, sound and touch (of the keyboard). The sound of one's name and a welcome message or the sound of 'YOU Have Mail' is often much more refreshing than just seeing the message on the screen. Perhaps we will get used to the novelty of voice messages but then there will be other multimedia approaches to make you feel welcome and happy working with a computer.

Multimedia systems include multiple sources of various media either spatially or temporally to create composite multimedia documents. Spatial composition links various multimedia objects

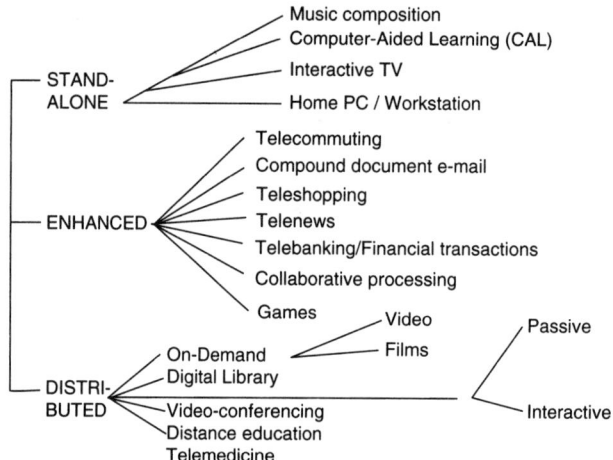

Figure 18.1 *Applications of multimedia*

into a single entity ... dealing with object size, rotation, and the placement within the entity. Temporal composition creates a multimedia presentation by arranging the multimedia objects according to temporal relationship ... (Furht, 1994: p. 53).

Multimedia can also be viewed as a set of objectives and data entities which are a multidimensional array of numbers derived from various sensors that record images, audio and videos. These are filtered and then available for browsing and sometimes even manipulation. Doing this using networks enables one to do collaborative processing using this vast and rich base of data and knowledge in all its many manifestations in different media. This enables richer video-conferencing, more meaningful learning and training even from a distance, and even better diagnostics by a doctor who is otherwise not available on-hand. It will contribute to the way we work, learn, communicate and even play. It could improve one's enjoyment of films and video, not only by increasing the range and variety, but by making them available whenever and wherever they were wanted. When interactive, multimedia could help in our work by improving communications and increasing dissemination of information and knowledge.

Requirements of multimedia

The distribution of multimedia makes many demands on computing resources. To better understand these demands and needs it is necessary to look at the basic characteristics of multimedia transport which creates these needs. One basic need is the timing of the multimedia components and the need for low latency, which is best appreciated by looking at Figure 18.2.

The reduction of delay is important in real-time applications such as in video or the showing of a film. Say that you are seeing a close-up of a singer. Would you not be unhappy if the sounds were not precisely synchronized with the corresponding movements of the lips? This synchronization is referred to isochronous processing and refers to real-time communication that ensures minimum delay by synchronizing to a real-time clock and establishing a virtual channel. Thus isochronous processing is similar to point-to-point circuits rather than packet switching where synchronization is not crucial in say e-mail delivery where the processing is actually not isochronous and data is moved in packets around possibly different routes at different times with small delays hardly perceptible. That is why packet switching is feasible for many text-based and applications-based information including e-mail say on the Internet that uses the TCP/IP protocol.

The second basic and important characteristic of multimedia is that it is very memory intensive. This has two consequences. One is that more storage space is required at the server and client receiver end; and secondly, large capacities are required for the transfer of the multimedia messages which means more bandwidth. Networks

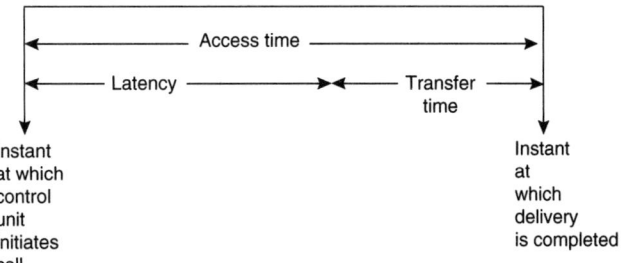

Figure 18.2 *Latency defined*

designed and optimized for data and textual traffic are not suited for multimedia traffic like video and films.

There are many techniques of overcoming the memory intensive problem. One is the use of techniques for economizing in memory. For example, in image processing of a film it is not necessary to capture and store all consecutive image stills in sequence but instead to capture and store a base image and then only the changes to that base making the computer processor do the construction of each changed image. Even then, storage is a problem and then one must resort to compression.

Compression over the years has greatly improved as shown in Table 18.1. However, despite compression and high compression ratios, storage can still be a serious problem as demonstrated in Table 18.2 where different types of multimedia are compared.

Once compressed, multimedia must be efficiently stored, transported, retrieved and manipulated in large quantities and at high speeds.

Table 18.1 *Progress in compression over the years for data rate acceptable for video*

Year	Bandwidth (Bits/s)	
1978	6000000	
1980	3000000	
1982	1500000	
1984	760000	A factor of
1986	224000	75 over
1988	112000	24 years
1990	112000	
1992	80000	

Source: Adapted from Joseph Braue (1992). The rise of enterprise computing. *Data Communications*, **21**(12), 27.

Table 18.2 *Storage requirement for typical applications (in megabytes)*

Text of 500 pages (normal standard)	1.0
100 Fax line images uncompressed	6.4
10 minute animation (15 : 1 compression)	100
100 colour images (15 : 1 compression)	500
10 minute digitized video (30 : 1 compression)	550
1 hour digital video (200 : 1 compression)	1000

Source: Adapted from Borko Furht (1995). Multimedia systems: an overview. *Multimedia*, **1**, (1), 48.

Different media have different needs of bandwidth depending on the nature of the application as shown in Figure 18.3. For example, video and audio stream playback and teleconferencing are both real-time applications but with different latency and delivery requirements with teleconferencing requiring very small latencies but playback applications requiring guaranteed delivery of real-time messages.

Resources needed for multimedia processing

Large bandwidth needs of multimedia require switching capacities for large megabyte and gigabyte transfers and will need equipment like ATM (Asynchronous Transfer Mode), SMDS (Switched Multimegabit Data Service) and SONET (Synchronous Optical Network) discussed in earlier chapters.

The need for large bandwidths, low latencies and real-time delivery makes many demands of computing resources. These resources have to be managed along all the paths between the end-user at a client device and the server, the

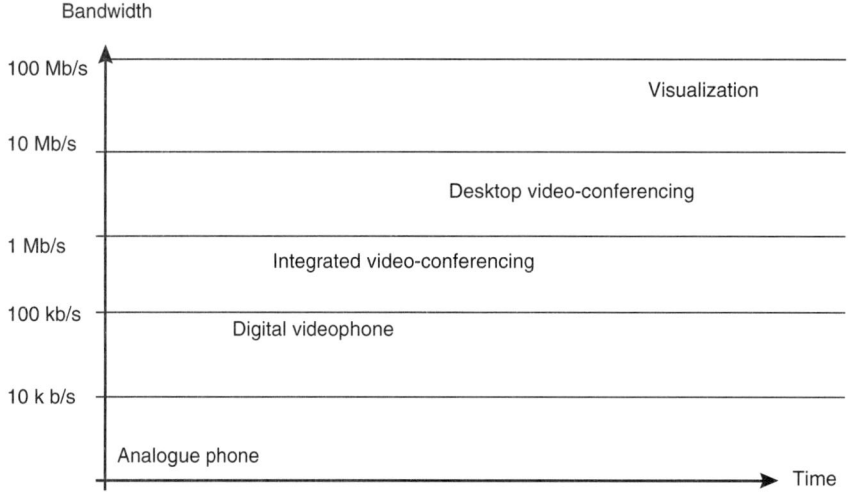

Figure 18.3 *Demands for bandwidth. Not to exact scale*

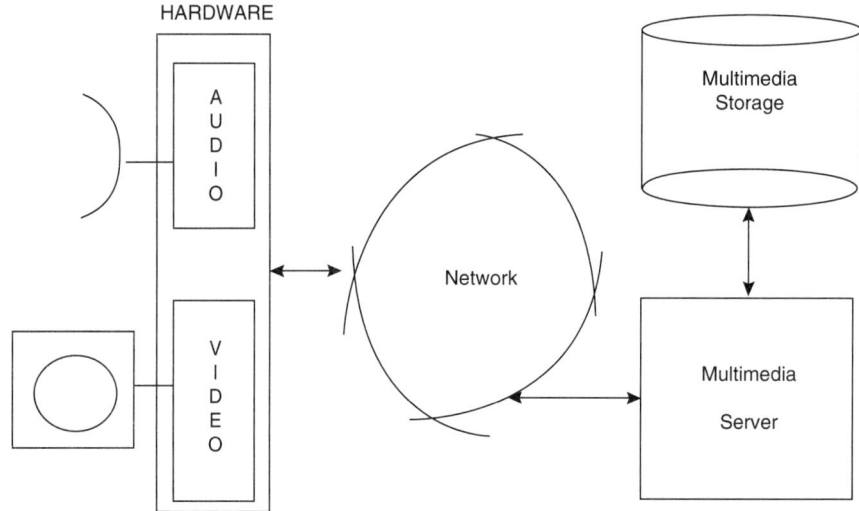

Figure 18.4 *Multimedia via telecommunications*

server itself, and the client. An overview of the main components of a multimedia system is shown in Figure 18.4. We will discuss these in turn starting with the server end.

Servers

The servers are specialized processors designed for handling large masses of data. They will vary with the type of multimedia being processed but whatever the media the server has to be fast

and powerful. It must also be supported by large and often very large storage capacity. The magnitude of storage space required for multimedia has been mentioned before. Now multiply the storage for one delivery with the large number that will soon be in demand. Even a modest choice of films and video can quickly run storage into gigabytes. These storage devices must be fast enough to serve the high volume of multimedia without being a bottleneck in processing. For simultaneous access, servers must also be equipped with multiple disk drives with video files for each

202

drive. Storage can be done by storing multimedia as objects in the object orientation methodology. Objects are also embedded in applications but can be retrieved by OLE, Object Linking and Embedding methodology from Microsoft, which is fast becoming the industry standard for large PCs and workstations. The objects could then be data from a spreadsheet, text from a word-processing program, or a fully fledged film or video.

Networks

We have mentioned the need for bandwidth earlier. One approach to the problem of bandwidth is microsegmentation where users are assigned to segments of the LAN. This reduces the users per segment and better serves the users in terms of bandwidth available to them. Microsegmentation, however, does rely on better switching as well as its integration with bridging and routing especially if a hub is being used. Intelligent hubs, it is hoped, will be able to do this and provide bandwidth-on-demand as needed and provide virtual switched circuits.

Besides bandwidth, there is a problem of architecture with multimedia. It corresponds to the OSI layer architecture discussed earlier and now compared in Figure 18.5. Multimedia needs a new applications layer to accommodate multimedia applications and the transport protocol needs an additional control capability to prevent congestion at the destination and at the routers. Also, the transport protocol needs the means to package timing traffic with audio and video to prepare them for transmission. Most upgrades to the transport and other layers required by multimedia can be handled by software upgrades.

Multimedia needs the addition of a base layer and a node layer. The base layer is added to the OSI model in order to coordinate the data access and make the data generated by these queries systems independent. The node layer, residing over the transport layer, represents the logical storage of multiple forms of data contained in the multimedia information ... By providing an organized method for computers to logically communicate, the layer model ensures data can be transferred independent of systems platform ... The interoperability ensures that different platforms can equally access multimedia information distributed over networks regardless of the incompatabilities in their operation. (Domet III *et al.*, 1994: pp. 42, 41).

Clients

The client processor has to be large and powerful though not as powerful as the servers. Early clients were the 386 or 486 processor. Nowadays, the Pentium is recommend for any respectable performance for multimedia processing. Besides the CPU requirement there is the

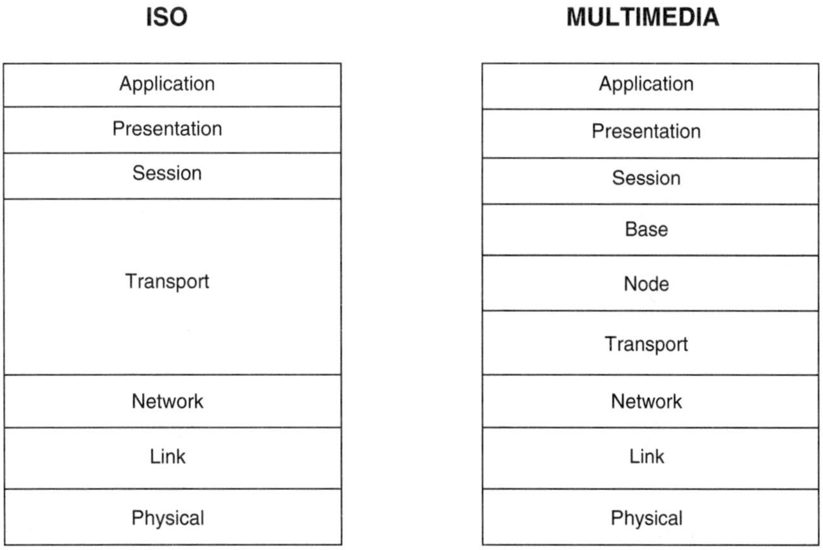

ISO
Application
Presentation
Session
Transport
Network
Link
Physical

MULTIMEDIA
Application
Presentation
Session
Base
Node
Transport
Network
Link
Physical

Figure 18.5 *ISO vs. multimedia architecture*

need for the audio and visual coder and decoder. Some systems come with a decoding board or a coder/decoder called the codec which needs to be equipped to both the client and the server. The codec is the multimedia equivalent of the modem (modulator/demodulator) we visited earlier. Many codecs also perform compression. Some codecs achieve 175 000 pixels/frame which comes close to high quality TV which is between 190 000−200 000 pixels/second.

The client could also have a set-top box, also called a home communication terminal, that has interactive program guides, movie-on-demand capability, robust graphics, full user interactivity and personalized information channels.

Standards

Multimedia needs its standards and there are plenty of them. The most prominent is the MPEG series, where MPEG is the acronym of the organization that developed the standards: Motion Picture Experts Group. **MPEG** deals with moving images while JPEG standards deal with still images. There are other standards organizations involved in developing multimedia standards, at least seven in the US alone. The MPEG-1 standard defines a video resolution of 352-by-240 at 30 frames per second. This gives a quality comparable to the videos on rent in many stores assuming a 150 Kbyte/ps bandwidth.

The MPEG standard is the best known since it is an international standard approved by the ISO though it does not carry the ISO prefix. The MPEG is being watched by the profession because it is an exception to the rule that you need a stable technology before a standard is defined. Here we have a standard before the technology has stabilized. MPEG will test the rule of which comes first: the chicken or the egg, the standards or the technology. The widely accepted standard may encourage the industry to bring out new products but it might also inhibit innovations and freeze the technology of multimedia.

There are also special standards being developed for codecs such as the H.261 by CCITT. It defines display formats with specific resolutions and provides processes for reading, decoding, video information and the organization into blocks and groups of blocks. H.243 defines exchanging a unique encryption key at start of compression. There are other standards like H.221, H.230, H.261, H.242 and the umbrella standard of H.320.

Applications of distributed multimedia

Even without discussing the more exotic multimedia application of telemedicine and digital libraries, we can see that the implementation of distributed multimedia is going to be neither easy nor cheap. It is therefore safe to assume that implementation will come in stages and will come first in the office rather than the home because the offices will be able to afford it and make good use of it. Also, implementation will be first in the areas where existing applications do not yet have multimedia and need it. Examples are telecommuting, e-mail and retrieval of multimedia information for decision-making and problem-solving. But even then there will be a high cost since these applications will be very storage-intensive. For example, an e-mail containing a one-minute full-motion video will range between 10 Mbytes for VHS quality. Other combinations of e-mail are also storage (and bandwidth) intensive like text augmented by audio, text with images, and images with audio. Sending video mail in the store-and-forward mode does not require isochronous processing but video mail does demand minimum delays and consume prodigious bytes of storage. Eventually we will have compound document e-mail with text, voice and video that can be sent in one form and read in another. It is already possible to send an image document and review it as text (using an optical character reader) and have text read aloud (using voice synthesis).

Video-conferencing

Another important application of multimedia in the office is video-conferencing. It may be considered an extension of existing teleconferencing but it has many important characteristics that make it complex to implement. These characteristics are:

1. It has bandwidth needs that are on-demand but an unknown size demand unlike say the downloading of a known sized full-feature film or video. There is some uncertainty about how many participants there will be and how much each will participate.
2. It is a two-way interactive communication with both ways possibly containing multimedia unlike the downloading of a film or video which is a one-way downloading.

3. It is real-time and hence isochronous processing.

Multimedia conferencing utilizes the concept of shared virtual workspace which describes the display at each client station and where each participant can send or receive data, voice or video as part of collaborative processing. Furthermore, the functions performed include real-time control of audio and video, dynamic allocation of network resources, synchronization of shared workspace, multiple call set-up, and graceful degradation when faults occur (Furht: 1994: p. 58).

The basic process of video-conferencing is shown in Figure 18.6 where the object to be transmitted over the networks is found stored at a server, compressed, and then communicated through the layers of transport, network and link to the network. On reaching its destination client, it flows through the layers of link, network and transport and is then decompressed before appearing on the screen as the desired image with its multimedia package. The process for the opposite direction is the process in reverse. For both cases we do not show all the layers nor all the components involved and stick to the basics to keep things simple.

One of the unresolved problems of multimedia conferencing is the optimum communication architecture for composite streams of audio, video and images in transmission. Other implementation problems concern hardware interrupts and interoperability problems and the global adoption of international standards on video-conferencing quality despite the existence of many standards for video-conferencing like the H standards from the CCITT concerning interoperability: H.261, H.221, H.230, H.242 and the umbrella standard H.320. What we need is a high resolution video and audio overlayered with spreadsheets, engineering drawings, text and other relevant multimedia information.

Despite these difficulties, there are some distinct directions that video-conferencing is taking. It will include the upgrading of telephones and the downgrading of meeting rooms leading to desk video-conferencing; improvements in size of image, quality, picture and colour support; the initial use of the PC with live video in a window of the screen; the ability of participants of videoconferencing to take notes, change memos and diagrams, jot annotations, and manipulate data on spreadsheets, all during the video-conferencing; and the development of a library of all relevant multimedia objects that are easily and quickly retrievable.

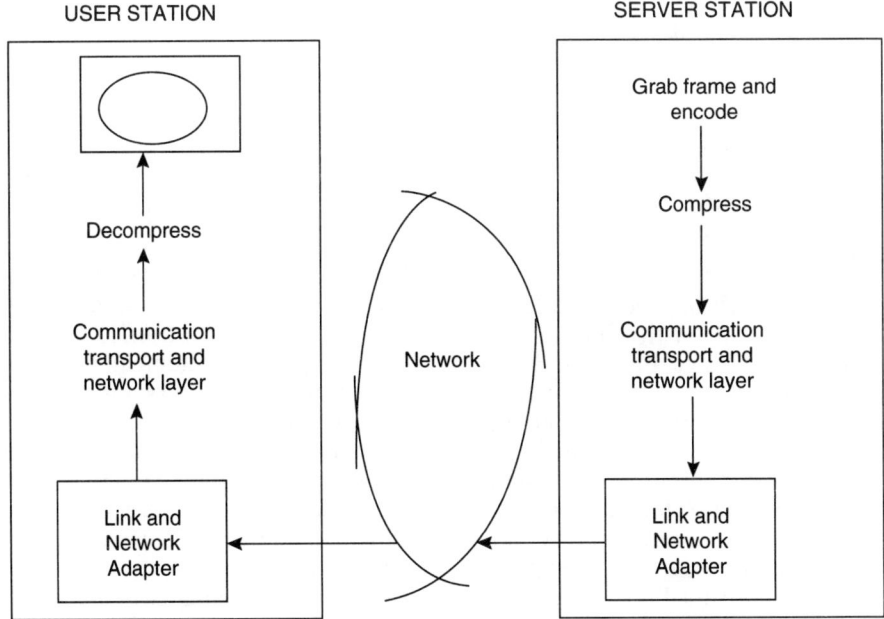

Figure 18.6 *Video transmission*

Video/film-on-demand

This is where full-length films or videos are available on demand and delivered to the office but most likely in the home and on the home entertainment device. Conceptually, there is little difference between the processing of video-conferencing and the very common teleconferencing. The main technological differences lie in that the multimedia in video/film-on-demand is transported only in one direction and that the storage at the server end has to be very large in capacity and very fast in its retrieval. The great advantage that both video and film-on-demand have is that they can be viewed, stopped and replayed, rolled-back, zoomed for any frame, run in slow motion or at fast speed, all under the control of the end-user while in the relaxed comfort of the library or bedroom at home.

We have already discussed the storage considerations but it is important to note that what is stored is mostly entertainment and that brings the entertainment industry into the picture. They become important players in the market for they control the content. However, they also want to control the distribution and so there is a confrontation between the holders of content and the distributors which is resolved in the market-place through mergers, alliances or takeovers. We shall discuss these relationships later but only after we identify another group of players, the educational sector, for they also generate and own content. This group includes the digital libraries and educational institutions that offer distance learning. Before we examine these applications, we look at another application which is somewhat similar to video-on-demand, that of telemedicine.

Telemedicine

Telemedicine is the delivery of medical care to a patient by an expert (or institution like a hospital) separated by distance but connected by network and a multimedia communication system. A package of information on the patient is multimedia in that it may contain CAT or MRIscans and other computer generated pictures; entire medical case history in text and graphics; and a verbal opinion or set of questions by the patient's doctor to the expert. The expert looks at the multimedia package and may ask questions and a dialogue may ensue after which a diagnosis is made. The multimedia package may be sent only one way but the dialogue is a two-way dialogue and maybe not between many people as in a video-conferencing situation but between at least two people on an on-line real-time basis for time could be crucial in many a situation. Thus telemedicine is not as complicated as video-conferencing in that multimedia is often not exchanged but transported one way. Also, this package is not grabbed from a server as in video/film-on-demand but is created in real-time with current pulse rate and heart beat transmitted as they occur. This may require a lot of sophisticated equipment for data acquisition at the patient's end but it may be the report of a nurse on a helicopter escorting a wounded person in a car crash (or in a war zone). This raises many important issues in telecommunications and networking along with moral and ethical issues of who has the priority in traffic congestion.

We will now define **telemedicine** as being the collection of data on a patient in digital form to be transmitted remotely to another point for analysis or diagnosis. It could be used for getting an expert opinion or any opinion because of the urgency involved such as in the case of injuries on a battlefield (incidentally, telemedicine was first supported by the US military) or for getting medical attention where it is not possible such as in rural areas or when travelling away from the family doctor.

The collection of data would include the recording of data such as pulse, blood pressure, etc.; the observation of physical conditions like colour of the eyes or face through imaging devices; observations made by special cameras that can go into orifices like the ear, nose and throat; testing of blood samples; collection of data from special instruments; and the past medical records of the patient. All this is sent by telecommunications using a wideband telecommunications system capable of handling multimedia traffic.

The impact of telemedicine is on both the patients and the doctors. One impact will be on the patient who will no longer have to go to a family doctor but a medical facility equipped for telemedicine and have a diagnosis made by a doctor or a specialist. In some cases, treatment may include an operation. Some operations have been performed remotely by a surgeon on a dummy with all the movements being transmitted electronically to a robot working on the patient. And what if the patient lives in a remote area of the country, or is remote from home, or just lives in a developing country without good medical facilities? Telemedicine may then be the only good answer available.

The impact on the family doctor will be access to expert opinion instantly but a loss of patients over time. The family doctor will no longer have a monopoly over the local patients but may have to start seeking remote patients. Many issues are raised by telemedicine, such as:

How is the cost of the visit for telemedicine split between the local doctor and the remote expert?

Given that medical practice has to be licensed by each country and even each state (or province or county) in each country, at what location should the doctor be licensed? At the location of the patient, the expert, or both?

Who is responsible if there is a misunderstanding on 'this' or 'that' artery, limb or organ resulting in the wrong actions being taken and an 'injury' to the patient.

What happens if a document or message for one patient gets switched to another patient? Who is responsible? The doctor with the patient, the telemedicine facility, the transmission equipment manufacturer, or the telecommunications carrier?

How is the telemedicine technology to be regulated?

What is the liability of the providers of the technology that makes telemedicine possible, including that of the telecommunications and network provider?

Digital library

In the 1960s and 1970s there was much talk of an electronic library. John Kemeny, the designer of the programming language BASIC, envisioned one library in all of the US receiving queries on a book or article. The end-user could browse through the index and even glance through the contents or the book itself on a terminal and finding what is wanted would have the article or book downloaded to the terminal. There was much talk about the distribution of royalties and then there was a long lull. Then in the late 1980s there was a rebirth of the idea the term digital libraries surfaced. A grant from the NSF in the US in 1993 to six universities for a digital library initiative did

not hurt. But the transformation to digital was perhaps largely the result of the emergence of the ISDN (Integrated Services Digital Network) and the promise of B-ISDN with high capacity bandwidth designed for multimedia transmission. The interest was also sparked by the falling prices of costs of digital storage relative to the costs of library shelf-space and the increasing costs of the labour-intensive activity of cataloguing, checking-out, checking-in and reshelving of books in a library. (The operating costs of the Bibliothèque Nationale de France costs approximately \$260 million for its 5 kilometres of computer controlled belts that deliver books to some 150 points.) And then there is a capital cost (the Bibliothèque Nationale that opened in 1995 had cost around \$2 billion). Also, advances in networking made transmission of book material feasible and even affordable.

But there are many unresolved problems and issues. One concerns the conversion of all the material in our libraries into tangible objects with electronic digital representations such that they can be easily accessed and retrieved, copied and then distributed. This will change the printing and publishing of books and magazines as we now know it. There is also the issue of the protection of intellectual property rights. This is an old issue and is not getting near to any solution. In fact, things are getting worse since networking and the availability of holdings in the digital libraries may soon be accessible anywhere in the world by countries (especially in the Far East and the Pacific Region) that have not agreed to the adoption or enforcement of international agreements to protect intellectual property rights. Also, there is some philosophical and legal argument about the relevance of intellectual property law to software protection. Some in the software industry believe that 'existing property laws are fundamentally ill-suited to software. The problems are rooted in the core assumptions in the law and the mismatch with what we take to be important about software'. (Davis *et al.*, 1996: p. 21).

Books in a library on some scientific and high tech subjects get outdated sometimes even before they are printed. One relies therefore on journals which also have a long lead-time for publication. Using a digital library (or the Internet), a document can be available immediately after it is written and put in machine readable form. Here books and articles can be searched by content, title or author. Special hyperlinked versions will facilitate search and use.

What will be the impact of a digital library on levels of education and the acquisition of knowledge? How will it affect the delivery of educational services? Do students need to go to the library or can retrieve what they want from a room in the dormitory or from their homes? In addition to books being delivered electronically, why not also have lectures delivered electronically? Why can a student not learn from the best instructor and researcher in the country or the world, just as a patient get the best advice from medical experts wherever they are? Why can any student not have quick and easy access to the lectures and research findings in Yale, Stanford, Harvard, Oxford and Cambridge? How will this affect international education? Will the digitization of information make it less accessible to a class of information-poor who must then compete with the information-rich class? How will digitization contribute to economic development? How will this affect productivity and competitiveness (nationally and internationally)? Will it improve our life-style and increase our standard of living?

Whatever the answer to all the above possible questions, there is a general acceptance that digital libraries will contribute to formal learning, informal learning and professional learning and may even contribute to digital schools. It will greatly support distance learning which is the subject of our next topic.

Distance learning

Distance learning is education in a place that is remote from the sources of learning whether this be instructors or teaching materials. This is achieved by transmitting educational material like courseware designed for remote learning in multimedia packages to the point where they are needed.

The process of distance learning can be visualized as in Figure 18.7. The instructor using a multimedia workstation will instruct courseware to be sent through a network to a pupil or student who also sits at a multimedia workstation to receive the material. Also accessed are materials from data/knowledge-bases (which may include multimedia dictionaries, encyclopedias, atlases, etc.) either by the student directly or on the recommendation by the instructor, or both. This mode of education that blends networking with educational materials (including appropriate courseware with varied media providing richer insights) not only implies different modes of instruction but also different modes of learning.

Distance education along with digital libraries if at all accepted as a viable concept will have a profound impact not only on the way we educate the next generation but also on the infrastructure needed for such changes.

In technological and social changes there is always an interplay, a tension, between the forces of conservation and innovation. Cultures and communities do not, and should not, let go

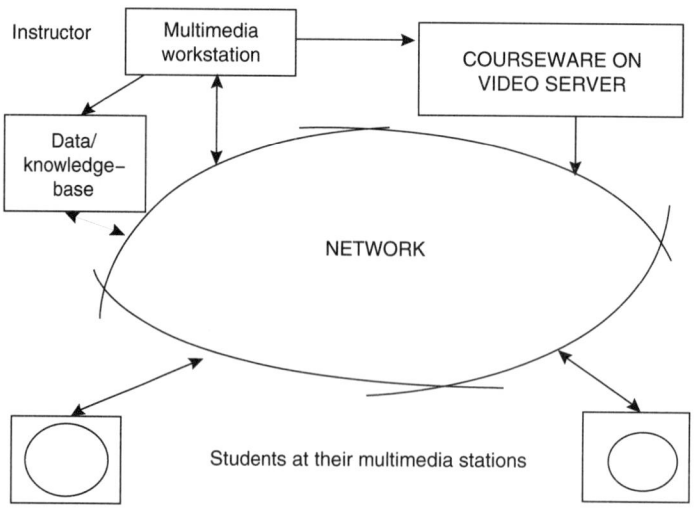

Figure 18.7   *Distance education*

lightly structures and practices in which they are invested heavily. The task in the years ahead will be to decide which existing practices and structures to let go and which to retain, and which innovations to reject or adopt. (Levy and Marshall, 1995: p. 83).

Multimedia electronic publishing

Electronic publishing is the publishing of material by electronic means like a computer with a fast laser printer. In the early 1980s we had a special computer program TeX that electronically published books for computing with symbols used in mathematics and science. The turn-round time from a finished manuscript to the bookshop was reduced from what was then around 1–2 years to 1–2 months. And yet electronic publishing is not very common. Partly because there is a high cost of capital and conversion and partly perhaps publishers are thinking of multimedia publishing. A contract for a book in 1990s often had an additional clause giving the publisher the copyright for use of the material for all multimedia publications to be done electronically if deemed profitable. Publishers are thinking in terms of making all publications multimedia and even interactive: a fiction book would have an appropriate background music for each chapter; a text in chemistry would have a film or video on an experiment being performed; and an encyclopedia entry on space travel will have the film clips of the actual space

launch. With the click of a mouse, the user, even a child, can get more information (or action) from any of the objects on the screen. Such interactive multimedia can add greatly to learning especially to electronic distance learning. It will also greatly reduce the costs of books once book publishing is integrated with distribution through telecommunications and networks. This could be done using a digital library or through communications media, be it cable TV or by telephone to a computer screen or to a combination device. This explains the great interest in communication carriers and the telecommunication industry by the publishing industry and the many alliances and mergers taking place not just in the publishing and entertainment industry but with carriers and the computer industry.

Integration of electronic publishing would have many advantages such as:

Reduction in current prices of publishing and distribution

Reduction in the time required for production and distribution

Ease and speed of updates, revisions, and later editions

Easier access of search and retrieval by different identifiers using computer indexing and hypertext methodology

The entire publishing industry will be environment friendly and save the cutting down of trees for making paper currently used in publishing.

Table 18.3 *Comparison of modes of delivery*

	Telephone	Cable TV	Networked (Internet)	Multimedia networked
Main users:	Everyone	Homeowners	Anyone	Everyone
Content:	Low	Low/medium	Medium/high	High
1/2 WAY:	2 way sequential	1 way only	2 way sequential	2 way interactive
Bandwidth needed:	Very low	High	Low/Medium	High to very high
Security:	Good	Not needed	Not good	Poor/good
Usage:	Very easy to use	Very easy to use	Not so easy	Easy
Access:	Excellent	So so	Not so good	Not so good
Topology:	Bus/star	Trunk & branch	Routed	Star
Switching:	Circuit switching	Unswitched	Packet switched	Switched/unswitched
Protocols:	POTS	Proprietary	TCP/IP	ISDN, analogue
Billing:	Per unit used	Per connection	Poor facility	Good facility

Source: Adapted from Furht (1995: pp. 28–9).

Integration requires the consideration of different modes of delivery which are compared in Table 18.3

We have thus far discussed the publishing of books and magazines and journals. But the publishing industry includes the publishing of newspapers too. If its publication and its distribution can be integrated then why not have one office do all the editorials, collect all the raw data, compose the news stories, add art work and coloured pictures, and then transmit this to the regional and local points where local or regional news can be added and printed off fast electronic printers? Wouldn't this be cheaper and more efficient? But would this also lead to a concentration of power assuming that information is power? If so, it that all that bad? If yes, then can the dangers of concentration of publishing power be contained and curtailed?

There is no question that telecommunications and networks make many an application now possible but along with it there arise many social and ethical questions and issues that we may not be ready to handle and cope with. There may be organizational problems too.

Organizational implications of distributed multimedia

For all the many applications of distributed multimedia, we have examined the technological implications but not the organizational implications. The important implication results of many large and important firms and industries in the distribution of multimedia. We mentioned earlier that the entertainment industry has become most interested in the distribution of multimedia films and videos-on-demand. This includes the publishing industry with Goliaths like Simon & Schuster in the UK. We have also mentioned the end-users implicitly as including libraries, the health-care institutions like hospitals, and the government who are interested in the welfare of their citizens and subsidize some organizations like libraries and hospitals. Another group interested is the computer industry which generates the products used for multimedia. This group include manufacturers of computers, software houses, and manufacturers of peripherals and supporting telecom equipment. And finally, there is the group of providers of communication services, which include the companies that offer telephone, cable TV, wireless, satellite and the Internet services. There is fierce competition amongst some of these firms especially in the communications provider industry. For example, the telephone and cable TV each think that they can produce one device that will do away with, if not marginalize, the other. This then could be a battle for survival. Each, however, has a comparative advantage as shown in Table 18.3. It is likely that all will survive but only after much restructuring and realignments.

The industries involved in multimedia communications face each other in the marketplace as depicted in Figure 18.8. Most of the

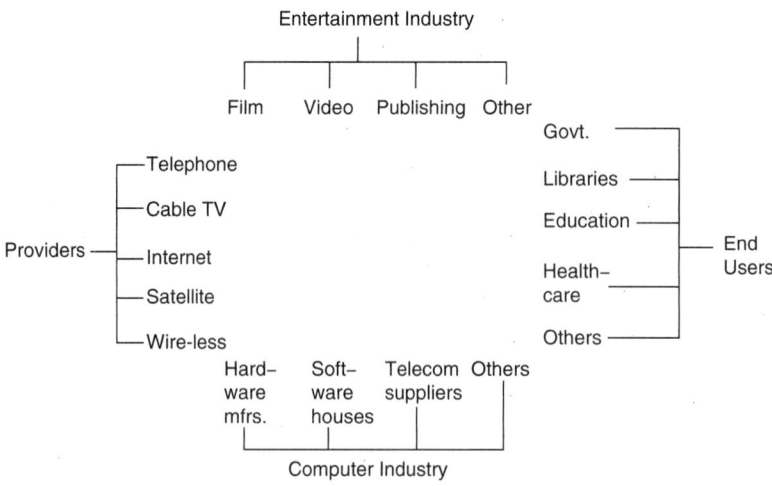

Figure 18.8 *Players in the multimedia market*

Table 18.4 *Evolution of multimedia computing.*

	1st Generation 1989—91	2nd Generation 1992—94	3rd Generation 1995—96	4th Generation 1996—
Media:	Text B/w graphics Bitmap images	Moving, still & colour-bit images Motion video	Full motion HDTV	On-demand video/film Multimedia video-conferencing
Technology:	JPEG	Motion JPEG MPEG-1	MPEG-2 MPEG-3 MPEG-4	Wavelets + ????
Base platform:	25 MHz 386 40 MB disk	50 MHz 486 240 MB disk	100 MHz Pentium 2600 MB disk	?????????? + Risc processor
Authoring:	Hypertext Hypermedia	OO multimedia	Integration of OO with systems operating system	Integration with corporate KBIS
Network:	Ethernet Token ring	FDDI (100 Mb/s)	FDDI (500 Mb/s) ATM	Multi-Megabyte ATM Isochronous Ethernet

interrelationships are between the four groups shown but some interrelationships exist within groups like the government supporting more than one library and hospital or more than one educational institution at the same time. These interrelationships if drawn on Figure 18.8 would be one jumble of lines with many being in the north-west corner between the communication service providers and the entertainment industry. This is where the large and very large multi-billion dollar companies are trying to stake out their share of the market. They will either do it through fair competition, through friendly mergers and alliances, or through unfriendly take-overs and buyouts. This will be decided in the near and far future and so will be discussed later in the chapter entitled 'What Lies Ahead?'. Whatever happens, it will certainly be exciting.

Summary and conclusions

Multimedia has come a long way since the first generation of applications in 1989—90. Technology is improving, especially with a trillion bps transmission, to the point that one can predict (with some caution) that we will soon be able to see any full feature film or video, or browse through any data/knowledge-base or surf anywhere on the Internet, all from the comfort of a home theatre. One could also do video-conferencing from the desktop and, like the home theatre, have access to knowledge wherever it

may be through personalized information channels. This evolution is summarized in Table 18.4.

There are few if any technological roadblocks for the above scenario. However, test beds show no great enthusiasm for such a scenario. There is the fear that costs of delivery will be so loaded with bells and whistles (features) that are not all needed that it may not be economically viable for many an office and family. The end-user also needs confidence in being able to control the content of what comes gushing down the information highway.

Multimedia applications can be classified in many ways. One way is shown in Figure 18.9.

There are two crucial characteristics of multimedia: one, high sensitivity to delays and lack of synchronization in transmission; and two, high volume and size of transmission. These factors vary with applications. Virtual reality applications will have high sensitivity as well as high volume running into the gigabit zone. On the other side of the spectrum of computer applications are the non-multimedia applications that have low sensitivity to delays and low to high volume. In between are the multimedia distributed applications especially video/films-on-demand, digital libraries, and telemetry and other data needed for telemedicine. They have medium to high sensitivity and medium to high volume sometimes going into the gigabyte zone. These applications are shown in Figure 18.10.

Multimedia applications are easier to use and even absorb than many traditional computer

	END–USER to END–USER	END–USER to COMPUTER	END–USER to DOCUMENT	COMPUTER to END–USER
INTER– ACTIVE ACCESS	• Conferencing • Training • Education	Graphical User Interface	Hypermedia access to documents	Real-time film/video/ shop/bank/ medicine
BROAD– CAST	Presentation	Information kiosk	Newsletter	Distance Learning
OBJECT– ORIENTED MANIPULATION	Collaborative processing	Data/Oo knowledge– base	Document e–mail	Tele–medicine Distance learning

Figure 18.9 *A classification of multimedia*

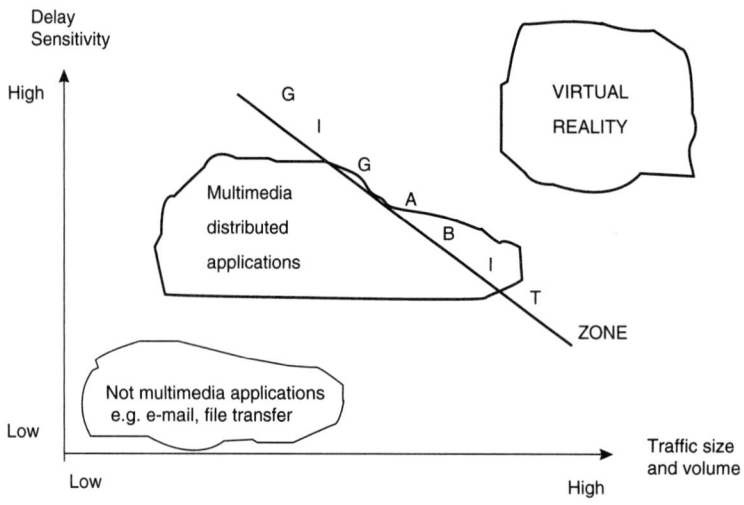

Figure 18.10 *Data sensitivity and traffic size*

applications but more difficult to produce. They have special resource requirements that are summarized in Figure 18.11.

The applications of multimedia examined in this chapter are no 'killer' applications like say e-mail was in the early 1990s. But the application of telemedicine may be a killer application in that its absence may kill people. Also important is the application of the digital library and distance learning for they may well save people from remaining illiterate and living on

the periphery of society. The value of spreading knowledge and increasing learning cannot be measured. Nor can you assign a monetary value to the saving of lives. In this sense the value of multimedia applications will be infinite even if they never become killer applications but become mainstream applications. And what is the value of mass entertainment on demand? This is somewhat questionable because that application has a downside of sometimes getting what you do not want either as violence or pornography or as being

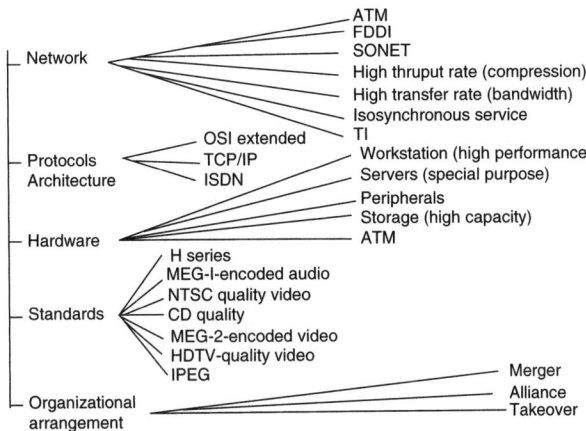

Figure 18.11 *Resources needed for multimedia*

undesirable for children. This poses a problem for technology: should technology be responsible for the control of content or should it be the owner of the system (parent in the case of children using the system)? The problem is less a technological that a moral and ethical problem and the response will vary greatly among societies around the world which will make the global use of multimedia all the more difficult to define, much less resolve. Businesses will benefit greatly from distributed multimedia applications. It is very possible that every worker will have the capability to video-conference from the desk without the elaborate equipment now available only in special rooms equipped for video-conferencing. Document processing through e-mail will improve business communications and teleshopping using multimedia may even increase sales and commerce. Businesses will have to devise new strategies for advertising and selling, while consumers will have to adjust to new ways of buying and receiving the products they want as well as the entertainment they desire.

Case 18.1: MedNet

MedNet is a medical collaborative and consultative system that was in its third phase of operations at the University of Pittsburgh Medical Center.

MedNet started in 1985 and is on-going on a daily basis at seven hospitals and multiple diagnostic and research laboratories. A major challenge in implementing MedNet was 'the development of techniques for processing and display of real-time multimodel medical information ... One area where a collaborative

system can substantially improve patient care is clinical neurophysiology, a consultative service of diagnostic and monitoring service of diagnostic and monitoring techniques used to assess nervous systems functions ... During brain surgery, monitoring techniques help prevent damage to nervous system structures by continuously measuring activity receded directly from the brain in real time. MedNet provides real-time monitoring and multiparty consultation and collaboration during brain surgery for approximately 1600 cases per year.'

MedNet has also developed communication control strategies for a wide variety of data including audio, video and neurophysicological data that arise in a medical environment.

Source: Robert Solomon. Multimedia MedNet, *IEEE Computer*, **28** (5), May 1995, p. 65.

Case 18.2: Electronic publishing at *Britannica*

Encyclopedia Britannica and Wide Area Information Services (WAIS) Inc. of the US agreed to jointly develop improved information retrieval for the Internet. The new search-and-retrieval engine optimizes the original WAIS engines producing efficiencies of up to 20% with easier and faster access to the 44 million word base of the *Encyclopedia Britannica*. *Systems Britannica*, the publishing platform that incorporates the new searching technology serves as an infrastructure for *Britannica's* electronic products including natural language searching which allows the end-user to enter a question like 'Why does the moon loom larger on the horizon?'. Also possible is an elaborate systems of hypertext links that make it easy to navigate among the related entries in the database.

The *Britannica's* Home-Page on the Internet is: http:/www.eb.com.

Source: *Information Today*, 1995, p. 2.

Case 18.3: The access projects at the British Library

The British Library, with a collection of over 18 million volumes, published a commitment in its *Strategic Objectives for the Year 2000* of 'providing maximum access to the collection using digital and networking technologies for onsite and remote users'. Some of the projects include:

> The Patent Express Jutebox holds over 34 million patents. Software searches and prints high-quality copies for users in under two minutes.
> The Electronic Beowolf holds the unique manuscript of the 11th-century Anglo-Saxon epic. 'Letters and words erased by the original scribes, damaged by the fire in 1731, or hidden by 19th-century restoration, are now discernible and test images have been mounted on the Internet to allow international scholars to see the progress of the project.'
> Electronic Photo Viewing System has a major photographic collection including Victorian spiritual photographs is accessible by subject and provides a hyperlink to the descriptive text which accompanies each image.
> The Network OPAC have over 6 million bibliographical records. 'Beta sites are now being selected in the US to work with the library on establishing communications links and usage requirements for a future international Network OPAC.'

Source: Jonathan Purday. The British Library's initiatives for access projects. *Communications of the ACM*, **38**(1), p. 65.

Note: For digital library projects at Stanford University, University of California at Berkeley, Alexandria in California, University of Illinois and at Michigan University, see *Ibid.*, pp. 59–64.

Case 18.4: Video-conferencing in telemedicine at Berlin

Video-conferencing is part of the Desktop Conferencing (DTC) used extensively in the Bermed Project at the German Heart Institute and the Rudolf Virchow University Hospital in Berlin. Advanced multimedia and communications technology support all areas of medicine from computer tomography and magnetic resonance imaging to intensive care monitoring units and administrative support systems. Diagnostic information comes from a variety of imaging techniques and formats as well as reports, graphs, charts, reference books, handwritten notes, film and video sequences, audio recordings, and conversations with patients and colleagues. However, systems do not operate optimally and many systems operate in isolation with little or no integration with other related systems.

A self-evaluation concludes: 'The use of multimedia DTC has shown the need for video, high quality audio, gesture support and overall simplicity. The future success of multimedia DTC in medicine critically depends on interoperability and solutions to security problems. Final acceptance in routine medical practice, however, will depend not only on technological factors but also on the many social and organizational issues, including changes in work practices to obtain maximum benefits.'

Source: Lutz Kleinholz (1994). Supporting cooperative medicine: the Bermed project. *IEEE Multimedia*, **1** 44–52.

Case 18.5: Telemedicine in Kansas

Increasing in popularity in the US is the use of telemedicine as in the case of the Kansas Medical Center where the patient is 300 miles away and the doctor, Gary Doolittle, a 37 year old oncologist, talks to the patient and uses two-way television along with stethoscopes, X-ray transmission and lab tests that are performed remotely using telecommunications. The cost of such an equipment configuration is currently $50 000 but is expected to drop by over 50% as technology improves and it becomes easier to transmit large blocks of data by wire and fibre-optic cable. Many insurers will not cover telemedicine of this type and many lawyers worry about malpractice and privacy. The important resistance comes from local physicians who say that they are afraid of out-of-state quacks that are not governed by state licences but actually fear the economic threat to their practice of clients that can now use telemedicine to go across the state borders (and maybe across national borders) to seek medical treatment or at least get a second opinion through a televideo telemedical appointment with a world famous medical specialist. Telemedicine is thus becoming another form of electronic commerce much like teleshopping, telebanking and securities trading, by eliminating the middlemen and bringing new competitors into previously insular markets.

Arthur Caplan, professor of bioethics at the University of Pennsylvania, argues that 'technologically, telemedicine is already here', but that overcoming the economic barriers 'may take 20 more years . . . as medicine moves into cyberspace, we need a new system of checks and balances'.

Source: Bill Richards, Doctors can Diagnose Illnesses Long Distance to the Dismay of Some, *Wall Street Journal*, Jan. 17, 1996, pp. Al and A10.

Bibliography

Akselsen, S., Eidsvik, A.K. and Folkow, T. (1993). Telemedicine and ISDN. *IEEE Communications Magazine*, **31**(1), 46–58.

Aldermeshain, H. Ninke, W.H. and Pile, R.J. (1993). The video communications decade. *AT&T Technical Journal*, **72**(1), 2–6.

Bleecker, S.E. (1995). The emerging meta-market. *The Futurist*, May/June, 17–19.

Chang, Y.H., Coggins, D., Pitt, D., Skellern, D. Thapar, M. and Venkatraman, C. (1994). An open-systems approach to video on demand. *IEEE Communications Magazine*, **32**(5), 68–79.

Coopock, K. and Cukor, P. (1995). International videoconferencing: a user survey. *Telecommunications*, **26**(4), 33–34.

Davis, R., Samuelson, P., Kapor, M. and Reichman, J. (1996). A new view of intellectual property and software. *Communications of the ACM*, **39**(3), 21–30.

Deloddere, D., Verbiest, W. and Verhille, H. (1994). Interactive video on demand. *IEEE Communication Magazine*, **32**(5), 82–88.

Domet III, J.J., Rajkumar, T.M and Yen, D. (1994). Multimedia networking. *Information Systems Management*, **11**(4), 39–45.

Furht, B. (1994). Multimedia systems: an overview. *Multimedia*, **1**(1), 47–58.

Furht, B., Kalia, D., Kitson, F.L., Rodriguez, A.A and Wall, W.E. (1995). Design issues for interactive television. *IEEE Computer*, **28**(5), 25–39.

Gemmel, D.J. Vin, H.M., Kandlur, D.D., Rangan, P.V. and Rowe, L.A. (1995). Multimedia storage servers: a tutorial. *IEEE Computer*, **28**(5), 40–49

Grosky, W.J. (1994). Multimedia information systems. *Multimedia* **1**(1), 12–23.

Levy, D.M. and Marshall, C.C. (1995). Going digital: a look at assumptions underlying digital libraries. *Communications of the ACM*, **38**(1), 77–84.

Lippis, N. (1993). Multimedia networking. *Data Communications*, **22**(3), 60–69.

Little, T.D. and Venkatash, D. (1994). Prospects for interactive video-on-demand. *IEEE Multimedia*, **1**(3), 14–23.

McQuillan, J.M. (1992). Multimedia networking: an applications portfolio. *Data Communications*, **21**(12), 85–94.

Nahrestedt, K. (1995). Resource management in networked multimedia s1562 ystems. *Computer*, **28**(5), 52–63.

Ozer, J. (1995). Shot by shot. *PC Magazine*, **14**(7), 104–163.

Peleg, A., Wilkie, S. and Weiser, U. (1997). Intel MMX for multimedia PCs. *communications of the ACM*, **40**(1), 24–30.

Reisman, S. (1994). Multimedia Computing: an overview. *Journal of End-User Computing*, **6**(4), 26–30.

Richardson, S. (1992). Videoconferencing: the bigger (and better) picture. *Data Communications*, **24**(9), 103–111.

Rabina, L.R. (1995). Towards vision 2001: voice and audio processing considerations. *AT&T Journal*, **74**(2), 4–13.

Strauss, P. (1994). Beyond talking heads: videoconferencing makes money. *Datamation*, **10**(19), 38–41.

Snider, J.H. (1996). Education wars: the battle over Information Age. *Futurist,*. **30**(3), 24–33.

Tobagi, (1993). Multimedia: the challenge behind the vision. *Data Communications*, **22**(2), 61–64.

Tolly, K. (1994). Networking multimedia: how much bandwidth is enough? *Data Communications*, **32**(13), 44–52.

Thuston, F. (1992). Video teleconferencing: the state of the art. *Telecommunications*, January, 63–64.

Wiederhold, G. (1995). Digital libraries, value, and productivity. *Communications of the ACM*, **38**(1), 85–96.

Woolfe, B.P. and Hall, W. (1995). Multimedia pedagogues. *Computer*, **28**(5), 74–80.

19

TELECOMMUTERS, E-MAIL AND INFORMATION SERVICES

Networks connect people to people and people to data.

Thomas A. Stewart

The general idea was that the industrial revolution had taken people out of their homes, and now the telecommuting revolution will allow them to return.

Tom Forester

Introduction

Knowledge workers handle knowledge, not necessarily in the sense of Artificial Intelligence where knowledge includes heuristics and inferences, but knowledge in the sense of the basic units of data and information. One subset of knowledge workers includes those that work at home and they are referred to as telecommuters.

With the coming of computers, knowledge workers have increased in number and importance. With the confluence of computers and telecommunications, telecommuters are increasing in importance and numbers. In this chapter we examine what knowledge workers and telecommuters do and how they do it.

All knowledge workers, and especially telecommuters, need interconnectivity; that is, their computers must be connected to other computer systems. This interconnectivity can be local within an organization, or regional within a community, or international connecting the whole world. International connectivity is currently achieved through the Internet and e-mail. The Internet we examine in the next chapter; in this chapter we discuss e-mail.

Telecommuting

Knowledge workers work mostly in offices but some work at home using telecommunications to keep in contact with their corporation instead of being in the office physically. These workers are called **teleworkers**, or **telecommuters**. There is a common perception that many teleworkers are mothers who stay at home and join the workforce to work around feeding their children and doing their household chores. But this perception does not hold for the teleworker, at least not at Rank Xerox, where all the teleworkers are men. Rank Xerox in the 1980s called it networking. There are other names for this type of worker at home: electronic homework, virtual work, distance workplace, flexiplace and electronic cottage. The emphasis is variously on work being independent of place and time. But why is teleworking a relatively new phenomena? Because it has at least four prerequisites that have to be satisfied: one, a cheap, fast and robust computing device that is satisfied by the PC or workstation; two, a connection of this device with the corporate office, customers, suppliers and other business players which is supplied by telecommunications and networks; and three, a worker at home who is at ease and comfortable with a computer and telecommunication. This third condition is largely satisfied with many in the workforce who are computer literate and comfortable with the PC which is end-user friendly, though perhaps not as end-user friendly as some would like it to be. The fourth condition for telecommuting is that there are professions where telecommuting is feasible, viable and desired by corporate management. This condition too is often satisfied, largely on economic grounds: corporations do not have to pay for the high overhead of an office building nor for parking and other facilities for its telecommuting employees. There is also greater flexibility in hiring because telecommuting adds to the labour supply. Telecommuter applications can be stored in the database and called upon when needed as part-time employees for the duration needed. Thus, telecommuting offers corporate management flexibility in the hiring of workers when needed, whilst dropping them when demand drops, all without any cost or firing or any hassle with the unions. This part-time arrangement has tax advantages in some countries and is an opportunity for people who want to stay at home and still be part of the productive workforce adding to the GNP (Gross National Product) of their country.

Telecommuting is the substitution of telecommunications and computers for transportation to the office. It was first talked about during the Arab oil embargo in the mid-1970s. It was then recognized that roughly 5% of the US oil was for commuting and that a savings of 190 000 barrels of oil a day could be gained if 20% of the commuters stayed at home. A few corporations did experiment with home-workers at the time, but not till the 1980s did the idea come of age. By that time reports on the success of experimental alternative work-site programs were in circulation and the benefits of telecommuting were better understood. In addition, the personal computer was on the market and at an affordable price, another factor contributing to the economic feasibility of working at home.

The percentage of time an employee telecommutes varies from one organization to another. Some companies have no main office at all. All work is done at dispersed workstations joined by a network. Informal meetings take place in low rent neighbourhood centres or at the homes of employees. Sometimes a company will establish a small satellite office with an electronic link to the main office. This type of alternative work style is classified as telecommuting since the time and cost of commuting to an office is avoided. Other organizations design work around assignments so that employees who operate out of their homes spend one or two days each week at the main office.

Telecommuting reduces the costs of carrying employees not needed all the time and the cost of fringe benefits that must be given to the full-time employees. When teleworkers are not union members, the unions are unhappy in losing membership and the clout and power associated with large memberships. Thus, the unions are unhappy with telecommuting and are often actively opposed to it. Unions argue that teleworkers are exploited with low wages and no fringe benefits compared to what is paid to full-time employees for the same work. The teleworker argues that in return for not being paid like full-time employees that work in the office, the teleworker does not have to dress up for work and drive to work. They can work when and where they please, though when not in the office they miss the socializing and time off they sometimes get in the office.

Implementation of teleworking

The equipment required by a telecommuter could be quite modest: a terminal with a modem. For complex tasks it may be necessary to have a workstation equipped with peripherals. Often though, the peripherals run by the corporate computer system can be used. Once connected to a corporate LAN, the telecommuter can not only have access to expensive peripherals but also to the main computer system, its servers and the corporate database. Such a system is depicted in Figure 19.1. If the corporate LAN is connected to the Internet (which is a net of other nets) then the telecommuter can access any customer or vendor (or friend and relative) anywhere in the world and at any time desired. Such an interconnection through the Internet is shown in Figure 19.2.

Careful planning is required for telecommuting to be successful. The people selected for telecommuting must work well on their own. Their assignments must be jobs that can be done away from the office. IT personnel that include data preparation clerks, programmers, and even some systems engineers, are good prospects. Technical or documentation writers and IT training specialists who must develop educational materials are also good candidates for telecommuting to do part of their jobs. Outside IT, telecommuting is viable for sales persons, architects, accountants, auditors, reservation clerks, some secretarial positions and other white collar professions.

Persons selected for telecommuting should be trained to cope with off-site employment while managers should be trained to supervise subordinates from a distance. Remote work does not mean all contact with co-workers is lost. On the contrary, training should include ways to remain in contact through electronic mail, phone, fax, regularly scheduled meetings, memos, routine

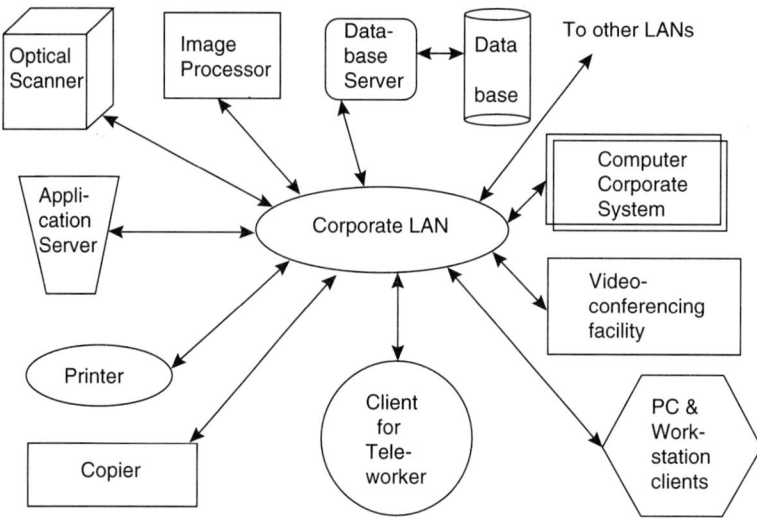

Figure 19.1 *Possible access to equipment by teleworker*

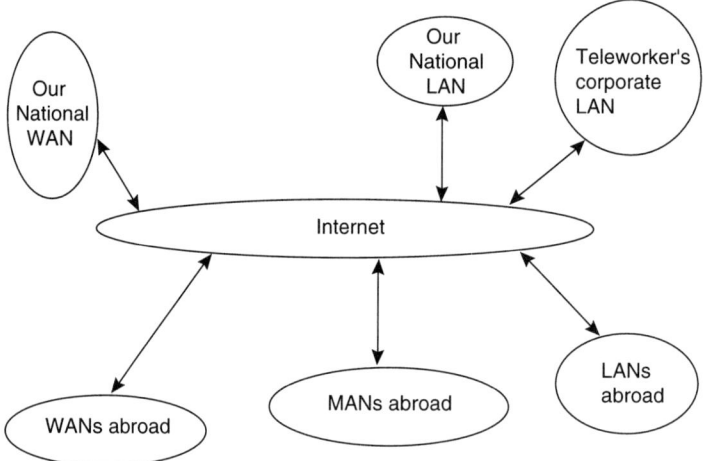

Figure 19.2 *The Internet as a net of other nets*

reporting, and so forth. There is some evidence that persons that opt for telecommuting look only at the short-term benefits (less commuting and reduced expenditures for petrol, meals, clothing, etc.) and do not realize the problems associated with working alone. A training period can help employers evaluate whether the right candidate for telecommuting has been chosen and help employees gain a realistic view of telecommuting as a work mode away from the social contacts of a city and an office.

Besides the economic, social and technical considerations of telecommuting, there are new problems for management. For one, how do you evaluate workers with whom you have no (or little) eye contact and do not observe while at work? Work performed at home by telecommuting can be evaluated quantitatively and perhaps, to an extent, qualitatively also. But many evaluations are often qualitative and subjectively judged by observing the workers attitude and behaviour at work. This is not possible in telecommuting and new ways of evaluation have to be devised appropriate for the tasks and personnel involved. This may involve some face-to-face meetings, but the rest of the evaluation must be based on measuring and monitoring productivity at home.

The benefits and limitations of telecommuting

One of the primary benefits of telecommuting is increased productivity. Workers claim that there are fewer distractions from the central office. Gains in productivity of 20 to 40% are not uncommon among professionals. However, it should be recognized that many persons work best in a structured environment or when they interact socially with others. Not every one has the work habits, discipline and planning skills to direct their own activities. Careful screening of candidates is necessary to ensure that productivity is not lost, instead gained by telecommuting.

In service-oriented organizations, telecommunications is a good way to extend hours of operations. For example, employees at home can take orders and answer customer queries. J. C. Penney, a large retailer in the US, uses telecommuting for some of its catalogue-order telephone centres. Companies that have seasonal peak volumes or certain hours of the day with heavy demand can handle the load by telecommuting without tying up office space.

When organizations are recruiting new employees, telecommuting can help attract applicants. Many persons like the idea of flexible scheduling instead of the standard 9–5, Monday to Friday work week. When working at home employees can choose their hours of work. Organizations trying to hire professionals when their demand exceeds supply, will find that the option of telecommuting gives them a recruiting edge.

Telecommuting is also attractive to persons with family responsibilities (a young child or aged parent) who may be unable to work at certain hours of the day or cannot take the time to commute, or who just do not like to commute. Persons with medical disabilities who have difficulty leaving home, can now enter the work-force through telecommuting.

Telecommuting can also be an alternative to corporate transfers. Many employees today resist relocation. A common reason is that a spouse is employed and unwilling (or unable) to uproot. Whatever the objection, a valued employee who might resign rather than move can be retained when telecommuting is an alternative to transfer. In addition, the employer avoids the cost of the relocation or the cost of hiring a replacement.

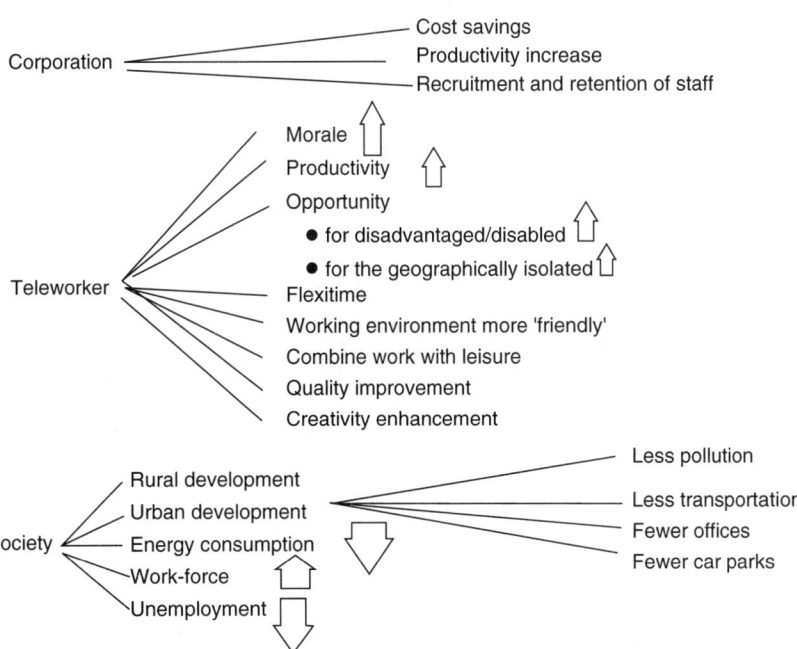

Figure 19.3 *Advantages of telecommuting*

Table 19.1 *Advantages and disadvantages of teleworking*

Advantages

Offers flexibility to corporations in the hiring of
    transient personnel for peak and unexpected loads.
Offers teleworkers the flexibility of working at their
    choice of time and place.
Provides a possible 'creative' environment of the
    home.
Provides access to 'piece-work' and part-time work.
Expands the workforce to:
    housewives,
    disabled people,
    people in remote regions and even farms.

Limitations and disadvantages

Raises problems of:
    providing proper access to information needed,
    integrity and security of data/information,
    evaluating teleworker.
Raises issues with unions over:
    membership,
    work conditions,
    payments, direct and indirect.
Psychological and social needs of teleworkers may
    not be met.
Conflicts and 'over-crowding' at home.
Inability to separate work and pleasure.
Possible 'burn-out' of working at home.

Finally, telecommuting is advantageous because it reduces office space and furniture
requirements. With telecommuting such overhead expenditures are minimized. Furthermore,
telecommuting means that space is not a constraint when changes in the workload are proposed. Organizations can expand rapidly without
worrying where to put people or retrench without being saddled with expenditures for unused
space and furnishings.

The advantages of telecommuting are summarized in Table 19.1 and Figure 19.3.

When telecommuting?

When the advantages can overcome the disadvantages and limitations of telecommuting, there are
many people who would want to telecommute.
These people (see Ursula Huws (1991: pp. 28–9)
include:

> people who wish to work at home in order to
> avoid the commute to work;

individualists who find the large corporation
so antipathetic an environment that they wish
to set up their own office at home;
women who cannot find child care arrange-
ments;
women who wish to stay with their children at
home and yet want or need to work;
people who feel that they are exploited and
underpaid at work;
people who use computers at work and might
as well use them at home;
workers who are 'loners' by temperament and
rather work at home than in an office;
people with disabilities but who can work at
home;
people who want an extra job;
firms and organizations that find telecommuting a way to downsize their organization and
find telecommuting of some of their employees
an economical thing to do.

Future of teleworkers

Before making predictions about teleworkers, one
must remember that predictions of the use of
computers and telecommunications in the home
have been made in the past and have been notori-
ously wrong. Witness the less than expected use of
home shopping and home banking through videotex both in the US and in Europe: Prestel in the
UK, Viewtron in Florida, Keyfax in Chicago, Gateway in California and Bildschirmtext in Germany.
There were some 50 videotex's operating in 1982
in 16 countries and all of them operated below
expectations and often below the breakeven point
of profitability. There was one exception: Minitel
in France. The success of Minitel was due partly
to high end-user acceptance but also to a combination of political intrigue and the deep pockets
of government support.

Most predictions in some countries went wrong
partly because the predictions were made by
vendors and vendor-biased commentators and
partly because the predictions were based on
poor assumptions of the behaviour of parties at
home and their ready acceptance of computer
technology.

Despite these reservations, there are some predictions that can perhaps be safely made. The
home-office of the teleworker will soon be the
confluence of computer technology, AI technology (including expert systems), networking (local

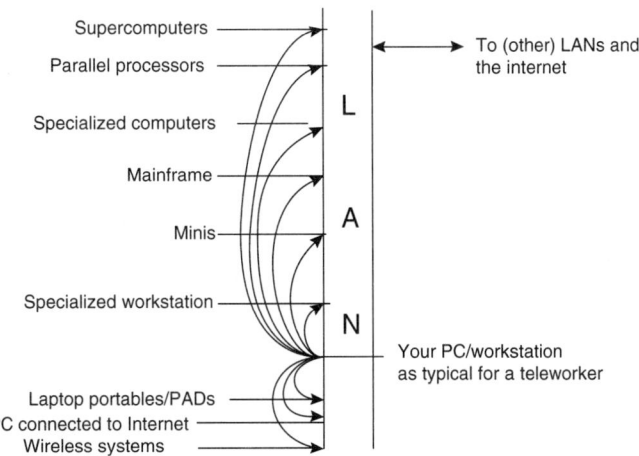

Figure 19.4 *Interconnectivity for a teleworker*

and global), wire-less technology, signalling over fibre optics, image processing and the fast processing of massive volumes of data and knowledge. There will be the capability of faxing documents and using palmtop computers when needed. All these, or at least most of these, devices will be 'smart' because of their use of AI programs. They will have the ability to process not just data but knowledge, not just discrete variables but 'fuzzy' variables, and not just on-line but in real time too. The telecommuter will have access to many different peripherals through a network that are not available at home. One such scenario is shown in Figure 19.4. These are no great predictive leaps into the future but merely a projection of what is technologically possible and what is being currently done (as, for example, PADs and the Internet computer).

Access by telecommuting may even extend to certain professions such as the salesperson who works out of a van or car. The electronic office on wheels will enable the salesperson to give video and animated graphic demonstrations to the clients at the client's site and respond to queries that will bring instant responses from the decision-maker in the corporate headquarters.

The changing technology and changing work environment of the 'electronic briefcase,' the electronic office on wheels or at home, will require organizational responses. Bureaucracies will be reduced and hierarchies will flatten out. In many cases bureaucracies and hierarchies will be replaced by adhocracies because there is little time for critical information to go up and down the rigid chains of hierarchical command and percolate slowly through the many layers of bureaucracy. There will be continuous contact between management and customer and back to managers making strategic decisions, often through the intermediary of the teleworker, sometimes not so. The worker is now liberalized and dispersed without set locations or schedules. Some workers may well be displaced from their long-held jobs if not unemployed. For many workers, the roles and tasks may require designing, restructuring and re-engineering; new standards for work and performance must be instituted and monitored by corporate management; supervisory roles and leadership roles redefined; and, most important perhaps, employees must be educated and trained for the new work environment. This requires a corporate plan for the migration of work that was done in traditional ways to the newer ways using the latest technology both efficiently and effectively.

An important resource for any knowledge worker and teleworker is to have the ability to receive and send messages quickly and easily. This is achieved by e-mail provided both parties have an e-mail connection.

E-mail

We start with a definition of e-mail and compare it with other means of communication like the telephone, the telegraph, the post office and face-to-face discussions. We then examine the resources needed for e-mail. For local communications a local information service provider may

221

be sufficient. To communicate across the world, there is the Internet. We shall briefly discuss the Internet as preparation for our discussion of cyberspace in the next chapter.

E-mail is electronic communication. You enter the message you wish to send on a computer, provide the 'address' you wish to send the message to, press a few buttons, and the message is sent to the desired 'address'. Of course you must follow the protocol (procedures) required for the transmission of the message and you must provide the address of the destination according to a prescribed format. The computer system uses a numeric address but that may be difficult for the average user to remember and so the address you provide is often alphabetic with some special symbols and the systems protocols does the translation. For example, you may use the DNS (Domain Name Service) format which translates the address provided by the user to the numeric address used by the machine. This address you provide starts with the 'screen' name of the receiver, then the network ID and working towards the finest locational information with the '@' sign and dots '.' separating different fields (parts) of the address. The e-mail address is somewhat like the addressing of an envelope in Germany. After the name, you start with the area code, name of the city, street and then the number of the house. Thus the author's e-mail address is:

KMHussain@AOL.com,

i.e. K. M. Hussain at (@) AOL (the information service provider) and com is the short for 'commercial service'). Some addresses have more subdivisions separated by dots and international addresses may even have a country alphabetic code at the very end. This is just one class of addressing. There are at least three other classes but we shall not be concerned with them except to recognize their existence in case they turn up sometime.

In the postal service of many countries, we go the opposite route to e-mail: we start with the detail name, street and then city. But in principle, e-mail is much like any post office. Both use the store-and-forward technology. Your e-mail messages, like your letters, are delivered to a local sorting centre where they are sorted for the address of the destination, and then forwarded through more sorting desks (like in post offices) before they are delivered at the address where they can be collected when convenient. The delivery point may be the home PC (or

workstation) or the office. If a printer is available, both incoming and outgoing messages can be printed in addition to the soft message on the screen. The transmission in e-mail is swift and very much faster especially for international communication. And it is much cheaper too, both locally and internationally.

Another commonality of e-mail with the post office is that both are asynchronous processes; that is, the two people involved, the sender and the receiver, do not have to be available at the same time, as is the case with a telephone call or a face-to-face meeting.

The great advantage of e-mail compared to face-to-face meetings and telephone calls is that you have time to think of an answer. You can break off at any time by simply not responding. You are not on the spot (as with a telephone call) and under duress to express your opinion instantly and seem intelligent and profound. You do not stand the risk of being misunderstood as in a fact-to-face relationship resulting in unintended emotional responses. You can take your time, and respond when you are ready. This is somewhat like the telegraph communication which is also swift once you have composed your message, but, unlike the telegraph, e-mail is often not charged by the length of the message, and, if it is so charged, the cost is not as prohibitive as that of the telegraph.

One advantage of e-mail is that when sent you know it is delivered and so if for some reason it cannot be delivered, you will get a message from the e-mail postmaster explaining the reason for non-delivery like an unknown computer address. There is no dead letter box for e-mail as there is with the post office. You are never sure of delivery in surface mail (unless you get a reply) especially with foreign mail for you do not know if the postman has taken a liking to your stamp.

E-mail increases secretarial performance by a factor of two to three. Retrieval and the file management capabilities are particular strengths. In addition, the system reduces the need of office storage since cabinets with files correspondence can now be replaced with computer memory.

It must also be recognized that e-mail is just one of the many modes of electronic message transfer. Mailgram and fax are alternate methods. Fax currently is slow in transmission, taking between $\frac{1}{2}$ to 6 minutes to transmit a document copy comparable in quality to a Xerox copier. The process is also costly. A document page in facsimile form requires 200 000 bits. That is 1000 times the number of bits required for a typical telegram, and 60 times that of a typical office memo.

In spite of its obvious advantages, corporate managers tend to be ambivalent in their attitudes towards electronic mail. They like the speed, retrieval and cross-indexing capabilities, being shielded from constant interruptions by jangling phone calls, as well as receiving and responding to messages when convenient. But e-mail demands a change in style of management. E-mail often eliminates the need of voice and personal contact in business. Business people, especially the gregarious ones, miss the human interplay of visual messages or a phone call or a visit to an associate.

E-mail has its drawbacks especially when compared to face-to-face meetings. You cannot hear the sounds of anger or pleasure; you cannot express joy or sadness through a change of tone or inflection; you cannot see emotions in a blush of embarrassment or of joy; and you cannot sense the atmosphere that may be tense or frivolous. You cannot interact. You are alone. You have no physical clues of how others are thinking or reacting to what you have just expressed. But you have total control over what you can say, how you say it and when you say it. But for these and the other advantages of e-mail, you must pay a price. You need computing resources.

Resources for e-mail

There are three important resources needed for e-mail: a computer, a modem and a connection to a e-mail provider (who provides you with the software necessary for the connection). Let us consider the modem though it has been mentioned earlier.

A modem stands for modulator and demodulator and is used for communication by a computer (digital) over a telephone line (analogue). It is a device that transforms a computer's electrical pulses into audible tones for transmission over the phone line to another computer. This is the modulator part. The modem also receives incoming tones and transforms them into electrical signals that can be processed by a computer. This is the demodulator part of the modem. Essentially, a modem allows us to convert the digital data of a computer into analogue data for the telephone and vice versa.

A modem often comes built into a computer. If not, then it can be added on. Do not let the computer buff confuse you with bandwidths and megahertz speeds for they are significant only if you have large volumes of data or are transmitting voice, images and sound which you will most likely not be doing if you need to know the basics about a modem.

Besides a computer and a modem, what is also needed is a connection and the software to be able to do e-mail. There are at least three possibilities:

1. Get an Internet connection through an Internet gateway which then enables you to do e-mail not just nationally and regionally but also internationally. In addition you have all the facilities of the Internet. We discuss such a connection and the Internet in general in Chapter 20.
2. Get connected to an Information Service or Business Communications Service provider that offers e-mail as just one of its services. These are LAN based systems and the most cost-effective for most users. We shall examine such providers and their services later in this chapter.
3. Build your own e-mail backbone with a dial-up access and also access to the Internet. This will be considered a private network as distinct from the other two alternatives which are public backbones. This private systems approach is more expensive but it allows customization, better administrative and management control, independence, and even better performance and scaling than the shared public systems. But it is such private systems that only the large organization can afford or indeed need.

Whatever the approach to connection, one must use a protocol for messaging. In the case of the Internet, it is TCP/IP (Transport Control Protocol/Internet Protocol); for the information service provider, there is PPP (Point-to-Point Protocol) or SLIP (Service Line Internet Protocol) connection; and for the private e-mail system, there are several alternatives: TCP/IP, PPP, SLIP, or STMP. STMP is Simple Mail Transfer Protocol. It is also used by the Internet and is simple in the sense that it handles simple messages of text and numeric data. If, however, one wants to transmit pictures, sound and video, then one needs an additional protocol, the MIME, Multiservice Internet Mail Extension. MIME is the protocol for multimedia e-mail and can be used also with STMP.

The protocols mentioned above are most common approaches for the US where there is a spectrum of vendors and corporations that are large enough to go their own way and to embrace a message handling system that includes e-mail as just one of the many message transport systems. In Europe, however, there is greater respect for the ISO standards. The X.400 was prepared by the ITU (International Telecommunications Union) and endorsed by the ISO. The X.400 is the alternative to STMP as a mail backbone and it is also the standard for EDI in both Europe and Japan.

Having just mentioned EDI, it may be the right time to establish the relationship of e-mail to EDI. They lie on top of what can be looked at as a cake with e-mail and EDI being the top layers; the X.25, switches and leased lines, the bottom most layer; with managed data on top of the lowest layer and information retrieval on top of that. This hierarchy is illustrated in Figure 19.5. What must be deduced from the cake structure is that you cannot always just simply choose one layer and not the bottom layers, at least the bottom most layer. And so the consumer is sold a package with layers that they may not want or need. They must thus seek the package that serves their needs best and is most cost-effective.

An irritation for the consumer and perhaps the greatest obstacle for an easier and cheaper system is that there is no common global directory or universally accepted standard for e-mail addresses. For example, in the US, you may have an alpha address (with an 'at' and 'dot' symbol) and yet a relative or best friend may have an alphanumeric and symbolic e-mail address like 76361.656@compuserve.com.

In Europe, there is the X.500 that is more widely used and is the directory standard for the X.400 (and X.435 for EDI on X.400). Creating and maintaining comprehensive lists poses great technical and organizational problems when the scope of the addressing is not just national and regional but also international. There is already an acknowledgment that a larger width of the fields for addressing will be necessary for any future global messaging system before we can link heterogeneous networks and synchronize distributed directories of networks and e-mail users. There is the need in e-mail standards for file formats, transports, directory services and e-mail APIs (Application Programming Interfaces), as well as better directory synchronization. This will greatly help in building the messaging infrastructure which would facilitate not just e-mail (and fax) routing but also database access, document sharing, group work and ultimately decision-making, as in the DSS and EIS. E-mail may yet become an important enabling technology for enterprise computing. For corporate management, this might result not just in better and faster communication but perhaps also in the flattening of the management, the improvement of control and tracking, faster offering of products to the market, changes in work structures, and increased productivity.

One annoying problem faced by e-mail users is the volume of junk mail that comes through. If you are on various lists for mail, it is easy to get 200–300 messages a day, which may well consume around two hours just in reading time. There are some programs around that will filter your mail. It could check the address of the origin against a list that you approve, or it could filter by subject matter according to key words that you provide. The filtering and screening of e-mail can

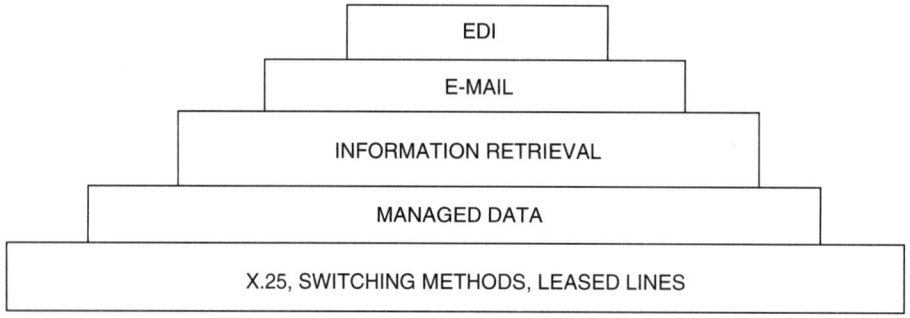

Figure 19.5 *E-mail in the hierarchy of layers*

control the stream of e-mail but that will greatly depend on the updating of the lists used for the filtering.

Despite problems with e-mail it is growing rapidly in volume and scope. In scope, it is extending from local to global communications; from sending mail to customers and suppliers to sending mail to friends and family; from access to databases to collaborative applications; from business correspondence to electronic form routing and process automation; from people to people to process to process (virtual users); from a formal correspondence vehicle to a e-mail transport in the enterprise-wide messaging infrastructure. In numbers, e-mail in the US has risen over 60% from 1992—1993 and from 5.9 million in 1992 to over 38 million in 1995. And this is with only 42% of the 47 million computer users in 1993. As the total number of computer users increases and the percentage of users increases, e-mail will also increase. Will it increase faster than our capacity to fully use it? Or will there be a backlash from the inefficiency and unfriendliness of the system? Time only will tell but the chances are that with the awareness of the problems and attempts to overcome them, there is a good chance that e-mail will continue to grow and prosper.

Information service providers

The tricky question now arising is to find an e-mail provider among the many information service providers. Such a provider is the equivalent of the post office that does the storing-sorting-and-forwarding of messages and provides you with an e-mail address and the software to receive and send messages. Early systems had a command driven language but nowadays you can get a GUI (Graphical User Interface) which is menu driven and has icons (graphical symbols that perform functions, for example a **scissors** icon to **cut** a selected passage). Different providers have interfaces with varying degrees of end-user friendliness. But e-mail is not the only service that the information service provider provides. Its primary function is to provide information on subjects like

- investing and finance,
- travel, business careers,
- news — local, national and international,
- weather,
- entertainment,

- health,
- sport,
- membership services,
- billing services,
- hobbies, leisure,
- games,
- shopping.

E-mail is only one of the services offered. Some are good at e-mail and others are not. Each should be evaluated for appropriateness to one's environment. Criteria for such a comparison and evaluation could be the following:

- content for areas of interest,
- ability of parents to control content of what children and teenagers may watch,
- type of 'community' in chat and discussion groups and forums,
- interface,
- GUI, graphical user interface,
- features of end-user friendliness,
- installation,
- e-mail,
- ease of use,
- features such as address book and its use in composing mail,
- ease of installation,
- Internet connectivity,
- cost, fixed or variable,
- e-mail fee, if any
- access,
- local/long distance affecting costs,
- speed of access affecting waiting-time in queue.

Summary and conclusions

E-mail is fast, reliable and cheap if you have a connection to some service provider. E-mail capability can be acquired through an information service provider, an Internet provider or a LAN. E-mail services are often offered through a GUI, graphical user interface, that is menu driven, often in colour, and end-user friendly.

E-mail is becoming increasingly central to the way we work in the office and at home as a teleworker. It is also central to a corporate messaging system. One study showed that, as the parties involved double in number, the messages increase fourfold. Thus the intracompany and intercompany messaging increases rapidly (Burns, 1995: p. 108).

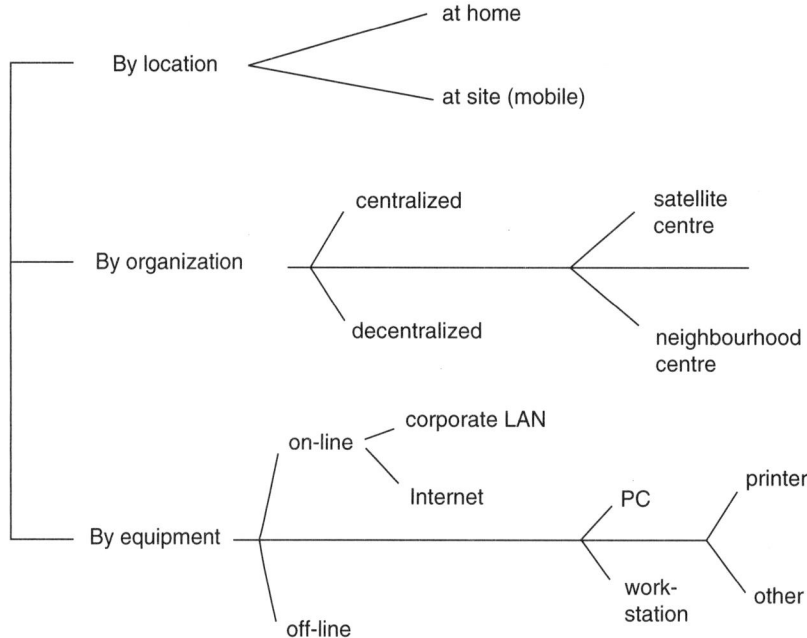

By location
— at home
— at site (mobile)

By organization
— centralized
— decentralized
— satellite centre
— neighbourhood centre

By equipment
on-line
— corporate LAN
— Internet
off-line
PC
— printer
— work-station
— other

Figure 19.6   *Classification of telecommuting*

As information processing and knowledge processing increases, there will be a great demand for the knowledge worker. Many of these knowledge workers will be teleworkers. There are many types of teleworkers. One classification is shown in Figure 19.6.

The impact of teleworking is summarized in Table 19.2. Not obvious are the very intangible benefits to society in the form of better and cheaper urban development with fewer high rise office buildings, fewer car parks, and fewer roads, motorways and highways. The greatest impact will be on society if it transforms working in the office to telecommuting and working at home. This will not only change our life-style, but could also affect town planning for banks, offices, car parks, transportation, and pollution.

Just as the industrial revolution has brought the worker out of the home, the new information era of telecommunications and computers have brought the worker back to the home. But this is true only of certain professions and certainly not true of the agricultural, manufacturing, and service sectors of the market economy. Despite its limited scope, telecommuting has raised many new issues and conflicts and will need new management strategies ways of resolution.

Table 19.2   *Impact of telecommuting*

ON THE INDIVIDUAL
Allows employment for part-time work
Allows employment for those who do not want to or cannot leave home easily, like mothers and the disabled
Requires a new discipline for working at home

ON THE CORPORATION
Enables transient and piece-time employment without the burden of fringe-benefits and working with unions
Reduces overhead for office
Requires new approaches to scheduling work, evaluation and control of remote work

ON SOCIETY
Reduces pollution from traffic by office workers
Reduces urban development and transportation infrastructure
Increases potential labor force and potential GNP
Makes unions unhappy by decreasing their potential membership.

We conclude with some thoughts by Ursula Huws, a long time commentator on telecommuting:

The extent to which the electronic cottage becomes a reality, and the specific forms which that reality takes, will depend on the decisions taken by a range of social actors — large employers, entrepreneurs, creative individualists, women with dependants, and planners. These decisions will not be unidirectional, nor will they be permanent. They will interact with each other to produce new and unexpected patterns; new ideas of conflict will arise and, in the resolution of these conflicts, new social forms will be negotiated.   (Huws, 1993: p. 29).

Case 19.1: Examples of teleworkers

Telecommuting at Bell Atlantic

In 1991, Bell Atlantic, a telephone company in the US, had 100 managers working on trial as teleworkers. In 1994, the number grew to 16 000 and Bell Atlantic is working with its unions to increase that number to over 50 000. (*The Futurist*, March–April, 1994, p. 13)

Ursula Huws

Ursula Huws is a freelance writer and writes on telecommuting including the book *Telework: Towards the Elusive Office*. She works out of a room at home that has, according to her, 'three computers, two telephones, a fax machine, two printers, five filing cabinets, six book cases, two desks, four chairs, ... clothes, shoes and household linen and, above it, my bed.' (Huws, 1991: pp. 19, 21)

Thomas Hubbs

Thomas Hubbs has been called a knowledge worker, a telecommuter and a virtual worker. He is the Vice-President of Verifone which dominates the market for credit-card verification systems used by merchants. Hubb's nominal headquarters is in California and his CEO lives over a thousand miles away in Santa Fe. They meet face-to-face every six weeks or so.

Hubbs spends 80% of his time on the road often getting behind the walls to link his Hewlett-Packard laptop computer to an outside line. He rarely uses his cellular phone to ship faxes or e-mail because 'right now the costs are so high it doesn't make sense, except in an emergency'. Verifone's e-mail system runs on a VAX mini-computer.

At Verifone, 'everyone from the chairman to the most junior clerk is issued a laptop computer and is expected to learn how to use it. Internally, paper memos are banned, secretaries — a rarity at the company's offices — are prohibited from handling an executive's e-mail.' (*Business Week*, Oct. 17, 1994: pp. 95–6)

Tom Bacon

Dr Tom Bacon, an instructor in Operating Systems in California, was offered a job with a software house in the East of the US. The two parties agreed on all contractual matters but there was one hitch: Mrs Bacon was not in agreement. She did not want to go to the winters of the East, nor did she want to pull her children out of schools in California. So the firm made another offer. Dr Bacon had to come three times a year to the main office for face-to-face meetings and the rest of the time he could stay at home and use teleprocessing for work. All parties agreed, but Mrs Bacon still had to make a sacrifice: she had to give up the bedroom of the fourth daughter for an office. Some fifteen years later, the Bacons are still happily married and Tom is still a teleworker. However, Dr Bacon now has a two-room office and lots of 'bells and whistles' on his office computer system.

Case 19.2: Comments from teleworkers

'My professional and home life have become so intertwined that I can never get away from work.'

'I've gained forty pounds. My terminal is too close to the refrigerator.'

'I feel isolated at home. There is no incentive to work hard, no sense of personal accomplishment, no social contact, no feeling that I am part of a team that is accomplishing something.'

227

'Everyone interrupts me – the dog, the kids, the postman, the neighbours, even my spouse who knows better – I am just too available.'

'There have been no raises, no promotions, since I've had a home office. I'm afraid that I'm jeopardizing career advancement by being a tele-worker.'

'I'm getting claustrophobia (we simply lack the space) and my wife doesn't want me home during work hours.'

Case 19.3: Telecommuting at American Express

In 1991, management at American Express Travel Services, faced a problem. Its many woman travel counsellors wanted to work at home instead of having to brave the driving (sometimes three hours) to work and back each day. To set them up at home would cost the company $1300 each but it would release $3400 in each office space. The big question was the effect in productivity. And so American Express started its Project Hearth, which eventually involved 1000 travel counsellors (10% of all employees) in order-entry at 15 locations.

In 1993, studies showed that a typical agent was able to handle more calls at home than in the office resulting in a 46% increase in revenue from booking. Supervisors retain ability to monitor agent's calls, ensuring tight control.

Source: *Fortune*, **128**(7), Autumn 1993, pp. 24–8.

Case 19.4: Advice from teleworkers

1. The first and maybe the hardest is to 'try and convince your boss to let you try it ... Don't stress how it will benefit you; emphasize how the company will profit from your time in the home office.'
2. 'Remind your boss that given fax machines, modems and communication technology, you can be as hooked in at home as you are in the office.'
3. For hardware, get a system for 'all home and office work ... Get a notebook with a docking station. The docking station stays on your desk at the office. When you come in to work you plug your notebook into the docking station, and it hooks up to the network, provides power, and gives you expansion capability.'
4. Get a network administrator.
5. Get remote control software that allows your modem to transfer files between computers using phone lines.
6. Have some basic troubleshooting software on hand.
7. During working hours, check and respond to your e-mail at least once every two hours. Send as many messages as makes sense and this will make your colleagues think that you are working in the office.
8. Close the door of your office at home with a sign 'Do not disturb'.
9. If you are an animal lover, then there is great potential for distraction and destruction. Cats have been known to pounce on the keyboards and dogs have chewed up many a floppy disk.

Source: Preston Gralla, Work @ Home, *ComputerLife*, Jan. 1995, pp. 106–10.

Case 19.5: Teleworking at AT&T

In 1994, AT&T declared a Teleworker Day for all its professional workers. Over 25 000 people worked at home. Of the people surveyed, 70% said that they were more productive working at home as opposed to working in the office.

According to plans of AT&T, half of its 123 000 US managers will try working from home by the end of this century.

The Link Research firm estimates that 43.2 million Americans will work at home at least part of the time (although 12.7 million of that are self-employed). 'The number is expected to grow by a whopping 15% a year, so by 1997, 56 million people will be working at home.'

Source: Preston Gralla, 'Work @ Home'. *ComputerLife*, January 1995, p. 107.

Case 19.6: Teleworking in Europe

As part of the RACE programme, the EU has been promoting a series of studies to stimulate transborder teleworking. One such study concluded that no single core of tools and platforms for teleworkers is realistic since there are different communication and network services required and as there are different applications for teleworking. Further, it was observed that:

> ... the degree of technical complexity used in teleworking applications depends on a number of factors, such as the nature of the work undertaken, pre-existing equipment within the organization, communication costs, and the availability of value added networks. Thus the communication requirements adopted by teleworkers will reflect both the nature of the task (is access to a host computer or LAN required? Is a private e-mail service available? Are voice communications required?) and the prevailing network conditions (the extent of national ISDN service, the cost of the service, or the availability of public e-mail or data services).

Source: Julian Bright, Teleworking: The Strategic Benefits, *Telecommunications*, Nov. 1994, p. 81.

Case 19.7: Telecommuting in the US (in 1994)

3.9 million Americans commute to their jobs either full-time or part-time through their PCs. 3.2 million Americans communicate through the Internet with 541 949 British, French and Germans, as well as 347 888 Australians and Canadians.
Average productivity increases per telecommuter (measured by employers) was 10−16%. Self-employed persons with a PC generate almost $70 000 in household incomes − 42% more than otherwise.
85% of telecommuters are equipped to communicate with their employer's system miles away.
The average work time spent in office per teleworker was one day a week.
The average work time increases per teleworker per day was 2 hours.
The annual savings in facility costs per commuter was $3000−$5000.
The typical system used in home-based systems cost $3000 including software bought at the time of purchase − about $1000 more than the average cost of a PC system purchased in the US. The average telecommuting equipment was a 386 processor with a fax modem and a dot-matrix printer.

Over 2 million self-employed professionals do not use any accounting software.

Source: Jonathan L. Yarmis, *Telecommuting Changes Work Habits*, 1994; *Inc. Technology*, **16**(13), 1994; and *U.S. News & World Report*, Feb. 27, 1995, p. 14.

Case 19.8: Holiday cheer by electronic mail

A Christmas cheer chain-letter sent through IBM's internal network illustrates the benefits and problems of e-mail. The computer program that forwarded this Christmas tree greeting was designed to rifle through each recipient's files in search of automatic routing lists, and then pass the greeting to each name on that list. The result: the card boomerangged through the network, clogging communication channels so that important business mail was delayed. No provision was included in the program for checking the duplication of names on different mailing lists so the global link was in knots for hours.

The problem was eventually resolved, according to a spokesman, by trapping the files and deleting them. The company was to review operational procedures to prevent future disruptions of its electronic mail service. For example, personnel have been told not to execute or store any messages if the sender's specific purpose is unknown.

Source: Patricia Keefe, Holiday Cheer Brings IBM Net to Knees, *Computerworld*, Dec. 2, 1987, p. 2.

Bibliography

Ayre, R. and Raskin, R. (1995). The changing face of on-line. *PC Magazine*, **14**(4), 108−175.

Barrett, T. and Wallace, C. (1994). Virtual encounters. *Internet World*, November/December, 45−47.

Beard, M. (1994). The mail must go through. *Personal Computer World*, September, 384−387.

Burns, N. (1995). E-mail beyond the LAN. *PC Magazine*, **14**(8), 102−175.

Caswell, S.A. (1988). Electronic mail − the state of the art. *Telecommunications*, **22**(8), 27−30.

Engler, N. (1994). Buying e-mail is harder than ever. *Open Computing*, **11**(9), 88−91.

Gasparro, D. (1993). Moving LAN e-mail onto the enterprise. *Data Communications*, **22**(18), 103−112.

Griesmer, S. Jesmajian, R.W. (1993). Evolution of messaging standards. *AT&T Technical Journal*, May/June 21–45.

Griffiths, M. Teleworking and Telecottages. *The Computer Bulletin*, **2**(9), 14–17.

Hurwicz, M. (1997). E-Mail: old meets new. *LAN*, **12**(2), 87–91.

Huws, U. (1991). Telework projections. *Futures*, **23**(1), 19–30.

Huws, U. Korte, W.B. and Robinsons, S. (1990). *Telework: Towards the Elusive Office*. Wiley.

Maritino, V. and Wirth, L. (1990). Telework: a new way of working and living. *International Labour Review*, **129**(5), 529–552.

Reichard, K. (1995). Mr.Postman@INTERNET. *PC Magazine*, **14**(8), 111–151.

Reinhardt, A. (1993). Smarter e-mail is coming. *Byte*, **18**(3), 90–108.

Richardson, R. (1997). E-Mail detail. *LAN*, **12**(1), 57–62.

Roberts, T. (1994). Who are the high tech homeworkers? *Inc. Technology*, **6**(13), 31–35.

Runge, L.D. (1994). The manager and the information worker of the 1990s. *Information Strategy: The Executive's Journal*, **10**(4), 7–14.

Sproull, L. and Kiesler, S. (1991). Computers, networks and work. *Scientific American*, **265**(3), 116–123.

Strauss, P. (1993). Secure e-mail cheaply with software encryption. *Datamation*, **39**(23), 48–50.

Trowbridge, D. (1993). Now remote computer users never have to leave home. *Computer Technology Review*, **13**(8), 1, 10–15.

20

INTERNET AND CYBERSPACE

Clearly, the Internet and its associated technologies and applications are taking us into new intellectual and business territory. There are powerful 'leveling' effects to be seen in which elementary scholars and senior statesmen correspond without necessarily being conscious of age and experience distinctions.

Vinton G. Cerf

In a few years, there may be more people talking to each other on Internet than on the telephone.
International Herald Tribune, 1994

INTRODUCTION

At the University of Canterbury in New Zealand, a professor in MIS (Management Information Systems) regularly assigns his students the reading of advertisements in professional magazines. He argues that commercials are the best predictor of future technology. He is perhaps right at least in the case of the commercial by IBM in 1995. In it, there are nuns walking in a monastery in Czechoslovakia when a Mother Superior confesses in a whisper: 'I'm dying to surf on the Net.'

Enter cybernuns. Welcome to the cyber-era, the cyberworld where you can find cyberpunks, cyberphobia, cyberwork, cyberbucks, cybersluts, and if you chose, cybersex. The term 'cyber' (according to cyberwatchers) has been used in publications at least 1205 times in January 1995, up by 621% over a period of two years. This cyber interest should raise questions like: Is the cyber-era a hype or a reality? Is cyberspace peopled by gearheads and technojunkies that oppose any intrusion by the commercial world and the government? Or, with its expansion, has cyberspace become more mainstream?

We shall explore cyberspace through its most ubiquitous manifestation: the Internet. Internet is an international net of local networks that connects lots of host computers where information and knowledge resides and can be readily accessed. It is the fastest growing sector of the economy. In 1994, Internet had over 60 billion packets transmitted per month starting with zero in 1988. By 1997, it is predicted that some 400 million people

in North America will use the Internet at least once a week with world-wide users approaching 80 million (Bayers, 1996: p. 128).

In 1994, Internet saw many changes: its financing changed hands from an agency of the US government to a firm in the information provider business; there were teenagers that were seduced and ran away receiving instructions on the Internet; Chinese dissidents flashed messages to China on the birthday of the Tiananmen Square massacre that would be banned in China; there was obscene and pornographic material on the Internet that prompted the US legislature to pass a bill fining and sending to jail those who put such objectionable material on the Internet. But the courts may have to decide what is and what is not objectionable without violating the freedom of speech that the US constitution guarantees. Even if the courts define what material is objectionable enough not to go on the Internet, how can the US government ensure that someone from outside the US will not put objectionable material on the Internet? And so the Internet is very much a political and social issue. The common citizenry, including computer professionals, have little say on how these disputes will be resolved. We can, however, look at the technological problems involved. This we shall do in this chapter. We examine the ways to connect to the Internet; the many uses of the Internet; the potential uses by businesses; and the problems of security on the Internet.

The Internet is part of the scene of what is called cyberspace. We then start with an introduction to cyberspace.

Cyberspace

We visited cyberspace implicitly when we discussed Internet and the information highway. In this section, we will briefly discuss its evolution and its possible future. In its past,

> ... there is a ghost in the machine: the traces of the people who created them ... where technology has been the medium for the expression of human creativity ... working within the structure that created and maintained the technology were individual, creative human beings ... This largely unofficial meeting ground of culture, populist politics and technology is called cyberculture ... some humanistic values flowed back into the society of hackers ... One thing that the hackers shared with the counterculture of the '60s and '70s' was a profound distrust of society ... In general, they cared almost nothing for the forces that moved their employers, namely money and power. (Evans, 1994: pp. 10–11)

One person who captured this thinking was William Gibson, a science fiction writer, who coined the term **cyberspace**. He viewed the free hacker in the future was of 'Bohemia with computers': counterculture outlaws surviving on the edge of the information highways ... 'the street will find its own use of technology' while cyberpunks will use technology to undermine the misuse of technology. The term **cyberpunk** was coined by Gardner Dozois to evoke the combination of punk rock anarchy and high tech. Gibson in *Neuromancer* (1984) talks about the cyberpeople he knew. 'They develop a belief that there's some kind of actual space behind the screen ... Some place that you can't see but you know is there ... place of unthinkable complexity ... with lines of light ranges in the nonspace of the mind, clusters and constellations of data.'

Elmer-Dewitt talks about the shadowy space as:

> great warehouses and skyscrapers of data ... By 1989 it had been borrowed by the on-line community to describe not some science-fiction fantasy but today's increasingly interconnected computer systems – especially the millions of computers jacked into the Internet ... Cyberspace ... encompasses the millions of personal computers connected by modems – via the telephone system – to commercial on-line services, as well as the millions more with high

speed links to local area networks, office E-Mail systems and the Internet ... wires and cables and microwaves are not really cyberspace. They are the means of conveyance, not the destination: the information superhighway. (Elmer-Dewitt, 1995: p. 4)

Internet

A network of networks is the **Internet**. Both public and private networks can be part of this loose confederation of networks. In 1993, it served almost 1 800 000 nodes in some 40 countries. It was then supported partially by the US government for its operating costs while the remaining costs were borne by subscribers who paid a fee for connection to local computer hosts that direct traffic to local access providers that tie into the hosts. Still, the costs of transmission by Internet in 1993 was well below the costs of surface mail and airmail and especially below the $2 costs of a comparable fax. Also, transmission by Internet is much faster, often a few seconds or minutes as compared to hours and sometimes more than a day by fax for large jobs.

Internet allows the transmission of mail, documents, basic digital services including e-mail, banking, video phone calls, movies and shopping. Eventually, Internet is expected to provide the combined integrated services of a computer and phone at the cost of a TV today.

Internet handles queue (waiting line) control and flow control automatically through its mail protocols. Its messages encapsulate fax, sound, video as well as many character sets of foreign languages and multiplexed messages across common links. Group information exchange of group news, mailing lists and other related information are delivered according to group specific profiles to the computers of the subscribers. Internet started as a communications service for researchers and academics of the Defense Department in the US. What was restricted to scientific and research communications was soon extended to libraries and many existing databases and accessed by information providers in both the private and the public sector.

Connecting to the Internet

There are at least three ways to connect to the Internet. One is to go through a corporate LAN connected through a router and requiring a TCP/IP

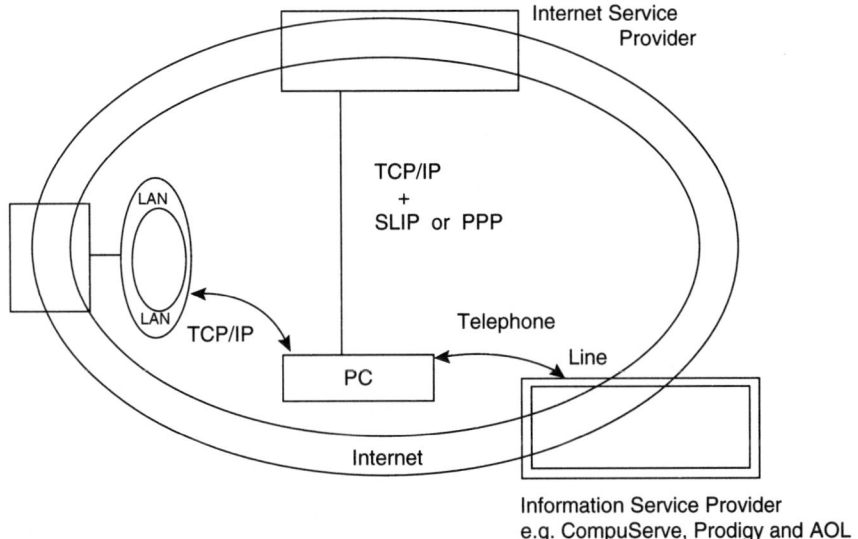

Figure 20.1 *The Internet connection alternatives*

connection. The second route is through a Internet provider and this requires not only a **TCP/IP** (Transmission Control Protocol/Internet Protocol) connection but also a **SLIP** (Serial Line Interface Protocol) or a **PPP** (Point-to-Point Protocol) connection. The third way is to use the telephone and go through a public information provider such as CompuServe, Prodigy or AOL. These three alternatives are shown in Figure 20.1.

Given alternatives, you have to make a choice. What is the best alternative for you? The answer depends on your consumption patterns and what you may have to pay. If the organization you are associated with has a LAN (and access to other LANs and the Internet) and you are an occasional user, then a LAN connection may suffice. There may be some constraints on when and how much you can use the Internet and you may be monitored but for an occasional user this may be no burden. However, if you do not have a LAN connection and need access to the Internet then you must explore other alternatives but you must

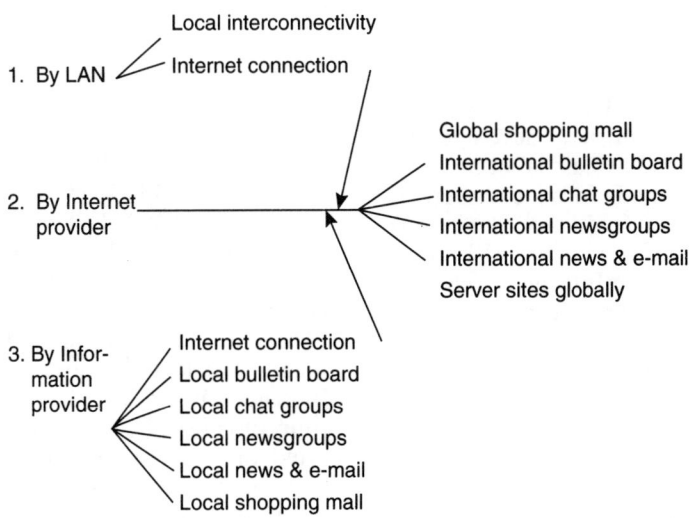

Figure 20.2 *Uses of the Internet*

233

now pay for the connection. One way may be connecting to an Information Services where e-mail is part of the services offered. This may be quite adequate and offer other services like news and discussion groups, shopping, and even the opportunity of playing computer games. However, if the services that you want are abroad, then you need the Internet. This connection will not only give you access to the international shopping malls and international discussion groups, etc., but it will give you access to a lot of material that is available on international sites only accessible from the Internet. The Internet connection is offered by some information service providers but for a cost which may be worth it if you have much intentional correspondence and a lot of global traffic **downloading** (receiving) and/or **uploading** (sending). There is a lot of freeware and shareware that is both of a technical and general nature on foreign sites especially if you live outside the US. Even the US firms benefit. One US firm is on record as downloading $64 000 worth of software and documentation in just one year. Downloading or uploading involves a **file transfer** and a special protocol called the TCP, Transfer Control Protocol.

The services offered by each alternative Internet connection is summarized in Figure 20.2.

Surfing on the Internet

Surfing (working or 'playing') on the Internet is not currently an easy or end-user friendly pursuit, but it is getting easier by the day. Browsing is difficult but easier with excellent software like the Gopher and the Web. The **Web** is the short form of WWW, World-Wide Web. The WWW is called the Web for short and was initially developed by CERN (European Particle Physics Laboratory) in Switzerland. It was later made into a commercial product by NCSA (National Center for Supercomputer Applications) at the University of Illinois.

As a navigating tool for the Internet, Web provides access to servers that are the repository of information and documents (including computer programs) that are of interest to computer scientists. The Web protocol is a superset of other protocols and embraces multimedia-based systems that are capable of delivering data, text, voice and images. The protocols are OS (operating system) and hardware independent, allowing Web to be used across different computer

systems all around the world. Web is based on **HTTP** (HyperText Transport Protocol) with 'hot links' to documents through another interface: the **HTML** (HyperText Markup Language). In contrast to Web which is hypertext oriented, **Gopher** has a hierarchical structure.

Web and Gopher are solutions to publishing information on the Internet. While many Gopher servers are limited to publishing text, Web servers can publish text and graphics, and in some cases even sound and video, as well as any combination of multimedia. Both solutions are client–server based and can link documents on the same server or remote servers.

To access and use Web or Gopher, one needs an interface. For Web, a GUI (Graphical User Interface) is **Mosaic** or **Cello**. Cello is restricted to Windows systems while Mosaic is open to Windows as well as Macintosh machines and X Windows. Mosaic is a very user friendly graphical interface for the Web allowing you to navigate freely by merely clicking the mouse.

There are other tools for the most popular Internet functions. **Archie**, **Veronica** and **Jughead** (named after comic strip characters) and **WAIS** (Wide Area Information Services) are software tools used to search archived files residing on FTP sites around the world, where **FTP** (File Transfer Protocol) allows large files and programs to be transferred between a remote server (computer) to your PC. There is also **NetScape** (an elegant and powerful interface to Web); **Telnet** allows logins to remote Internet servers often anonymously; and **Usenet** News, one of the many newsreader services that is like a large bulletin board specializing in topics where group members can post and reply to messages. Usenet also has discussion groups (10 000 of them). All the tools and services have a **home-page** that gives you information of the services offered. The home-page, when well designed, has 'hot' links that allows for easy access to related services that are relevant and accessible. Access to the Internet is summarized in Figure 20.3.

A lot of software on the Internet is free. This is not too surprising given the fact that much of Internet was developed by the free contribution of time and effort by a large number of people interested in wanting such a system. Such free software on the Internet is called **freeware** (such as the WSGopher configured to serve most Gopher servers) as opposed to **shareware** that requires you

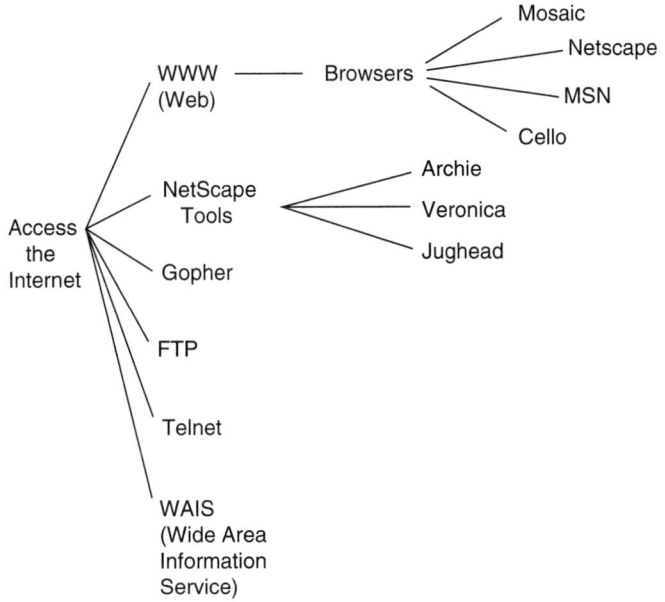

Figure 20.3 *Access in the Internet*

to pay a registration fee to the developer. You typically would use a combination of software, one to grab your Internet mail and another to read the postings on the Usenet newsgroups. Many of the Internet services are available through commercial on-line services like AOL (America on Line), Delphi ComputServe and Prodigy.

One of the problems with Internet is that it is an informal organization growing very rapidly (in 1994 it grew by 95%, adding 22 new nations to the net making a total of 159 countries). The growth of the Internet is often very *ad hoc*. What is needed are standards that allows for an orderly growth and yet maintains the security and privacy of data transmitted. This is partly being done by standards like the S-HTTP (Secured HTTP) that will secure (largely through encryption) data transmitted to-and-fro on networks and through protocols for security of monetary transactions to be discussed later in this chapter.

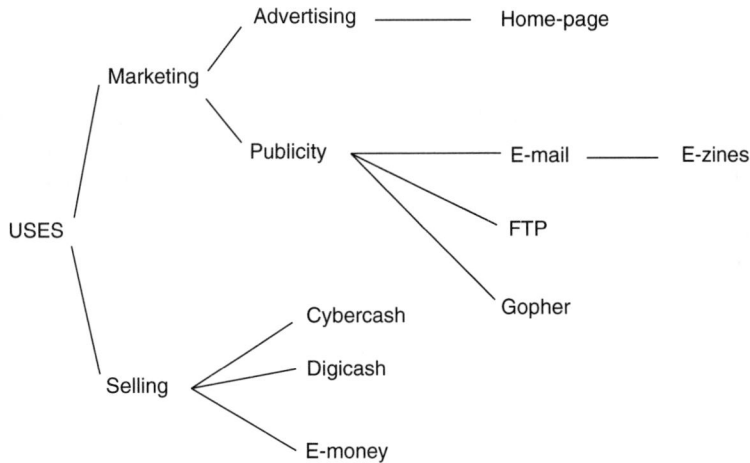

Figure 20.4 *Uses of Internet in bussiness*

Internet and businesses

The use of the Internet by business was discussed earlier in Chapter 16. The uses of Internet in business are summarized in Figure 20.4. Internet is of interest to all users of the old ARPANET that was developed by scientist, academics, researchers and the defence agency in the US. It was not specifically designed for conducting business for a profit, but cyberspace is of great interest to business and commerce. In 1994, 56 000 businesses were interconnected world-wide (*US News & World Report*, Feb. 27, 1995, p. 14). Studies on the profile of users of Internet show that it is filled with bright, well educated, upward mobile people, a demographic population most attractive to businesses that sell to such people.

Security on the Internet

An important consideration for businesses, and even for the communications of e-mail and all documents with confidential or proprietary data, is security. Some of the considerations of security that are applicable to systems other than those using telecommunications are applicable to the Internet as discussed in Chapter 12 earlier. Other considerations are importants to the Internet and we now examine them. For transactions requiring security on networks, there are at least five conditions that must be satisfied. These are:

1. Authentication, where one verifies the two trading parties involved. The authentication may be of the ID of each party which will vary with countries but in many countries it is a driving licence, a national registration card or a credit card. Authentication constitutes 'what you have' like a badge, token or card; 'what you are' such as your personal physical characteristics; 'what you know' like your PIN (Personal Identification Number); and 'where you are' such as the terminal or client identification which can be checked for legitimate access by you.
2. Certification of the authenticity of the parties involved. A common source of authority would be a governmental agency or the credit card issuing authority. In the latter case, there may well be some specification of the issuing authority such as the requirement

that it be a VISA card, a Citibank card, Eurocard or an American Express card.
3. Non-repudiation is the documentation of the agreement between the seller and the buyer specifying the transaction in enough detail so that there is no ambiguity and misinterpretation on the agreement.
4. Confirmation is where there is documentation on the seller receiving the order and the buyer receiving the goods sold.
5. Encryption, where the data on authentication, certification, non-repudiation and confirmation are all coded, so that it cannot be tampered with and altered.

Satisfying all the above five conditions should greatly ensure safety of any transaction though a hundred percent security is never possible. Unfortunately, these five conditions do not come neatly packed in a protocol package. Various strategies and protocols for security are available but they satisfy only a few of the five desired conditions.

Strategies for security can be classified into two main types: a **channel-based security**, or **perimeter-based security**, is where the machine ('where you are') is secured as distinct from the **document-based security** where the document of transaction is secured. In this latter approach, it is the person making the transaction who is authenticated by what they know (name or password), or what they are or what they possess (some biometric identification like fingerprint, hand shape, voice-print or eye retina scan print).

The most common are the document-based security measures which includes passwords, badges or cards like credit cards. None of these approaches are inviolate and their presence is no guarantee of the owner since these identifications can be lost, stolen or even forged. Some are more difficult to forge like a driving licence with a photograph but then passports come with photographs and elaborate seals and watermarks (some). What is more difficult to forge is a combination of biometric features that are naturally unique. Some methods of verification of identification require special equipment seen mostly in movies and not in everyday business operations. Instead they have **smart cards** and special cards like the **PCMICA card** that contain a set of unique information that is difficult to copy or duplicate. The PCMCIA Type II card has a broad array of data elements (including PIN, digital signature and proof of ability to pay)

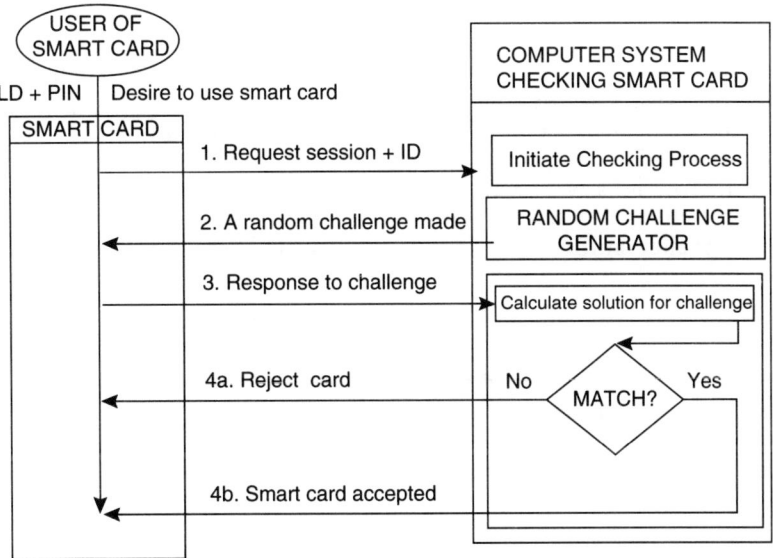

Figure 20.5 *Security for the smart card*

that will allow for privacy and non-repudiation protection. This card can be used not only for validating POS (Point-of-Sale) transactions but also for logging on to the Internet through a PCMICA slot in many PCs.

The process of checking for smart cards is shown in Figure 20.5. The steps are as follows:

1. A user enters the smart cards in special equipment that initiates the process.
2. The system offers a challenge generated randomly which makes the system dynamic and not static where the user can guess.
3. The user then responds to the challenge but can only do so with the knowledge of his/her PIN (Personal Identification Number). The system compares this response with a solution that it calculates internally knowing the PIN from the ID. This internally calculated solution is compared with the solution of the user.
4. If there is no match then the card is rejected.
5. If there is a match between the two solutions, then the card is accepted and further transactions are performed.

Some smart cards can also recognize the owner, log user activity and also do privilege mapping, i.e. map user to specified data, process or program only. The costs of such smart cards in 1995 were high, around £180 each. As international standards

develop and these cards follow those standards, the costs will definitely drop. The standards for document security now being tested is the **SSL**, the Securities Sockets Layer. It is far ahead of the other standard being developed, the **SHTTP**, the Secure HyperText Transport Protocol, which is designed for channel-based security. What the world is waiting for is the integrated SSL and SHTTP.

Encryption is a powerful security measure for the Internet and telecommunications in general. The two basic systems are Kerberps (developed by MIT (Massachusetts Institute of Technology) and RSA (named after the initials of the three founders R. Rivest, A. Shamir and L. Alderman). In the public sector, the US has developed the **Clipper Chip** which codes and decodes messages including e-mail and is protected from snooping by anyone except the government itself. The government claims that it needs a 'back-door' key so that it can intercept messages from terrorists, drug dealers and mobsters. There are many who cry foul and argue that such measures violate privacy and are merely a way to catch those who avoid paying their taxes.

The US has such good systems of identifying both the sender and the receiver that the US government bars its easy export, although they are available abroad. However, people outside the US cannot use it, even though it is based on public key algorithms, without the risk of an infringement suit. Encryption software in the US is a political tussle between its governmental agencies that want access to data for security reasons and those who consider it a violation of privacy rights. A summary of requirements for security and some of the solutions are shown in Figure 20.6

While waiting for internationally recognized standards there are some channel-based security measures. One is the **packet sniffer** that is a computer program that runs on a computer site that needs protection and watches all data passing by and records names and passwords that make transactions. The packet sniffer does not stop or even mildly restrict unauthorized intruders but collects data which when analysed can identify the loophole if not the criminal.

One approach that does control access is the **firewall**. There are many approaches to building

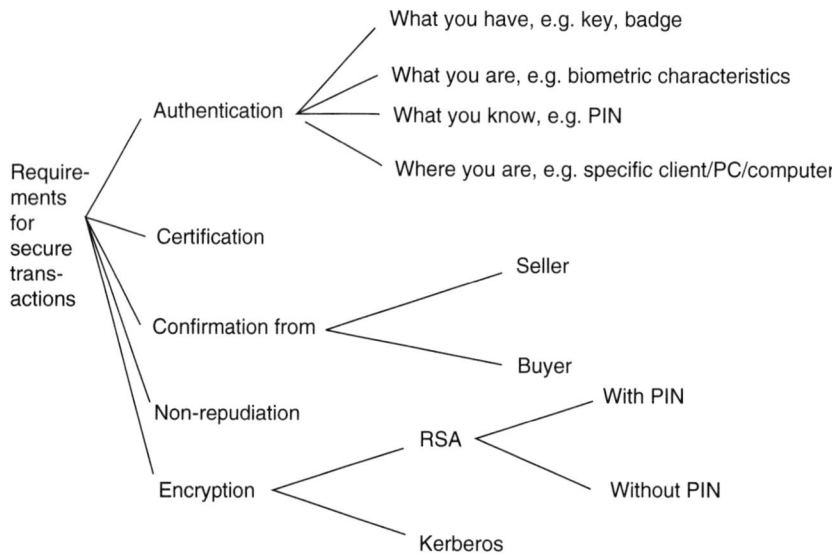

Figure 20.6 *Requirements for secure transactions*

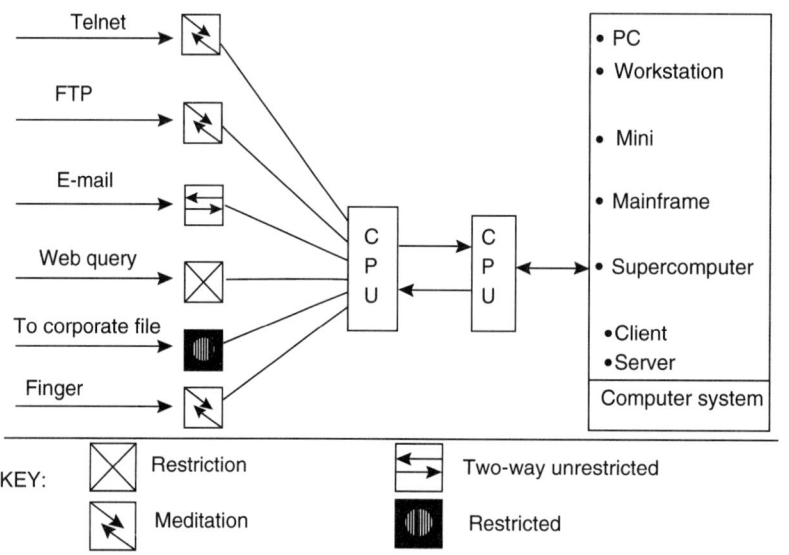

Figure 20.7 *Firewall for security*

a firewall. One is packet filtering and the other uses gateways to isolate intruders. There is also the hybrid approach of combining the filter with the gateway.

One approach is to have a dual set of computers as shown in Figure 20.7 that funnels all the incoming and outgoing traffic to a system. Some of the traffic, like traffic from a browser seeking information on a server site, may be restricted but allowed if the request comes from a known and 'safe' site; traffic like **FTP** (file transfer protocol) or that from **finger** (a service that provides data on users) or **Telnet** (a tool of interactive communication) is mediated, allowing only certain types of access and processing to be done; e-mail may go uninterrupted; and some traffic is blocked altogether, like attempts to enter a sensitive corporate database.

Whether channel-based or document-based, there are some simple common-sense rules that can be taught to users in training for Internet security, for example some simple rules on choosing a password. One study showed that almost 7% of passwords are names of the user. One system protects against this by outlawing all common names and even words in a dictionary.

Some of the types of security on the Internet are summarized in Figure 20.8

One approach to security of monetary transactions of the Internet is to use the EDI and EDI VAN. The great disadvantage of the EDI VAN is

Table 20.1   *EDI vs. Internet*

	EDI	*Internet*
Experience:	Extensive	Little/None
Support:	Good	None
Standards:	Many	None
Security:	Relatively secure	Not secure
Reliability:	Relatively reliable	No data available
Software costs:	Expensive software	Inexpensive software
Other costs:	Substantial/finite	Almost none
Response time:	High	Very fast (real-time)
Relationship required?	Yes	No
Open or closed?	Closed	Open
Complex?	Yes	No

that it does not provide instant confirmation. The transactional settlement in the EDI VAN involves a store-and-forward batch processing that is complex, costly and slow. In contrast, the Internet is real-time, quick in transaction confirmation and is the ultimate in impulse shopping. The Internet is also an open system, it is free (other than the Internet connection), and it is widely used. These and other comparisons of the EDI with the Internet (and the discussion of the EDI in Chapter 17) are summarized in Table 20.1.

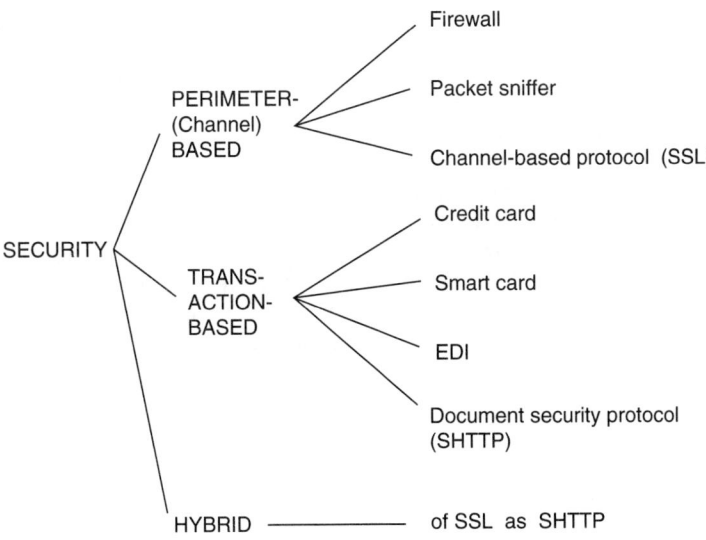

Figure 20.8   *Types of security on Internet*

Despite the many good things that can be said for Internet security, there is still great concern for its state of security. For one thing, there are many weaknesses (perhaps half to three-quarters of the holes in Internet security) that are known to hackers that are not yet openly acknowledged. No wonder that Weista Venema of the Eindhoven University of Technology suspects that the most respected domains on the Internet contain computers that are effectively wide open to all comers, the equivalent of 'a car left unattended with the engine running.' (Wallach, 1994: p. 94). Another commentator on security, cautions:

> ... e-mail and other communications can be almost tracelessly forged – virtually no one receiving a message over the net can be sure it came from the ostensible sender. Electronic impersonators can commit slander or solicit criminal acts in someone else's name; they can even masquerade as a trusted colleague to convince someone to reveal sensitive personal or business information. (Wallach, 1994: pp. 90–1)

It can be said of Internet security, as for the security of all information systems, that no absolute security is possible. We may not be able to eliminate all intrusions of the Internet but we can try to minimize the risks involved. All security measures can be thwarted by a person who is determined and clever. The motivation may not be money but merely the demonstration of beating the systems. And there is plenty of such motivation around the world. The better the security measures, the greater the challenge. What makes security on the Internet so difficult is that the Internet belongs to no organization or government that can control it. The Internet is autonomous (despite some change of financing to an information provider in 1995) and dedicated computer people who volunteer their time and passion. Disciplining them and imposing a regime of protocols and control on them is very difficult.

Organization of the Internet

Security of the Internet is partly due to the nature of its organization structure, or lack of it. There is no central authority or command to impose a discipline or a standard. No body owns the Internet, at least till 1995, when the NSF in the US withdrew its no-strings-attached financial support and at least one information provider bought into it. It is still a resolute grass-roots structure. It is open and non-proprietary. It is rabidly democratic. It is almost lawless. It crosses national borders and answers to no government. There is not even a master switch to turn it off if so desired. Most of the work is done by dedicated volunteers who pride their work on the Internet, in what they have to display and share, and in providing a community with free communication.

There is a 'netiquette' on how the Internet should be used. It is self-policed by spamming (jamming angrily) the offender. In one case, they so deluged the Siegels couple that the Internet provider cut the Siegels off. Siegels sued the company for loss of sales and vowed to keep advertising. The Internet users took the threat seriously and included a Norwegian programmer who has written a program to keep the Siegels off the Internet. But the case illustrates that there is no organizational entity responsible for what goes on the Internet. Some call it anarchy. Some say that the organization has tenets:

1. All information should be free.
2. Access to computers should be total and absolutely free.
3. Decentralization should be promoted.

These tenets and ethic was easy to maintain and police as long as the membership was a small and homogeneous group dedicated to the Internet. Now the membership is larger than the population of many countries in our world. And every year, when the universities open in the US, there is a new breed claiming membership. They are the students armed with a free LAN account number and have time to show their worth. There is often a clash between these 'newbies' and the old guard. There is a clash of culture and a difference on how the Internet should be used and what its content should be. How much cyberporn, cybersluts and cyberculture is healthy and is an expression of the freedom of speech and thought? Why is this the concern of college authorities or even the government? Why is the Internet organized the way it is?

The Internet does offer a change in the paradigm of how information is collected. Traditionally, information was top-down with the editor (of a newspaper or TV programme) deciding on what to include and what to exclude. The journalists at the bottom followed the lead of the top and dish out information to the reader. For the Internet, the flow of information is not unidirectional nor is it a one-to-one relationship. It is a two-way many-to-many relationship.

> The magic of the Net is that it thrusts people together in a strange new world, one in which they get to rub shoulders with characters they might otherwise never meet. The challenge for the citizens of cyberspace — as the battles to control the Internet are joined and waged — will be to carve out safe, pleasant places to work, play and raise their kids without losing touch with the freewheeling, untamable soul that attracted them to the Net in the first place. (Elmer-Dewitt, 1995: p. 46).

The Internet and information services

The Internet is accessed by many information service providers but there are many differences between the two. These differences are summarized in the Table 20.2.

Table 20.2 *Comparing information service providers with the Internet*

	INFORMATION SERVICE PROVIDER	Internet PROVIDER
Audience:	Household	Researcher Educationalist
	National	International
Services:	Information on news, shopping, weather etc.	Net of databases
	E-mail	E-mail
Delivery:	Prompt	Can be delayed
Cost:	Fixed + Variable	No variable cost
		Provider is paid
Control:	By business	Indirectly by governments
Directory of users:	Available in most cases	Planned

Summary and conclusions

The Internet can be viewed as an evolution in our modes of computing. In the 1970s, we had shared

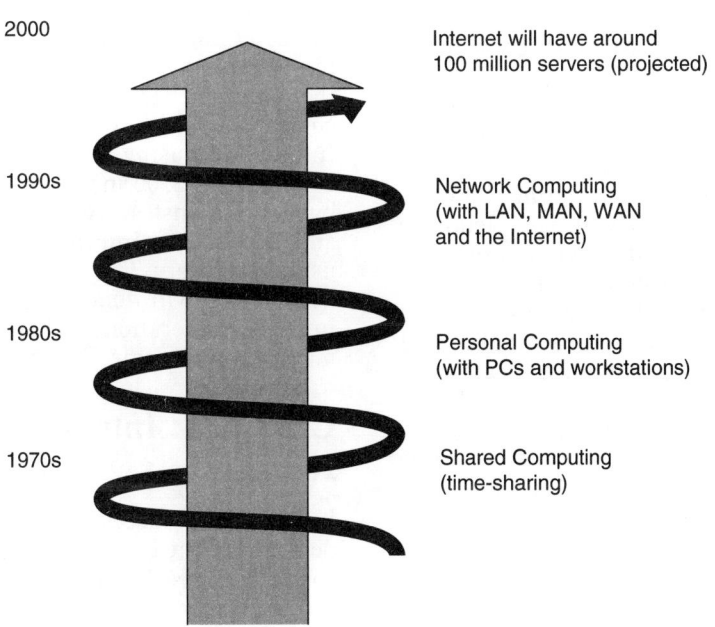

Figure 20.9 *Evolution of the Internet*

computing and time-sharing; in the 1980s, we had personal computing with the ubiquitous PC; and in the 1970s we had network computing and the Internet. By the end of this decade, it is projected that the Internet may still remain the 'Greatest Show on Earth' and have over 100 million servers. This evolution is depicted in Figure 20.9.

On the Internet, informal information can also be transmitted quickly between scientists and researchers as was the original purpose of ARPANET, the precursor to Internet in 1969. It is in this spirit that on 30 October 1994 the mathematician Dr Niecly discovered mistakes in his calculations by the new Intel Pentium chip and checked with colleagues on the Internet. This is how the world soon got to know about the flaw. At first, Intel argued that the flaw would not affect most users and was willing to replace the chip when it determined that this was necessary. There were delays and much paperwork and formalities involved. These experiences were transmitted rapidly on the Internet including to stockbrokers. The price of Intel's stock dropped. Under increasing pressure, Intel agreed to replace all faulty chips. The Internet in this case was the user's communication device to put pressure on the vendor, in this case Intel, the manufacturer of the pentium chip.

The future role of Internet is very much in the area of speculation. Will it remain a large global on-line real-time bulletin board and interactive information service? If so, who will control the content? Could the content include pornographic material? If there is to be pornography, who is to define pornography. Or how do you define the broader concept of 'obscene' for children and for teenagers, and for adults? Can we define allowable content on the Internet that will not offend the many national, cultural and religious interests? Currently, the Internet 'is so fragmented, in fact, that some fear that it will ultimately serve to further divide a society that is already splintered by race, politics, and sexual prejudice. That would be an ironic fate for a system designed to enhance communication.' (Elmer-Dewitt, 1995: p. 10).

Predicting the Internet is a common game among computer scientists as well as futurists. One set of questions relates to publishing. Will the Internet go into electronic publishing? If it does, how will the problem of royalties be resolved? There is also the problem of the protection of intellectual property that is now being perhaps addressed by ITO, successor to GATT. This issue

was the subject of conferences in the US in the 1970s and is still being debated in international forums. Or will the Internet be a resource centre for telecommuting and assist the growth and viability of telecommuting? Or will Internet be a channel for marketing and distribution for businesses? How can Internet be managed so that it controls unwanted mail and blatant advertising? How can the Internet be maintained and enhanced, for example, by going ATM (Asynchronous Transfer Mode) to make it faster and cheaper to use? How could Internet protect against violations of privacy, security and misuse? We do not know the answers to many of these questions, any more than we could have predicted the consequences of the first telephone or the first computer coming to town. We cannot predict the future for the Internet and the confluence of telecommunications and computers for the next ten or twenty years, let alone the next fifty years, except that the experience will be breathtaking and exciting.

It is difficult to predict the future of computing and much more difficult to predict the future of the Internet. It is difficult to predict the technology and its acceptance in the market-place, except to say that computing and telecommunications must be very end-user friendly, robust, tolerant of human errors, as well as secure from viruses and outside interference before they become as common or comfortable as a telephone. What is more difficult, if not impossible, is to anticipate correctly the response of the common person (especially the student) who has yet to accept the Internet. The common person must feel comfortable with cruising and surfing on the Internet and using a PC (or a Internet computer available for $500 in 1996); prefer corresponding by e-mail instead of using the post office; and browse the Internet server sites instead of using the library. The human−computer interface boundaries will tend to disappear. This may involve a generation of change in life-style. That may, however, come. Slowly but surely.

Case 20.1: Intrusions into Internet

Internet is a successor of ARPANET that was designed by the US government to help researchers connect with each other. Internet was not designed for general and extensive use by individuals and business organizations and hence no security measures for such use are in place. Internet gateway suppliers have maintained that anti-virus scanning is the responsibility of the end-user through security measures are now being implemented and include 'firewalls', electronic tokens and the PersonaCard. Meanwhile, numerous intrusions into the Internet have taken place:

- In 1986, a worm on the Internet adversely affected 6000 host computers.
- In 1988, a self-perpetuating worm virus, appropriately called the Internet worm, found its way into the University of California campus system and wriggled out of control for many days infecting and corrupting thousands of computers on the Internet.
- In 1993, the Lawrence-Livermore National Laboratory in the US, conceded that an employee had used its computers to distribute pornography on the Internet.
- In 1993, CERT (Computer Emergency Response Team) at the Carnegie-Mellon University in the US warned network administrators that tens of thousands of Internet passwords had been stolen.
- CERT estimated that, in 1993, there were 1300 'incidents' on the Internet compared with 50 a few years earlier.

Source: *Business Week*, Nov. 14, 1994, p. 88.

Case 20.2: Bits and bytes from cyberspace

Netiquette

Canter and Siegel, a law firm in Phoenix, Arizona, blatantly advertised one of its services on the Internet. It was soon 'flamed' with letters for breaching the 'netiquette' of the Internet. The rush of letters caused the server through which the firm accessed the Internet to crash and it was eventually kicked off the net.

Source: *IHT*, No. 34, 787, Jan. 2, 1995, p. 9.

Thomas site

In 1995, a Web site, Thomas, became the repository of all legislative information available to anyone on the Internet, putting them on par with legislators and lobbyists in having equally quick access to future legislation in the US.

Thomas is named after Thomas Jefferson, an early champion of democracy, as well as of freedom of speech and expression. Some commentators call the Thomas Web site the first step towards a virtual democracy and cybergovernment. Others call the Thomas site a rich potential for cyberspace anarchy.

Marriage in cyberspace

Cybermen and cyberwomen can match themselves for potential marriage by advertising themselves on the Internet by using: *alt.personal.ads*. The advantages are that it costs nothing; the ad is unrestricted in size so that one can advertise oneself at any length; and perhaps most important, one can get immediate responses. The one big problem is that most best matches live on the other side of the world.

Children in cyberspace

There is much concern for children in cyberspace being exposed to pornography and violence. This was the subject of research of Leslie Shade, a mother of two children and a doctoral student at the University of Montreal. She comments: 'Even though I might be logged in on the Internet an average of four hours a day, we would have to actively seek out and deliberately find something offensive.'

Experts advise children (and adults) on the following rules for working on the Internet: never give out any personal or family information, such as numbers or addresses, and never respond to abusive or suggestive messages.

A good primer for children in cyberspace is: *Child Safety on the Information Highway*. For a free copy, call +1 (800) 843 5678.

Source: *US News World Report*, Jan. 23, 1995, p. 60.

Suicide hotline

'The Samaritans, a British non-religous charity, offers emotional support to the suicidal and despairing through HELP by E-mail service ... Callers are guaranteed absolute confidentiality and retain the right to make their own decisions, including the decision to end their life.'

Source: *Internet World*, Jan. 1995, p. 16.

Censorship in cyberspace

In October 1994, Martin Rimm, a research associate at the Carnegie-Mellon University (CMU) informed the university administration that he was soon going to release his study on on-line pornography. The study was based on a collection of 917 000 images. The study tracked the usage (6.4 million downloads) and the frequency of retrieval of different types of images including pictures of men and women having sex with animals. The administration sensed the delicacy of the situation because pornographic images were declared obscene by the State Courts just a few months previously.

The CMU administration determined that the collection in question was part of the Usenet newsgroups, with functional titles like alt.sex.rec.arts.erotica, and alt.binaries.pictures. erotica. The administration decided to 'pull the plug on all major "sex" newsgroups and their subsidiary sections — more than 50 groups altogether.'

The battle lines soon got drawn over the preservation of free speech in the new cyberspace interactive media. Opponents of the university's action organized over the LAN network a 'Protest for Freedom in Cyberspace'. The core issues were: 'to what extent can the operators of interactive media be held responsible for the material that moves through their systems? ... Publishers who venture into international networks like the Internet are particularly concerned about libel and slander ... Unless computer users exercise some self-restraint, control could be imposed from the outside.' Whether this control would be at the local level, the state level, or a combination of the two, is still a matter to be decided.

Source: *Time*, Nov. 21, 1994, pp. 102–4.

Cyber weapon in takeover war

Canfor Corp. made a hostile bid to takeover Slocan Forest Products, a lumber and pulp company in Canada. Slocan fought with the usual defence tactics of newspaper ads and press releases. In addition, Slocan went on the Internet and issued detailed arguments as to why shareholders should withhold stock from Canfor. Slocan also invited questions on e-mail and answered all of them. It is possible that going on the Internet gave Slocan an image of a leading-edge company and it is also possible that the e-mail helped. It is not clear what made the difference but Canfor conceded defeat.

Source: *International Herald Tribune*, Feb. 9, 1995.

Cybercafés

The first cybercafé, Cyberia, appeared in London, but they can now be found in cities like Hyderabad in India. These cybercafés differ in the food and drink that they offer but the services offered always include time on the Internet. The price is charged hourly or half hourly at a rate of about £5 per hour. In the UK there are discounts and commissions given to students, OAPs, the unwaged and the 'scrounging journos'. For more information on cybercafés, access the home-page on the Internet using the URL: http://www.hub.co.uk/intercafe/

Source: Ed Ricketts, A Day in the Life of Cybercafé, *Net*, No. 10, Sept. 10, 1995, pp. 58–63.

In the US, there is no tradition of cafés like those in Europe. Internet acccess is available (also for a fee) in coffee houses in many US cities.

Case 20.3: Business on the Internet

- Federal Express Corp. delivers over 2 million packages each day and had a serious problem of tracking the packages. Now much of this is done on the Internet being a very cost-effective way of communicating with its many customers. In May 1995 alone, it tracked 90 000 packages through its Web site.
- Hewlett Packard, a computer systems manufacturer in the US uses its Web site to distribute revisions to its Unix operating system and additions to its printer software. Customers can even get software revisions through the Internet.
- Apple Computer makes software updates through its "home-page". It often responses to a request by downloading the necessary program needed.

- Well Fargo & Co. of San Francisco gives customers access to their transaction histories as well as their current balance. It was used by 2000 customers each day in 1995 who would otherwise have called the customer service representative to discuss their accounts.

Case 20.4: Home-page for Hersheys

Hersheys, the manufacturer of sweets and chocolates, has a 50-page home-page on Web, accessible by keying in http://www.hersheys.com, that provides marketing information on Hershey products as well as corporate information including the history of the popular product 'Kisses'.

The cost of the home-page for the first year was $4200 composed of the following components:

Design @ $12/hour	$2400
Internet account	$1200
Ongoing charges	$12/page

Source: *Computerworld*, Nov. 27, 1995, p. 24.

Case 20.5: Court in France defied in cyberspace

Soon after the death of François Mitterand, ex-President of France, his personal physician, Dr Claude Gubler published a book *Le Grand Secret* in 1996. Dr Gubler asserts that Mitterand was diagnosed with cancer in 1981 and suppressed this information despite the promise to keep the nation informed of his health in national bulletins. Gubler further argues that the President of France was unfit to rule in the final months in power.

Mitterand's family sought a ban on the publication of the book on grounds that it violated their privacy and that it had contravened the right to medical secrecy. The gag order specified a 1000 French franc fine for every copy sold illegally. By this time, however, 40 000 copies of the book had been sold and there was a hot market for the book. It was then put on the Internet by a cybercafé in Beasonçon, available to anyone using http://www.le.web.fr/secret/, and was soon being accessed at a rate of 1000 calls per hour.

Pascal Barbraud, owner of the cybercafé, was assured by his lawyers that the ban applied to only the printed version of the book, which has a 70-year copyright protection. Barbraud says that he had produced the book electronically because 'the spirit of the Internet is against censorship'. He said that if there was any attempt to take legal action against him, he would immediately transfer the book to a server in the United States, where anyone including the French would have access to the book.

Plon, the publisher of *Le Grand Secret* decided to take no action because the book was banned. Christian Hassenfratz, the public prosecutor at Beasançon, acknowledged that Dr Gubler's intellectual property rights had been violated, but that these rights had been placed in doubt by the decision to ban the book.

This case raises the important issue of the breaching of book copyrights on the Internet and whether courts have legal jurisdiction in cyberspace for material published in their earthly jurisdiction. Some say that this case may be potentially as significant as Gutenberg's development of movable type. Meanwhile, publishers are moving cautiously. John Wiley & Sons is posting its *Journal of Image-Guided Surgery* on the Internet; Time Warner Electronic Publishing is running a serial novel on the Internet with no printed edition planned; while Simon & Schuster has hired a team of computer cybercops to prowl the Internet for new ideas.

Source: *International Herald Tribune*, Jan. 25, 1995, p. 7, and March 19, 1996, pp. 1, 7.

Case 20.6: Internet in Singapore

Mr Yao, Minister of Information and Arts in Singapore, announced that henceforth all Internet providers and operators must be licensed by the Singapore Broadcasting Authority. Under the new regulations, all operators – from main providers to outlets such as cybercafés, as well as organizations putting political and religious information on the Internet – must now register with the Broadcasting Authority.

The Broadcasting Authority will require service operators to take 'reasonable measures' against the broadcast of objectionable material, and will insist that providers block pornographic Web sites. 'It's kind of an anti-pollution measure in cyberspace,' said Yao.

Source: *International Herald Tribune*, March 6, 1996, p. 4.

Case 20.7: English as a lingua franca for computing?

The explosive global use of the Internet has raised the question as to which natural language should be used for global computing. By default, English is the language of computing, and even countries like Malaysia that are very nationalistic about their mother tongue are offering English as a language in order to prepare their citizens for the information age.

The Americans are developing a universal digital code known as **Unicode** that will allow computers to represent the letters and characters of virtually all the world's languages. However, there is some resistance to English as a universal language of computing. One Korean official states: 'It's not only English you have to understand but the American culture, even slang. All in all, there are many people who just give up.'

In France and the Franch speaking part of Canada, people are concerned about cybernauts not being able to use the Internet if they do not know English. In February 1996, a group of French researchers put up an all French search engine called Locklace (http://www.iplus.fr/locklace) that enables Francophones to find information in any of the thousands of French language sites using French only.

The Japanese, too, are concerned. They fear 'that if the language of computers remains English, it will be more difficult for them to compete in the information industries of the future.'

Source: *International Herald Tribune*, March 11, 1996, p. 13.

Supplement 20.1: Growth of the Internet

Year	No. of hosts
1971	23
1974	62
1976	235
1983	500
1984	1000
1986	5000
1987	20000
1989	100000
1991	617000
1992	1000000
1993	2000000
1994	3000000
1995	5000000

Supplement 20.2: Growth in Internet hosts around the world

Region	Jan. 1994	Jan. 1995
North American	1685715	3372551
Western Europe	550593	1039551
Pacific Rim	113482	192390
Asia	81355	151773
Eastern Europe	19867	46125
Central & S. American	7392	n.a.
Middle East	6946	13776

Source: Internet Society.

Supplement 20.3: Computers connected to the Internet in 1994

USA	3200000 (rounded)
UK	241191
Germany	207717
Canada	186722
Australia	161166
Japan	96632
France	93041

Source: *US News and World Report*, Feb. 27, 1995.

Supplement 20.4: Users of the Internet in 1994

Type of user	In Europe	In U.S.A.
Education	36%	22%
Computer	33%	31%
Professional	16%	23%
Management	9%	13%
Other	5%	12%

Source: Georgia Institute of Technology, *Computer Weekly*, July 13, 1995.

Supplement 20.5: Build or rent a Web site?

One can either develop a Web site in-house or rent one. The relative costs in the US in 1995 were as follows:

	Developing in-house	Renting
One time costs	Software: $500−1000 Hardware: $20 000−40 000	Design and programming: $5000−30 000
Ongoing costs (annually)	One full-time Web Master: $60 000−120 000	Rental fee: $2400−3600

Source: *Computerworld*, Oct. 16, 1995, p. 70.

Supplement 20.6: Users of the Web for business

A total of 161 businesses in the US and Canada were interviewed about their uses of the Web for business purposes. Multiple responses were allowed. The results are as follows:

Gathering information	77%
Collaborating with co-workers	54%
Researching the competition	46%
Communicating internally	44%
Providing customer support	38%
Publishing information	33%
Buying products information	23%
Selling products or services	13%

Source: CommerceNet, Menlo Park, US; and *Computerworld*, Nov. 6, 1995, p. 12

Supplement 20.7: Milestones in the life of Internet

1969 US Department of Defense commissions ARPANET for networking research with its first node at University of California at Los Angeles.

1974 Robert Metcalfe's Harvard Ph.D. thesis outlines the Ethernet which then became the technology adopted by ARPANET.

1974 Vinton Cerf and Bob Kahn detail the TCP for packet network intercommunications.

1976 First use of Usenet establishing connection between two universities, Duke and UNC.

1982 Eunet (European UNIX Network) begins.

1983 University at Berkeley releases UNIX 4.2 incorporating TCP/IP.

1984 JUNET (Japan UNIX Network) is established.

1986 Internet worm burrows through Net affecting 6000 hosts.

1986 The NSF in the US establishes the super-computing centre which results in an explosion of network interconnections.

1990 ARPANET ceases to exist.

1991 University of Minnesota introduces Gopher, named after its football mascot.

1992 WWW (Web) is released by CERN in Europe.

1993 Businesses and media discover the Internet.

1994 Shopping malls arrive on the Internet.

1994 Mosaic takes the Internet by storm while WWW and Gopher proliferate.

1995 Emergence of Java and Intranets.

Bibliography

Abernathy, J. (1995). The Internet. *PC World*, **13**(1), 131−146.

Ayer, R. and Reichard, K. (1995). Web browsers: the web untangled. *PC Magazine*, **14**(3), 176−196.

Bayers, C. (1996). The great Web wipeout. *Wired*, **4**(4), 126−128.

Baran, N. (1995). The Greatest Show on Earth. *Byte*, **20**(7), 69−86.

Berners-Lee, R.C., Luotonen, A. Nielsen, F. and Secret, A. (1994). The World-Wide Web. *Communications of the ACM*, **37**(8), 76−82.

Bright, R. (1988). *Smart Cards*. Ellis Horwood.

Brinkley, M. and Burke, M. (1995). Information retrieval from the Internet: an evaluation of the tools. *Internet Research: Electronic Networking Applications and Policy*, **5**(3), 3−10.

Bryan, J. (1995). Firewalls for sale. *Byte*, **20**(4), 99−104. *Data Communications*, **16**(1), S2−S29. Editorial supplement on The Internetwork Decade.

Chapin, A.L. (1995). The State of the Internet. *Telecommunications*, **29**(1), 24−27.

Elmer-Dewitt, P. (1995). Welcome in Cyberspace. *Time*, Spring, pp. 4−11.

Elmer-Dewitt, P. (1995). Battle for the Internet. *Time*, Spring, 40−46.

Evans, J. (1994). Where the hackers meet the rockers. *Computing Now!*, **12**(3), 10–13.

Fassett, A.M. (1995). Building a corporate Web site: advice for the hesitant. *Telecommunications*, **29**(11), 33–40.

Hackman Jr., G. and Montgomery, J. (1995). One-click Internet. *PC Computing*, **8**(7), 114–125.

Hurwicz, M. (1997). Netscape's bridge to the Intranet. *Internet*, **2**(1), 103–7.

Ivine, D. (1995). Internet infrastructure. *NetUser*, Issue 3, September, 7–19.

James, G. (1996). Intranets. *Datamation*, **42**(18), 38–40.

Kosiur, D. (1997). Electronic commerce: Building business security. *Internet*, **2**(2), 82–90.

Krol, E. (1992). *The Whole Internet: User's Guide and Catalogue*. O'Reilly.

Levy, S. (1994). E-money: that's what I want. *Wired*, **2**(12), 174–177, 213–215.

Lipschutz, R.P. (1995). Extending e-mail. *PC Magazine*, **14**(8), 157–175.

Marion, L. (1995). Who's guarding the till at the cyber mall? *Datamation*, **11**(3), 38–41.

Montague, A. and Snyder, S. (1972). *Man and the Computer*. Auerbach.

Obraezka, K. Danzig, P.R. and Li, S.-H. Internet resource discovery services. *Computer*, **26**(9), 8–24.

Reichard, K. (1995). A site of your own. *PC Magazine*, **14**(17), 227–271.

Reichard, K. (1995). Mr. Postman@Internet. *PC Magazine*, **14** (8), 111–137.

Rheingold, H. (1991). *Virtual Reality: The Revolution of Computer Generated Artificial World – and How it Promises and Threatens to Transform Business and Society*. Summit Books.

Schmidt, R. (1994). Internetworking: future directions and evolutionary paths. *Telecommunications*, **28**(1), 55–74.

Vetter, R.J. (1994). Mosaic and the World-Wide Web. *IEEE Computer*, **27**(6), 49–57.

Wallach, P. (1994). Wire pirates. *Scientific American*, March 90–101.

21

WHAT LIES AHEAD?

As we move from an industrial to an information society, we will use our brain power to create instead of our physical power and the technology of the day will extend and enhance our mental ability.

Naisbitt in Megatrends 2000 (1984)

In a sense, we have automated the process of gathering information without enhancing our ability to absorb its meaning ... Our challenge is to process data into information, refine information into knowledge, extract from knowledge understanding and then let understanding ferment into wisdom.

Al Gore (1991)

At the end of the century, the use of words and general educated opinion will have changed so much that one is able to speak of 'machines thinking' without expecting to be contradicted.

A. Turing (1950)

We can expect changes and improvements in computing technology. Even if the rate of technological growth decreases some, there will still be faster and more cost-effective computers. The smaller end of PCs will run at supercomputer speeds and will be simple to use. Some will be light and hand-held, battery-operated (with longer battery life), and capable of communicating across the land and even across the world with digital micro-cellular technology. But some computers will be wire-less which will mean that they may face the vested interests of companies that have investments in over 200 million lines of copper wire (in America alone) currently used for transmission. International standards for such cellular technology and mobile telephone networks such as the PCS (Personal Communications Services) will be needed. However, predictions of computer technology are always dangerous. As recently as 1984, AT&T predicted that there will be 900 000 wireless telephones by year 2000. In 1993 with seven years to go there are already 12 million subscribers in America alone. And the number will grow as powerful companies merge and consolidate in the US and abroad. The problem will be to find an IT platform that allows the convergence of the critical technologies in IS, such as personal computing, office and factory automation, and networking. These technologies will facilitate the overlay of information and knowledge which will enable corporate managers to pose questions that are fundamental and relevant to them. We need a corporate infrastructure of information that is cross-functional and integrated to provide a careful balance between centralized coordination and decentralized use while at the same time being end-user friendly.

In this chapter we look at the future of telecommunications but only in so far as it is relevant and important to corporate managers. One concern is the growth and trends in the increasing power of computers and telecommunications. To access such power at all levels of end-users we need better network management, including the development and acceptance of international standards for telecommunications and a rational management of the Internet. These are essential to applications like home-shopping, telenews, telebanking, the automated factory, the electronic office, etc. In integrating such systems across all functions and across all levels of end-users, we may approach a telematic society. The richness of such a society will also depend on advanced applications now under development such as the digital library, virtual reality, and other multimedia services.

We will conclude this final chapter with many predictions of the future with some reflections and caution on why predictions in computing often go wrong and what one may do to minimize its dysfunctional effects if not to eliminate them.

The future in the context of the past

Before making any predictions one should remember the caution of Niels Bohr, the eminent Danish physicist: It is hard to predict — especially the future. But there are some things that one can predict with some confidence. One such prediction is that we will soon have a three-giga machine (where giga is a measure in billions): giga-instructions per second, gigabyte bus and gigabytes of memory.

There will be more voice processing, biometric input (e.g. hand recognition devices) and interactive 3D graphics. We will also see the use of helmets, goggles, booms and data-gloves for achieving virtual reality, which will enable us to leave the computer screen and use our body to interact with a rich variety of virtual objects to replace the physical world with a computer generated one.

With mass ownership of computers and with interconnected computers, there will be computers used in every place imaginable: in automobiles, in aeroplanes (already so in the Boeing 777), in games rooms, doctor's offices and waiting rooms, libraries, offices, factories, and so on. There may even be many computers that will be locationless.

Locationless selling — selling that requires no retail store. Locationless inventory — inventory that needs no warehouse. Locationless training — training without a classroom. Locationless conferences — conferences without meeting places. Locationless management — management without headquarters (von Simson, 1993: p. 106)

Locationless implies interconnectivity of computers by networks which will continue to grow ... Between 1989 and 1993, the proportion of computers in America connected in networks rose from below 10% to over 60%. Telecommunications will also see great strides in performance and the law of microcosm will soon merge with the law of telecosm. George Gilder, states this well:

> Just as the law of microcosm essentially showed that linking any number n of transistors on a single chip leads to n^2 gains in computer efficiency, the law of telecosm finds the same kind of exponential gains in linking computers: connect any number of n computers and their total value rises in proportion to n^2 ... in a peer-to-peer computer arrangement, each new device is a resource for the system, expanding its capabilities and potential bandwidth. The larger the network grows, the more efficient and powerful are its parts. (Gilder, 1993: p. 78).

The laws of microcosm and telecosm, when combined with end-to-end digitization, may result in what Professor Solomon of MIT predicts: 'the public switched network will be transformed into one large processor'. The computer paradigm with its high bandwidth in both fibre and the air may replace the telephone, cable and TV. We may well see the merging of some telephone, cable, TV and computer companies in alliances with firms in the publishing and entertainment industry. Such merging, joint ventures and alliances, in parallel with the development of an electronic information highway and a NII (National Information Infrastructure), will give us access to anything in computer storage and do so interactively. This advanced communication environment will not only facilitate applications like video-conferencing, teleshopping and teleconferencing; but will also provide access to knowledge on entertainment, libraries, medical consultation, and public knowledge-bases where knowledge can be interacted, selected and manipulated at will.

As a preview of such a world, consider that in 1994 there were on-going feasibility studies of storing thousands of pages of text and thousands of films (full feature films not 10 minute documentaries) under the control of supercomputers and accessible to a homeowner. This will reshape our current distribution channels and will give access to the end-user that will be not just quick and easy but also comfortable. And distribution will be much cheaper. Currently, some 30 cents of every dollar in the US goes to distribution. With the use of fibre optics, the share of the entertainment dollar going to distribution may be five cents. With the use of a centralized supercomputer the costs are expected to be lower still.

Teleprocessing has made control by remote means possible. This capability in conjunction with visual processing has greatly increased the possibilities of monitoring personnel. This could have the advantage of providing more information to personnel about their own performance and be used as a coaching device. It would also enable salaries to be closely correlated to performance. This may be good for the employer but not necessarily for the employee because of the danger of personal privacy being violated. There may well be need for legislation to regulate monitoring.

Trends in telecommunications technology

In the 1990s, telecommunications will continue to be concerned with the distribution of information (and knowledge). This distribution will be greatly facilitated by advances in the implementation of the top layers of the ISO/OSI architecture and by digitizing information. In digitizing, we translate information (data, text, audio and video) into 0's and 1's to facilitate the efficient storage, processing and distribution of information.

We shall see greater use of ISDN (Integrated Services Digital Network) and enhancements of ISDN such as B-ISDN (which uses a fixed cell size with asynchronous transfer mode cell switching technology). The transmission will be over high speed electronic highways. Such a highway infrastructure in some countries (like the US) will not be owned and maintained by the government but developed, owned and managed by the private sector.

The transformation of computing by telecommunications and networks is summarized in Figure 21.1. It shows trends from the past, to the present and to the future. It is difficult to slice time into neat dimensions of past, present, and future. There is a long gestation period for telecommunications technology. There is a great overlap between developments of the past, present and future. Hence, the time horizon for Figure 21.1 is very approximate, yet can be quite revealing. What is needed is the movement towards systems that have an open architecture and are digital, broadband, integrated and end-user friendly. Also needed is that the vendor providers, who may well be vertically integrated, do not unduly restrict consumer choice and flexibility, but instead allow for a network system that is efficient and effective, and enables a seamless flow of network traffic. This seamless flow should not only be nation-wide through a national highway infrastructure (also called the superhighway or the Infobahn) but also globally through a GII, Global International Infrastructure. For this evolution from ARPANET and through Internet, and for the inclusion of business traffic on the GII, it would be necessary for many countries (especially the large trading partners) to change their national laws as they relate to the regulatory environment (for example, low telecommunication costs and better privacy protection) and change their labour laws (such as those allowing for flexible time and telecommuting).

We need systems that support the possibility of applications including telebanking, telemedicine, digital libraries, telenews and telecommuting. This can be achieved partly by integrating existing islands of computerization. It may be necessary, however, to restructure which may have to be a shift from a vertically integrated industry, in which a single entity provided

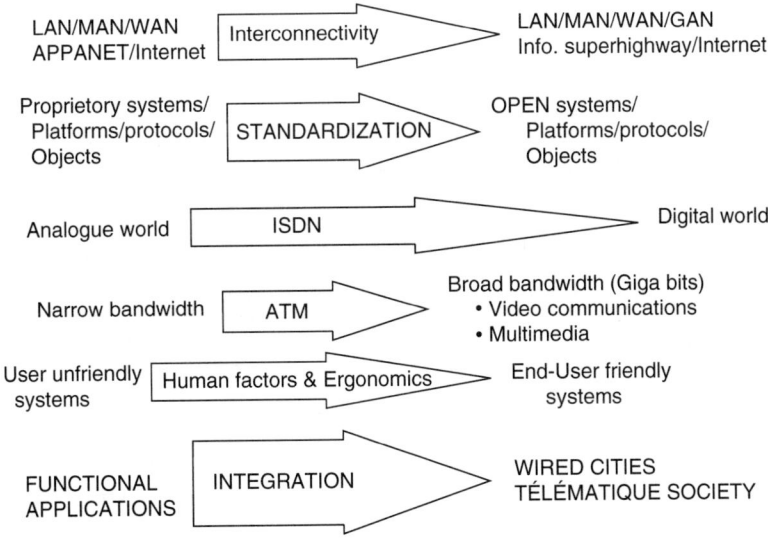

Figure 21.1 *Trends from past — present to present — future*

not only a telecommunications service but everything behind it in a network and on the customer's premises, to a richly diverse, horizontal structure ... Customers are not locked to a relationship established with a single provider of goods and services ... The new diversity of telecommunications markets ... forces essentially every player to harmonize its products and services with those of other players in other layers at least to the extent of assuring that products and services from the different layers can work together. (Heilmeier, 1993: p. 31).

Restructured and integrated or not, systems could use supporting technologies. These support technologies for telecommunications include software intelligent agents (agents are programs that when **passive** will monitor, when **active** will perform specific tasks, and when a **master agent** will customize applications like summarizing information from a newspaper), **intranets**, private networks using the Internet, **applets** (programs with specialized functions in **Java**), and **knowbots** (robots working on a knowledge-base). There are other support technologies that are not new and have been discussed earlier but they will appear in the future with robust enhancements and greater functionality. These are all listed in Figure 21.2.

We will also see a restructuring of the telecommunications industry. In the US, the telephone, TV, cable and computer industries are no longer forbidden to compete on each other's territory. Outside the US, many national PT&T (Post Telephone and Telegraph) companies are being privatized. Traditionally, telecommunications has been a public company, but because of its high cost and low responsiveness of technology and consumer demand, firms started their own private networks. The private networks have the added advantage of strategic control over their operations, they are faster to implement because of the lack of government bureaucracy found in the public sector, and they have unique features and services to offer. (The private networks do not always belong to one firm but could include hardware manufacturers, software houses suppliers and even public carriers). A summary of the comparison of private and public networks is summarized in Table 21.1.

Given the history of strong PT&Ts there is resistance to privatize by governments who are fearful of losing control.

The fearful ones will try to impose the dead hand of regulation (in the name of protecting privacy, or culture, or of clamping down on pornography or crime). Others will say that it is the job of governments to build and shape (and so, again to control) what is variously known as the 'information superhighway' or 'infrastructure' ... the changes are neither a big threat to culture or decency, nor a panacea for jobs ... Apart from imposing a few familiar safeguards, the cleverest thing that governments can do about all these changes is to stand back and let them happen. (*Economist*, Feb. 25, 1995: p. 13)

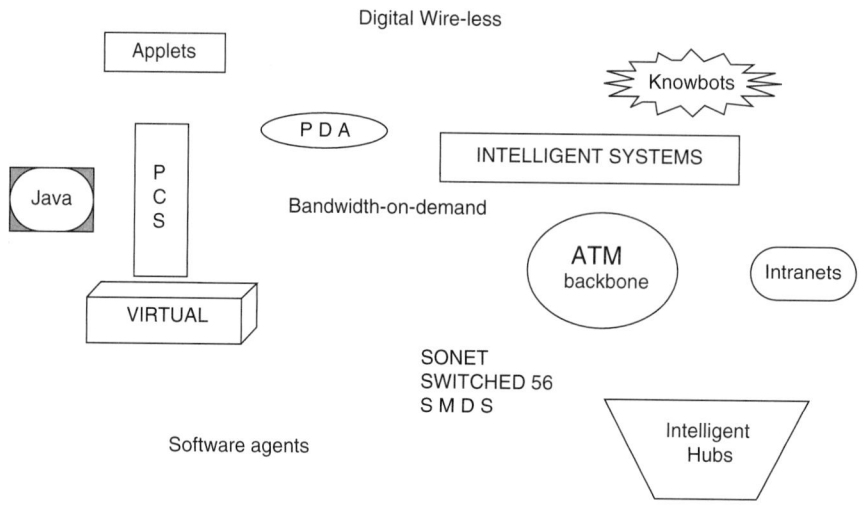

Figure 21.2 *Supporting technologies*

Table 21.1 *Public versus private networks*

	Public	*Private*
Cost:	Higher	Lower
Funding:	Public	By firm
Quality:	High	Adequate
Services:	Generalized	Customized
Security:	Good to high	Adequate to high
Control:	By government agency	Strategic control
Implementation:	Slow to adapt and learn	Quick to adapt, learn and innovate
Standards:	Tendency to wait for standards	Consider if any, or else go it alone
Features	Generalized	Uniquely configured for organization

There is a strong trend towards deregulation of telecommunications in industrialized countries. The US broke up the large monopoly of AT&T in telecommunications and allowed all communications companies (telephone and cable) as well as computer companies to compete with each other. It is generally agreed that the high rate of innovation and integration of such companies is largely due to their deregulation. We see deregulation of PT&T coming in Europe too though at varying rates among its telecommunications receptive nations. The UK is very committed, with Germany not far behind, and France still far behind. The value-added information related industries will soon be open to competition though there is still a strong feeling in many countries that their telephone and computer industries are too important to be left to the private sector. (France supported its ailing computer firm with 10 billion French francs in 1992—93).

Privatization has added more alternatives to public, value added, shared and even intelligent networks in the future. The firms involved use different media like fibre optics, telephone, TV, cable, or satellite, or a combination of these. Each firm may have a set of suppliers where some are multimillion dollar firm themselves.

The two main players in this game are the telephone and the cable companies. Both hope to have one device for both the telephone and TV.

Some technical experts think that this may not be possible because a TV that can produce a quality picture may not be able to produce good quality text which has the greater demand. What may happen is that we shall have two devices, one for text and e-mail and one TV pictures and films.

All this is just one dimension. The other dimension is the scope of the network: a LAN, MAN, WAN or a global network. And there is also a third dimension: that of media which includes data, voice, graphics, video and multimedia. Some of these media can be offered in one place including an access to the Internet instead of separately through a combination of hardware and software which makes hardware manufacturers and software houses (some being billion dollar companies) important players in the market.

Each cell in this three-dimensional matrix is represented by one or more company, some companies encompassing more than one cell. This matrix is shown in Figure 21.3.

There is actually a fourth dimension: content. We do not show the fourth dimension because of the difficulty in displaying it on the two-dimensional space of this book. The dimension of content brings other large oligopolies into play that include not just film companies in Hollywood, but video companies and book publishers. We all know about the bigness of Hollywood studios but there is less visibility of the many video companies and publishing houses that are large and powerful. These companies that have the content want an alliance with a carrier to carry their content just as much as carriers are looking around for content to carry. They are all dashing around trying to capture the market in as many cells of the market as possible. These firms are not small dwarfs but are often giants of industry that will invest billions in just testing the market (like Bell Atlantic that invested $11 billion into testing a fibre optic and two-way TV system in 11 million homes by the year 2000). The market though is too large for any one firm (even titans in large industries) and so the firms are prancing around making alliances and mergers. Their strategists are debating whether the most profits are in owning the content, or distributing it, or both. Each wants to capture the market where it has a comparative advantage which may be a row or column in our matrix of Figure 21.3. This results in intersections of interests and so there are strange alliances between some firms that are in joint ventures in one market but in fierce competition in another.

253

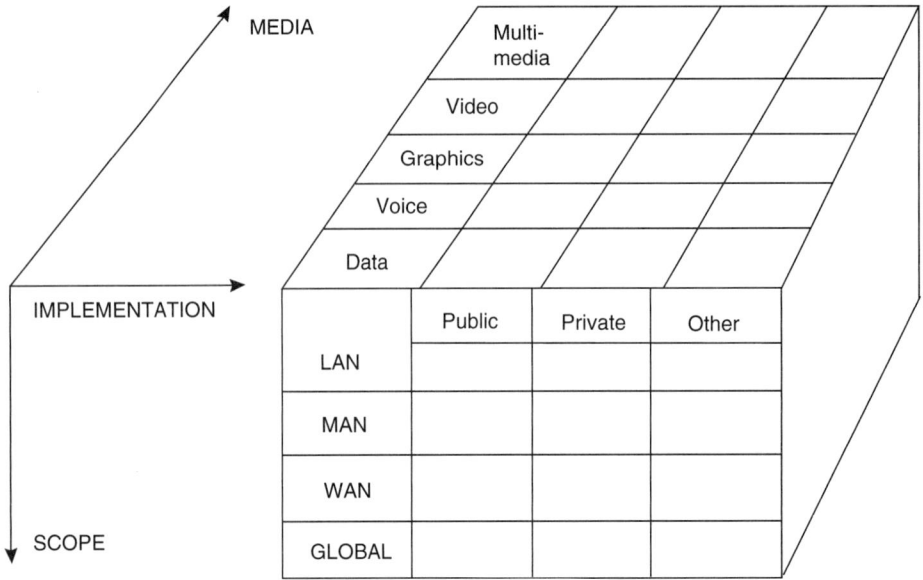

Figure 21.3   *Dimensions of networks*

This new mix of competition and cooperation is so important that it has been given a special name: **coopetion**.

In this battle for the market which is large and growing at a prodigious rate there may well be a shakeout with the strongest alliances surviving, many most likely being oligopolies. They may then get a foothold into what may soon (around the turn of the century) become the Infobahn or superhighway of the future offering a combination of services for different distances (local, long distance and wire-less) to a population of over 50 million in the US alone. Some firms and even industries may have to exit, like the video retailing industry, for soon it may be cheaper to store a video in computer memory and deliver on demand and store it on magnetic tape or optical disk.

The liberalization of the information industry and a change in attitudes in some countries toward freedom of transfer of information across national borders (such as in the Eastern European countries that are no longer tied to Russia) will greatly increase the growth of demand for information and knowledge and their related services. This will facilitate our approaching cyberspace, an on-line computer service galaxy. The growth in demand for information services will also increase because of an information superhighway and NII (National Information Infrastructure), which in conjunction with services such as the Internet will help

us towards a global information network and a télématique society.

Meanwhile, both Internet and any telecommunications infrastructure face common problems of defining and then providing universal access; offering security and protection against invasion of privacy; controlling 'obscene' material; establishing standards that are globally acceptable; and metering users and usage so that any charging system is fair and equitable. The problem of protecting intellectual property rights globally also is especially difficult when the traditional paradigm of author–publisher–library may now be disrupted.

There are also the cautions of authors like Postman in *Technopology* about the surrender of culture to technology like the Internet. Postman cautions that we may soon have a diverse and rich set of information without knowing how to handle it. He has the fear that we well lose our community responsibility and instead rely heavily on chat sessions and downloaded information from the Internet and the Web.

Steve Jobs, one founder of the Apple PC and later the father of the NeXT computer, correctly predicted in the early days of the PC that a PC will soon be on every office desktop. In 1996, Jobs asserted, 'The desktop computer industry is dead.' He then predicted that 'people are going to stop going to a lot of stores. And they're going to buy stuff all over the Web ... The Web's not going to capture everybody. If the Web got up to 10 per cent of the goods and services in the country, it would be phenomenal. I think it'll go much higher than that. Eventually, it will become a huge part of the economy... We're already in information overload ... We live in an information economy, but I don't believe we live in an information society.' (Jobs, 1996: pp. 103, 104—5).

Network management and the future

There is much that network management must respond to in terms of the changing environment of telecommunications and networks. We look at these developments from two points of view: the demand side, that is, the demand for more network services; and the supply side, the probable availability of new technology relevant to network management.

Networks are growing continuously and at an explosive rate. By the year 2000, it is estimated that there will be over 100 million microcomputers tied into corporate networks. Many of them will be using Internet. In 1994, there were over 60 million users of Microsoft Windows that could use networks for electronic shopping through credit cards and access information from public on-line information services like CompuServe and Prodigy which in 1994 had over 2 million subscribers each. In addition, there are minis and mainframes that use networks for daily transactions like reservations, financial services, electronic publishing, EFT (Electronic Fund Transfer), and other commercial applications including interactive ones. There is a great need for open systems in hardware and software components of networks allowing them to be interchangeable and interoperable. This then allows plug-and-play computing with products from different international vendors to be plugged into different platforms of integrated systems.

Another demand-driven development is the trend towards a GAN, Global Area Network. This will be a response to the need for a better architecture suited to commercial operations, the increasing globalization of our economies, and the increase in multinational corporations and multinational cooperation between national governments and regions. This does not mean the dismantling of the LAN, MAN and WAN, but rather their expansion and extension. On the way to a GAN, one will perhaps see a different architecture, more advanced network technology and a national infrastructure like the telecommunications superhighway (or information highway) in the US. The five channels of communication (telephone, TV, cable, satellite and wire-less cellular phones) are now being opened to competition by being auctioned rather than being licensed by the government. The desire for competition and the 'levelling of the playing field' for all players in telecommunications is also being pursued in the European Union, though through a different route: the Open Network Provision, planned for initiation on 1 January 1998.

It has been argued that we already have a GAN in the form of Internet, which is a confederation of networks with each having its own opinion on how things should work. Each net pays its own share with the backbones being supported on a national or regional basis. Thus, in the US, the National Science Foundation was the payer; in France, it is the EASInet partly funded by IBM; and for the 18 European countries, it is CERN, the Corporation for Research and Educational Networking, in Geneva, Switzerland. However, there is the counterargument that Internet is too *ad hoc* and informal. It is the outgrowth of ARPANET that was designed in the US to allow researchers and educationists to 'talk' to each other electronically. It was never designed for commercial use and is inadequate for the business and commercial needs of our modern world. Furthermore, its sources of income are drying out and the Internet needs reform and new management.

Management of the Internet

An important and urgent question for network management concerns the Internet. There are technological problems like insufficient bandwidth but improvement in performance of transmission media and component technologies will soon overcome the bandwidth scare. The problems of the future are not technological but organizational and financial. The Internet has matured, and so it is having financial problems. Its early development was the ARPANET, designed by the US Department of Defense as a communications system for the cold war. Since the cold war has cooled off (and because much of the network architecture had stabilized), the Defense Department withheld its support for network development but the National Science Foundation (NSF) continued to support operational expenses in order to help the Internet build its market. Then, in 1995, the NSF said: 'the market can stand on its own – without our seed money,' and withdrew its financial support. And so there are no longer any 'free lunches', free e-mail and downloading from all round the world. The Internet has to be self-supporting.

The Internet is so successful that recently there was talk of the danger of a meltdown. It almost occurred after the bubble telescope sent tons of data and swamped many users on the Internet. Melting of the Internet could occur once monetary transactions on the Internet are secure and businesses start using it not just for advertising and marketing but for operational traffic.

The nature of demand is also posing unique problems. The use of telemedicine raises the problem of assigning priorities to real-time processing by, say, a doctor wanting a CAT scan of a patient who is dying. Can such messages barrel down the Internet with the highest priority much as ambulances have priority on our roads? And how about a business wanting an important teleconferencing session but is behind a massive downloading of a computer program or a film? Who has priority? Should businesses get priority if they have paid for it? And, if so, how much should they have to pay?

It has been said that the Internet is autonomous and self-policing. That is largely true but there is a governing body of representatives from government as well as educational and research institutions that determine prices that can be charged and standards that must be followed besides the informal netiquette. They may have to assign priorities and fix prices. And they may have the company of others as organizations start subsidizing the Internet and replace the NSF benefactor, these organizations (like information service providers) may demand a say on how the Internet is to be organized and who pays what. The actual prices will have to be negotiated with the providers of local, regional and backbone services as well as the access providers.

Actually, the Internet was not free for everyone in recent years. Businesses, telecommuters and others working at home paid a fixed fee, usually a fixed fee plus a variable fee for the time beyond the 'free' time. The fee is also based on the bandwidth and so it is lower for a home user of 9 600 bps

as opposed to a 56 kbps by business users. There are other schedules of fee and this emphasizes the point that some income is collected by fees. However, the withdrawal of support by the NSF will hurt the non-profit-making organizations and perhaps even academia. How can we reduce the burden and pain for users without dampening their enthusiasm and bona fide use? Will pricing more on the Internet make it unaffordable and thereby destroy the diversity and populous nature of the Internet? How can we protect the freewheeling exploration and experimentation that takes place in the on-line society that the Internet is? How can we let researchers and libraries searching for information get what they want and yet control excesses and misuse? What is the value of public discourse and community services as opposed to commercial opportunity on the Internet? How can we prevent businesses from taking advantage of the system without limiting their right to make legitimate profits on the Internet? How can we prevent large businesses from swamping smaller ones?

Pricing is crucial to usage of the Internet. A recent increase in the price of the Internet usage in Australia resulted in a sharp decline of usage by keen students. The price elasticity of demand is expected across a broad spectrum of users though hard statistics are hard to come by.

One solution to pricing may be to charge for each unit consumed just as do the utilities. But this can raise unique problems on the Internet. Herb Brody gives the following example. 'User A sends a 100 byte request to User B, who responds by transmitting a 1 megabyte program. A naïve billing system would charge User B 10 000 times more than User A, even though User A initiated the transaction and received all the benefits.' (Brody, 1995: p. 29). This approach is very inequitable, and more so if the charges would depend not just on the size of the message but also on the distance travelled. This would dissuade users from getting what they want from wherever it is best available. Anyway, this may not be financially feasible. The TCP/IP protocol used on the Internet was developed for the ARPANET switching 'packets' of data and has no way of providing the detailed information on transactions needed for a charging system. The information can be collected and protocols adapted but the cost of doing so will require passing the extra cost to the customer and may not be worth the trouble.

Another approach is to keep the customer fee nominal but to charge businesses for their advertisements and for setting up shop in the cybermall.

A variation of this approach is to tax each business and use the tax money for operations and to subsidize libraries and community centres.

Standards

Another of the uncertainties of the future in network management concern standards. One example would be the differences in standards for EDI. In the US, the X.23 developed by ANSI (American National Standards Institute) is in vogue. In Europe, there is the EDIFACT, the EDI for the Administration, Commerce and Transport. Another example would be the existence of two parallel standards in messaging. We have the international X.400 adopted by many national governments in both public as well as private sectors around the world, but not in the US where the Internet standards are in vogue and fashion. Hopefully these two standards will converge over time on important considerations of addressing and naming, asynchronous access, security infrastructure, and messaging management. The danger of course is that there will be no merging and common standardization of messaging architectures and that the two systems will continue to develop along separate and parallel paths. We may have a similar problem with the network systems architecture. Europe and Japan and many countries have adopted the ISO model of OSI but the US and the large vendors of telecommunications systems support SNA by IBM and the TCP/IP used by the Internet.

The problem of international standards harmonization is a difficult one. You do not want to standardize too early in a fast moving technology like telecommunications or else you run the risk of freezing technology. At the same time you need international standards accepted by the developed countries or else vendors around the world with supporting applications will be inhibited less their product does not sell and becomes obsolete. That is not good for the consumer nor for world trade. Also, the absence of standards leads to uncertainty in the market which breaks the cycle of innovation and creative productivity.

A télématique society

Telecommunications provides the glue to integrated applications especially for dispersed computing whether this be internationally or locally. At the corporate level this would include the electronic office and the automated factory. At the city level it would include e-mail, electronic home, the electronic newspaper, home-shopping (teleshopping), home banking, on-line databases, and interactive TV and movies. At the regional level it would include the Technopolis Strategy in Japan where a number of wired cities (technological metropolises) are connected by fibre optics. At the world level, once we have many wired countries and technopolis integrated with each other, we approach a télématique society. It is the integration of telecommunications and computers that led to the term **télématique society** (telematic society, in English), a concept originally coined by the French administrators Nora and Minc. They were asked by their President to assess the potential of telecommunications and computers and the danger of the American monopoly affecting their culture and turning it into a 'McDonald's society' ...

Some of the world-wide integration may be in the far future but integration and telecommunication make them a viable possibility. A glimpse of the future is described in the book *Technopolis Strategy* (Tatsuno, 1986). The subtitle of this book is worth noting: *Japan, High Technology, the Control of the Twenty-First Century*. The technopolis strategy in Japan is to integrate regionally with 65% being fibre used by corporations and intercity communications. A national 'Next Generation Communications Infrastructure' is planned for 2015 at a cost of $410 billion. The Japanese have not planned fibre to the curbs of homes. Such integration of electronic homes into wired cities and then into regional integration is currently the approach taken by the US. The hierarchy of applications that could lead to a télématique society through wired cities is summarized in Figure 21.4.

The problems of a télématique society is less technological and more social and political. The questions we must also ask is: Will the quality of life in a télématique society be improved?

Optimists see freedom from drudgery, intelligent management of natural resources, and the elimination of war and poverty in a télématique society. They predict a new Renaissance since more time will be available for leisure and cultural pursuits. Access to the world's knowledge will contribute to mankind's understanding of the future, they say. Interactive communication tools will stimulate empathy with others and help bind human ties. The home will become the focus of daily life, promoting family togetherness

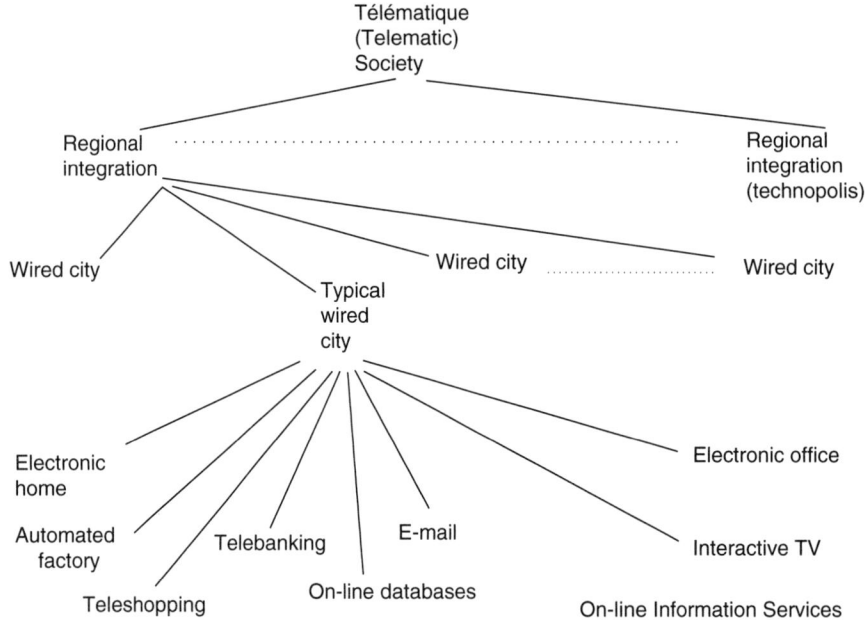

Figure 21.4 *Hierarchy for a télématique society*

and family values. National boundaries will lose importance as the world community is born.

Pessimists see illiterates glued to game shows on video screens in the world of tomorrow, waited on hand and foot by domestic robots. They see the responsibility for managing cities, running factories, growing crops, and distributing goods and services delegated to intelligent computers. Surveillance systems with little regard for personal privacy will be the norm. Individuals, they say, will sever all ties with human, in favour of computer companionship, which will indulge every individual selfish wish. With self-selective media, people will filter out unwelcome news, preferring to live with virtual realities – that is, with imaginary constructions of the world – rather than the physical realities of every day life.

In the Third World, Alvin Toffler sees a middle course for society, a 'practopia' that is neither the best nor the worst of all possible worlds. He sees:

A civilization no longer required to put its best energies into marketization. A civilization capable of directing great passion into art. A civilization facing unprecedented historical choices – about genetics and evolution, to chose a single example – and inventing new ethical or moral standards to deal with such complex issues.

A civilization, finally, that is at least potentially democratic and humane, in better balance with the biosphere and no longer dangerously dependent on exploitive subsidies from the rest of the world. (Toffler, 1989: p. 375).

Clearly, advances and innovations in information technology itself will not decide our future. It is the way technology is used that will decide the quality of life in the years ahead. We must begin now to construct a society that can meet the revolutionary changes of advanced technology. We have a destiny to create in the télématique society.

A télématique society, from the view point of technology, represents no great revolutionary trends on the immediate time horizon. There will be more of the same in many network technologies, but more enhanced versions. For example, we have heard of ATM (Asynchronous Transfer Mode) and ISDN (Integrated Services Digital Network) throughout the 1980s, with implementations in the late 1980s and early 1990s. In the future we will see advanced versions of both ATM and ISDN and likely in the open system 'plug and play' mode. ISDN in a historical perspective may be revolutionary in the sense that it will transform an analogue world of telephony to a digital world of computers and telecommunications.

258

The replacement of telephones will be accelerated by the ATM. Greater use of ATM and ATM-LAN switches will change the narrowband world to a broadband world (with gigabits per second), have wide-areas connections across the enterprise, reduce delays in telecommunications, and increase applications such as interactive video and multimedia. Such multimedia will improve applications such as those using 3D graphics in CAD (Computer Aided Design), medical applications, video-conferencing and entertainment.

Standardization will also help in more and better applications, more LANs, more users of network computing and more integrated systems. Standardization will also help in the shift of a proprietary platform to a standard platform and proprietary protocols and objects to standard protocols and objects. This is part of the shift from proprietary to open systems.

Advanced applications

Many of the applications shown in Figure 21.4 will survive in the future and may well be enhanced. There will be other applications that are innovative and advanced. These may include the digital library, which is not new in the conceptual sense but new in that it has never been fully implemented. Adaptation of new technology often takes long. Witness the case of the fax: the first fax message was sent from Lyon to Paris. It took over 125 years for it to be universally accepted. (During 1987—94, in the US alone, there were over 10 million fax messages delivered. This was in addition to the 11.9 billion voice messages in 1994.)

Back to the digital library. One view of the organization of a digital library is shown in Figure 21.5. It shows that the author no longer sends the publisher a typed manuscript but sends it electronically (Path 1—2 in Figure 21.5). The review management process in the publishing house may not change much though the roles of the author, the publisher, the librarian and the marketing department of a publisher will change. The process of publishing will also change and will no longer be a mechanical process. Instead, once a manuscript is accepted, the publishable material can be available within days (Path 2—4) greatly compressing the time now required for publishing (including a possible integration of multimedia material) and distribution, which in the mid-1990s was anywhere between 6 and 18 months.

The publishing house will be spared much of the publication headaches but will have new ones such as royalty management (Box 5 in Figure 21.5). This involves not just assessing royalties, but collecting them and distributing them. Collecting royalties will be a problem if the collection has to be done across national borders and the receiving nation is not willing to pay royalties. This is to be expected and is related to the problems of copyright management. Even in 1996, the US had problems with the acknowledgement of copyrights

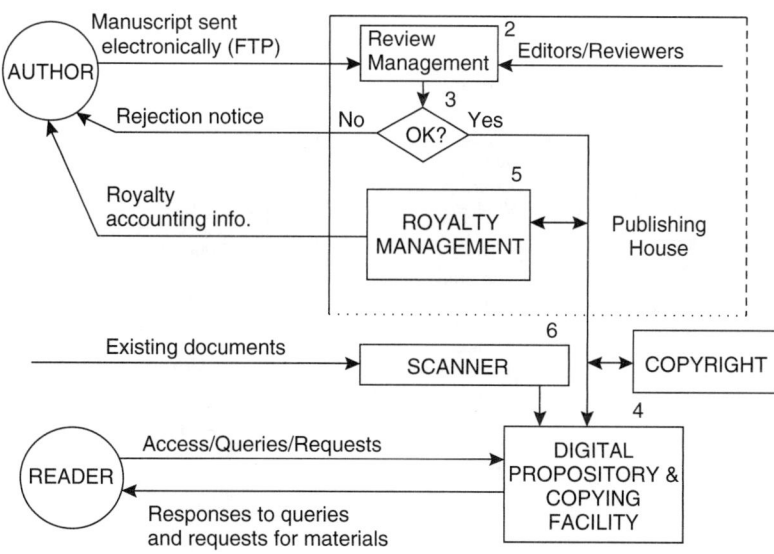

Figure 21.5 *A view of a digital library*

and the collection of royalties for software and music created by US citizens. The problem is not technological but legal and political. It may well have to be decided through international treaties or through international organizations like the WTO (World Trade Organization).

Meanwhile, there are technical problems with the digital representation of materials available in the library. This will be no problem for future publications because most of the input of new materials will be in machine readable form. The problem lies with existing materials that are not in digital form. They have to be converted through scanning (Box 6) that can be a manual process and hence slow and expensive. But libraries are converting their holdings into machine readable form (Path 1—2). Even some producers of Hollywood films are storing all their new films in digital form so that they can be easily manipulated and transmitted electronically.

Meanwhile there are technological problems to be resolved at the reader's end. Reading off a screen is not as good as reading from a printed page. The screen is slower to read (20—30% slower), has flickering that may be uncomfortable or injurious to the eyesight, and certainty poses constraints of location and posture of reader.

The reader may wish to project the pages of the book from a digital TV on the wall (or ceiling) and read while lying in bed! What if the reader wants a 600 page book and does not have a fast laser printer at home? That reader may well have to go to the local distribution centre that may replace what was once a local library. Or may be there is a flat screen display that is lighter than a heavy book and can be read while lying in bed.

These alternative for the representation of material are relevant not only to the person at home but also to the student in school, college and university. Students, teachers and reference librarians will have access through the Internet to materials and people (through chat sessions) around the world. Video monitors will replace blackboards. This empowers the student with control over what can be accessed. This may increase motivation for self-directed learning. Such access will also 'level the playing field' between the rich urban students and the poorer rural population offering equal access to all educational resources (materials and teachers). There will, however, be the problem of teachers having to learn about the potential of the new technologies made possible through telecommunications and having easy access to such technologies. Also, technologies may well change and become cheaper. The popularity of the Internet has spurred the computer industry to produce computers that will allow easy access to the Internet (including e-mail) in addition to word processing. Such computer systems could cost half of what they cost in the mid-1990s, thereby greatly increasing access to networks.

Many of the media currently available, their uses and their users (customers) are listed in Figure 21.6. Other service (and media) not listed may well emerge through the creativity of the

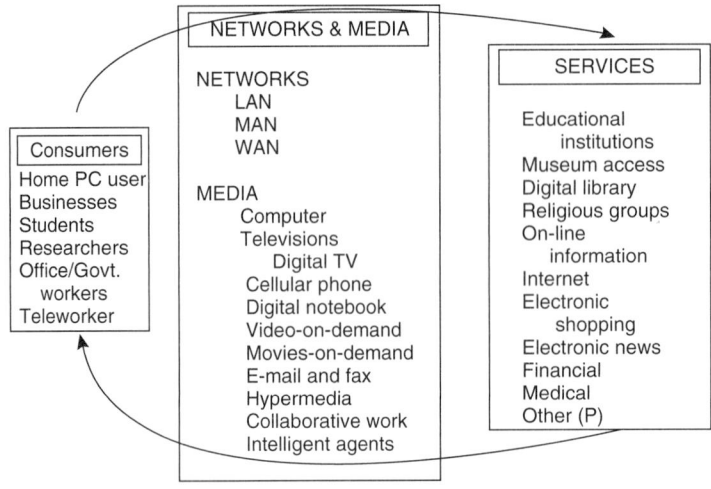

Figure 21.6 *Consumers, networks, media and services*

vendor, the user or the entertainment industry. But there seems to be a distinct trends that future services will be more robust, digital, user friendly and multimedia. We will see computer technology and telecommunications merging information into a digital stream of sound, images and words. In addition, there will be use of animation. This can be very useful in both industrial and educational institutions, as for example in the simulation of conditions of environment or a model.

The mapping of the media to the end-user and its use will be determined by the end-user, the industries involved (especially the computer and entertainment industries) and the national government that have to provide the telecommunications infrastructure. For the exchange of information across national borders we need the telecommunications infrastructure in other countries. Some countries will try to mix economic and information openness with authoritarian policies of restriction and control. Such policies may enjoy short term success, but in the long run free flowing information nurtures democracy as in Taiwan and Chile. Sometimes, a country may restrict information as was the case in 1995 when the German government prohibited CompuServe from transmitting information with pornographic content. CompuServe was then (December 1995) unable to control access to selected countries and so had to pull-off some 200 Internet sites from its approximately two million users. This made the users in the US very irate for they felt that their freedom of expression was being controlled by another country. Such issues of freedom of speech and expression across networks (national and global) are important and remain controversial as are the issues of security and privacy that remain unresolved.

Future systems will be more OLRT (On-Line-Real-Time) and interactive. Systems will be intelligent using techniques of AI (Artificial Intelligence). Systems will be more integrated, not just within an organization and corporation, but also within a city, a geographic region and maybe even in an entire society.

Perhaps the most exciting and controversial development in the future of telecommunications and networking is the information superhighway which in the US is designed to connect every home, office and school with high-speed data and multimedia links. Notice that it is an infrastructure that will bring the links to the nodes rather than the nodes having to go looking for the link. The linking may well be done by cable or telephone carriers and paid for by liberal depreciation allowances, other tax breaks and payment by advertisers. The corporate network manager can assemble a network of bulk rates and offer internal services at a lower rate than the public provider, but within the cross-subsidies allowed by governments. This should drop the cost of communications and increase not only corporate communications, but also make outsourcing of computing services more economically feasible. In the home, there will be two-way video terminals with multimedia capabilities and interactive services including movies and video-on-demand.

The superhighway will not raise many technological problems but social, ethical, moral and ethical ones. A taste of such problems arose in 1994 at the Carnegie-Mellon University where there were over 6.4 million downloads of material that had sexual content including pictures of humans having sex with animals. This type of question raises many questions like: How does one balance openness with good taste? If pornography is allowed, can we also allow say a manual on suicide? Who is responsible for slander and libel being transmitted on the network? Who is responsible for the content of what is transmitted? If a bookshop is not responsible for the content of all the books sold and the telephone carrier not responsible for the content of the telephone message, is the carrier of transmission responsible for what (pornography, libel or slander) is transmitted? Or is the sender responsible? Or, is the receiver responsible? Also, what if the message were to cross national borders like from the liberal Netherlands to the conservative Britain? Should governments intervene by legislation or subsidize technology that gives parents and managers better control of 'content' being transmitted? These are questions that will eventually have to be addressed if telecommunications is to be free and unfettered. One attempt at this was the Telecommunications Act of 1996 in the US which is being contested in court. Even if the Act is considered legal, the question arises: can the US enforce its laws over the international Internet?

Predictions often go wrong!

In evaluating the predictions made above, the reader should remember that IT is notorious for its overoptimistic predictions especially the macro predictions of 'megatrends', 'future shocks', 'third waves', a 'leisure society' and an 'information society'. Computers have permeated social and business life but have not transformed it as greatly as predicted. We do not have the 'electronic cottages', the 'cashless society', or even the 'automated factories' and 'electronic offices' that many had predicted for the 1990s. Some predictions are too long range to be evaluated today but the dates of some predictions have already passed without the predictions being achieved. For example, the 32 hour work week in the US resulting from computerization and automation by 1985 and a retirement age of 68 with more time for leisure predicted in 1967 is far from being a reality. In the 1990s, the average work-week was above 40 hours a week (including travel time) and the average worker in the US (and in many other countries) appeared to be working longer (and harder) than ever.

In the computer applications area we also have been overoptimistic in our predictions. Thus in manufacturing, there were predictions that in the US there would be 250 000 robots or more by 1990 when a survey shows that in 1990 there were only 37 000 robots. Also, only 11% of machine tools in the metalworking industry were NC (Numerically Controlled) and only 53% of the factories surveyed did not even have one automated machine. FMS (Flexible Manufacturing Systems) and MAP (Manufacturing Automation Protocol) have been slow to catch on and CIM (Computer Integrated Manufacturing) is still only a dream.

Given that many IT predictions have gone wrong, one should forgive those who are sceptical of the latest in IT predictions: the coming of the information highway. Is this more hype or is it the coming of a multimedia revolution and the integration of computers and communications? Will the information highway offer interactive programming of over 500 channels and facilitate home-shopping? Who will take the risks involved and pay for all the telecommunications infrastructure necessary? Who will assure the consumer that there will be no open season for viruses, software glitches, hardware 'crashes', systems incompatabilities, computer theft and the invasion of privacy? Will there be consumer acceptance and a willingness to pay for all the new additional services offered? The proponents of the information highway argue that all problems and obstacles will be overcome in good time. They argue that the real question is not whether we will have an information highway but when will we have it. They remind us that most revolutionary innovations took years before they

gained consumer acceptance. As examples, they recall that the radio took 11 years, colour TV took 20 years and cable television took 39 years before they became part of our daily lives.

Predictions of a 'cashless' society and a 'paperless' office have been Utopian rather than realistic. Despite the increase in the use of EFT (Electronic Fund Transfer), ATM (Automated Teller Machines) and the use of credit cards, there has been a rise in the use of paper. In the US, paper consumption rose 320% over the last three decades. A study *Business Week* (June 3, 1991) estimated that 95% of the information in business enterprises is still in paper form and that only 1% of all information in the world is stored on computers.

Why have our predictions for IT been so wrong, and are there any guidelines to help us prevent such errors in prediction? One problem is that we have often relied heavily on estimates made by the vested interests of vendors and inventors and on simple straight-line trend extrapolation and not paid adequate attention to human resistance, customer acceptance of small marginal changes in known products rather than large changes in new products, and the long time that it takes for IT to diffuse among customers and society.

Another problem is that we do not always realistically estimate the complexity of information systems. In the UK, the cost of failed software in the early 1990s is conservatively estimated at $900 million per year (Forester, 1992: p. 7). In the US, at General Motors, millions of dollars were lost in premature automation and robotization of the production line. The Bank of America abandoned one computer system after an investment of 5 years and $60 million. All State Insurance had the cost of its information system increase from $8 to $100 million and the time of completion extend from 1987 to 1993.

It is also difficult to predict the unintended and unplanned consequences of computing such as the high level computer piracy, the large number of successful 'hackers', and the extensive misuse and unauthorized access of electronic databases resulting in computer theft and the invasion of privacy.

One business strategy for avoiding or at least minimizing the danger of unexpected impact is to have an IT watcher. This may be a full-time person or someone assigned the duties of reading the literature, attending professional conferences and of evaluating IT as it may impact the organization. In doing so, the technology watcher must distinguish between technological forecasts and market trends, evaluate the sources of information, identify opportunities of potentially successful applications in the organization, give innovation time to diffuse and get consumer acceptance, anticipate problems of implementation and the availability of adequate infrastructure, and, finally, assess (with the help of other analysts, management and end-users) consumer and organizational acceptance of innovation. Only by evaluating innovation in terms of both its technological potential and its human implications can an organization exploit IT and eliminate or at least reduce the consequences of incorrect predictions.

Summary and conclusions

Networking in telecommunications can be seen as a recent stage in the evolution of computer technology. This is shown in Figure 21.7. Another view of the evolution is to look at it as another stage in computing (Figure 21.8). We started with centralized computing in the 1960s, went to shared and distributed computing in the 1970s, to personal computing in the 1980s, and we had plug-and-play, cooperative processing and network computing in the 1990s. This evolution is shown in Figure 21.8 with the corresponding hardware in use at the time. In either the technology view or the computing evolution view, we have not reached the end. We can expect more innovations in technologies and their application in collaborative computing and plug-and-play computing, where we are no longer concerned about compatibility of hardware and operating software or even programming languages. Whether the message be in French or Arabic, whether it be in text or pictures, as long as it is enveloped in a simple language like HTML it will be recognized by all computer systems around the world as long as they are connected to the Internet.

The Internet in recent years has been called the 'throbbing new center of the computing universe ... fostering the illusion that all of the Net's computers have been stitched together into one' through hyperlinks, making related information all around the world available at the click of a mouse. Businesses around the world are waiting to use the 'universal portal' and use the Internet platform if only it were secure for monetary transactions of international commerce. But there is a bigger problem, that of financing it, managing

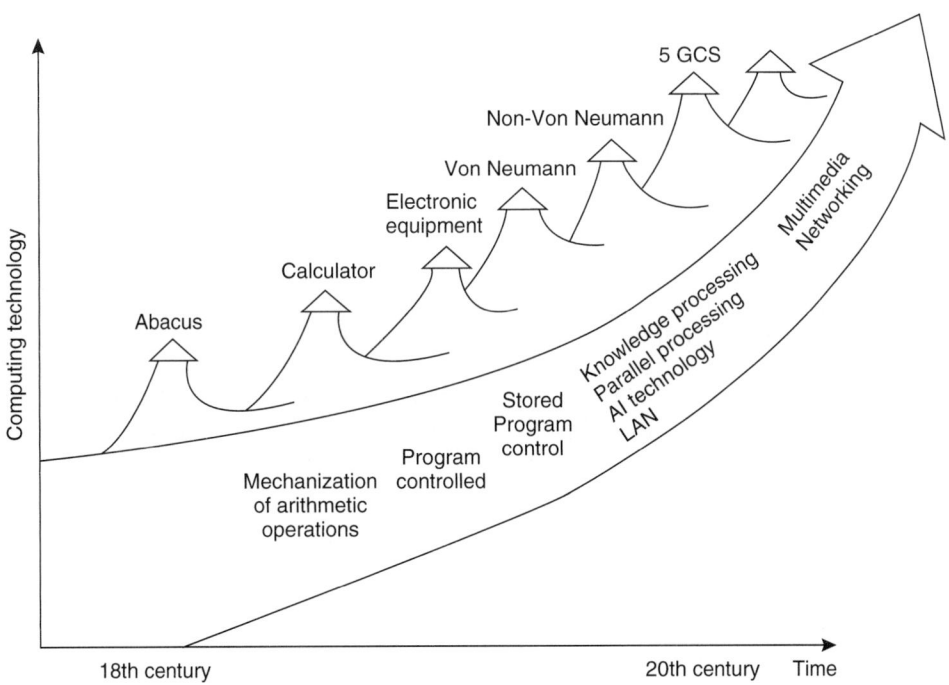

Figure 21.7 *Evolution of computing*

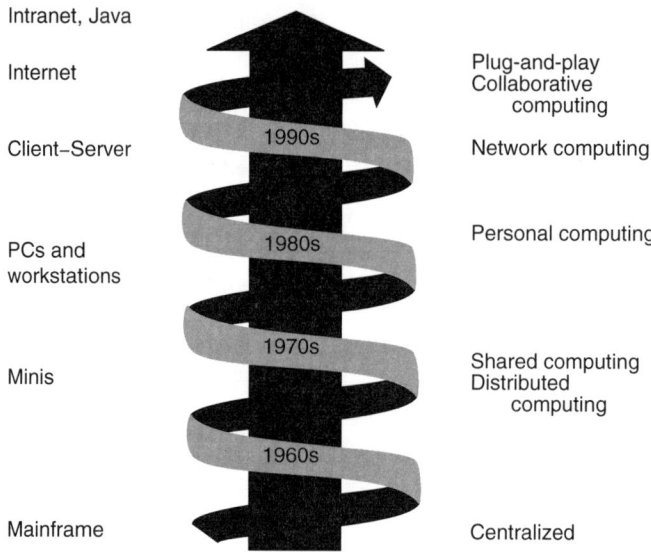

Figure 21.8   *Evolution in computing*

it and controlling it once the demand increases and the financial support from America dries up. In the resulting reorganization, the Internet may lose some of its character of being an electronic democracy, a freewheeling on-line public discussion place, of a workplace, playground and a social club all at once. This may be the time when we say good-byte to the freedom of the flat fee rate and the downloading of free software and unrestricted and fast e-mail. The use of computers in many offices and homes around the world may never be the same.

What is currently in use cautiously or experimentally may soon become necessary and a way of life. From a technological point of view, this will include the common use of hand-held computers connected by international networks to anyone around the world, high resolution flat screens, continuous speech recognition, speech synthesis, machine translation, use of natural language for computer interfaces, intelligent peripherals and secure user identification, as well as efficient and friendly end-user computing. In terms of applications, the future may well see our shopping, banking and reading the newspaper from our computer at home; replacing the library visits by using an interactive videotext; substituting distance learning and digital libraries for education and training at fixed sites; using e-mail instead of visiting friends; using EFT (electronic fund transfer) instead of cash, and even travelling to far-off

places by virtual reality. And doing much of this in cyberspace.

The future of IT will see much integration. The integration will be partly technological: the integration of text, sound, video, graphics and animation. This deployment of multimedia technology will be on a platform of interactive processing. The other dimension of integration is largely organizational and may come from strategic partnerships, joint ventures, acquisitions, or by hostile takeovers. Such conglomerates are designed to bring within one firm's jurisdiction the necessary informational infrastructure, which includes technologies (like digital telecommunications), very large client base such as television (including cable in the US, which is one-way and high capacity); and telephone (which is two-way but low capacity) subscribers, content programming (including education and movies); software for program navigation; capacity (including satellite and fibre optics transmission) and switching abilities to deliver full information services to all customers.

Communications will take place in cyberspace on an electronic highway initially with an NII (National Information Infrastructure) and later on an III (International Information Infrastructure). Access of information will be to businesses, individuals, libraries, educational and training programs, and databases. Furthermore, this access may soon become global. Access to the convergence of computer, communication structure and programming media will no longer depend on location but can take place anywhere, from one of many devices, and may be even in a visual environment and in virtual reality. Computing will not only be democratized but it will offer the consumer more choices and will be end-user friendly. The speed of technological and organizational changes will require adjustment of consumer behaviour and still possibly be pro-competitive. If there are important restraints (actual or perceived) to competition, then there will most likely be regulation by governments. In any event, our télématique future will require a redefinition of work, leisure, home and community. Along the way, we may well edge towards a cashless and paperless society. We may well see a blend of national, regional (like the governmental agencies technopolis strategy) and private networks with governments agreeing to the interoperability of national networks, provide open access to corporate players, universal service to consumers, and promotion of cultural and linguistic diversity in traffic. Also desirable would be that national governments provide ramps to the information superhighway and enforce international standards for the protection of intellectual property and data, as well as enhance the transborder flow of data and information.

Not all the predictions made in this chapter will come to pass. Some of the goals of the fifth and sixth generation computers may prove to be technologically unfeasible, or too expensive to implement. Limited resources, government regulatory schemes or interference (for example, unfavourable tax policies), high risk of capital investments, product development priorities in other fields, lack of markets, legal restrictions and user resistance are factors that may affect the speed and direction of change.

A high rate of changes in our information age and their many reasons and dismaying consequences confuse our ability to predict the future. However, we must be aware of the caution by the Librarian Emeritus of the US Congress Library, Daniel Boorstin: 'We have created and mastered machines before realizing how they may master us.'

Since business and IT managers often serve in a leadership role in their communities as well as at work, they will have a prominent role in shaping our telematic society. Planning must begin now for the computerized society to come. As stated by Montague and Snyder:

Many deplore the computer and some even fear it as more monster than machine. Whatever we think of it, however, we must adjust to it. This does not imply resignation, but rather that we must understand the true nature of this latest of man's inventions and learn how its powers can be combined with our own abilities to be used to the best advantage for humanity. (Montague and Snyder, 1972, pp. 1–2).

The author hopes that this book will help you to prepare for the challenges and opportunities ahead.

Case 21.1: Amsterdam's digital city

The digital city of Amsterdam is a city within a city replete with cafés, kiosks, town-squares, billboards, house rentals, an information centre and offices. Founded in 1993, 'the digital city has become a model virtual community for the digitally and politically active, inspiring analogous projects across the Netherlands and Europe' and has a 'population' of 30 000.

The city's manifesto includes the simple philosophy: 'Every plugged-in Amsterdammer should be able to click into the political domain to receive and exchange information on the latest governmental developments, be it about local or national elections, political party platforms, or interest-group agendas. The idea was to shed some light on the minutes of city council meetings and official policy papers, which up until then had been buried in the cabinets.'

Source: *International Herald Tribune*, Oct. 9, 1995, p. 15.

Case 21.2: Spamming on the Internet

Spamming is the flooding e-mail on the Internet with undesired and unauthorized information. Early spamming was when advertisers sent thousands of advertisements to addresses that did not want those messages. This type of spamming was controlled by the informal 'netiquette' on the Internet and was policed by members who spammed the advertiser to the point that the

Internet provider for the advertiser would cut the advertiser off the Internet.

A new twist to spamming is where people are targeted largely for political purposes by being put on mailing lists which then spam the targeted person with unexpected and unwanted messages. This crowds legitimate e-mail messages and is a great inconvenience. One such targeted person was Elmer-Dewitt, Senior Editor of *Time* magazine. He found himself being put on 106 mailing lists that generated about 50 messages a day each. Mr Elmer-Dewitt painstakingly unsubscribed himself from 106 of these mailing lists only to find that the next day he was on 1700 more lists.

The *Time* editor is not the only one spammed. The President of the US was spammed at his White-House e-mail address. The White House called in the Secret Service that managed to reduce the flow to 1200 messages a day.

Surely there will be a solution to such spamming problems but just as surely there may well be other twists to spamming the interruption of a valuable e-mail service rendered on the Internet. Every innovative application of telecommunication seems to pose a problems that calls for an innovative response.

Source: *Time*, March 18, 1996, p. 77.

Case 21.3: Minitel — its past and its future

Around 1980, France introduced Minitel terminals in Paris and other selected regions. The Minitel was connected to message-switching network called the Transpac which was integrated to the telephone system. Analysts in Paris calculated that it would be cheaper to give every household a free computer terminal with real-time access to the telephone directory than to annually update the printed telephone directory. Over the years, the price of the terminal went up and other related investments came to over 56 billion francs. This allowed over 6.5 million homes with Minitel terminals to be connected to the nationwide teletext system. There were 25 000 providers of on-line services with 2 billion calls logging over 110 million hours of connect-time. France seemed poised to take the lead in the race of developing the Information Society but never capitalized on its technology. It was unable to sell its system to other countries. 'We tried', said the France Telecom

spokesman. 'But we could never find the same ingredients in other countries that we had here.' These ingredients were a centralized government, with the political will to introduce a conceptually advanced technology, as well as a monopolistic PT&T for all telecommunications.

France has concentrated on local services which included a credit-card reader that makes it easy and safe for teleshopping because the verification system is at the terminal and does not have to be sent down a telecommunications line. Besides teleshopping, users were able to access stockmarket prices and perform banking transactions both nationally and internationally. The system also has the 'Minitel Rose' which is the equivalent of 'alt.sex' on the Internet. But meanwhile, the Internet was growing beyond anyone's expectations and France was being left behind. Between 1992 and 1995, the Internet grew 50 000 times faster in the US than in France. Minitel's technology design was rooted in the 1980s though Minitel's concept of letting the program and data reside in cyberspace and offering end-users a cheap terminal interface was conceptually advanced compared to other approaches at the time. Henri Gourald, Digital Corporation's Internet Technology Manager reminds us: 'The Minitel is both a brake and a boon. Don't forget that it has installed an electronic commerce ethos in 25% of French households.'

Today, however, Minitel's hardware is slow and monochrome text-based with rudimentary graphics as compared to the faster coloured multimedia ability to click around the world afforded by the Internet.

France has now recognized the potential of the Internet and is planning to catch up. According to François Fillon, Minister of Technology, the principal objective is 'to make access to the Internet possible for all French citizens, at a price which is attractive and the same anywhere in France.' It is expected that France Telecom will be able to meet the objective by routing Internet traffic along the same Tanspac network developed for the Minitel. Terminals will still come free or at a nominal rent and access to databases like a newspaper database will be charged at around 9 francs a minute and the bill simply added to the telephone account retaining the simplicity and reliability of the Minitel system.

Source: *International Herald Tribune*, Jan. 6, 1996, pp. 1, 8; and *Information Week*, October 2, 1995, p. 56.

Supplement 21.1: Percentage growth of phone lines connected to digital exchanges in the 1990s in Europe

Country	End of 1994	End of 1999 (predicated)
France	88	100
Germany	40	73
Italy	64	93
Netherlands	62	84
Spain	50	74
Sweden	63	96
UK	81	88
Rest of Europe	53	74
Western Europe	62	86

Source: Dataquest, May 1995.

Supplement 21.2: World-wide predications for 2010 compared to 1994

	1994	2010
Wired telephone lines:	607 million	1.4 billion
Wireless telephone lines:	34 million	1.3 billion
Number of PCs:	150 million	278 million
Number of desktop computers:	132 million	230 million
Number of mobile computers:	18 million	47 million

Source: *Business Week*, special Issue on 21st Century Capitalism, 1994, p. 194.

Bibliography

Brusseis, J.O.J. (1995). It's a wired, wired world. *Time*, Spring, pp. 80ff.

Brody, H. (1995). Internet@Crossroads.$$$. *Technology Review*, **98**(3), 24–31.

Carlyle, R. M. (1987). Towards 2017. *Datamation*, **33**(18), 142–154.

Coleman, K. (1993). The AI market in the year 2000. *AI Expert*, **8**(1), 34–44.

Evans, J. (1994). Where the hackers meet the rockers. *Computing Now!*, **12**(3), 10–13.

Forester, T. (1992). Megatrends or megamistakes? Whatever happened to the information society? *Computers and Society*, **22**(1–4), 2–11.

Fortune, **128**(7), 1–162. Special issue on '1994 Information Technology Guide'.

Galliers, R. (1992). Key information systems management issues for the 1990s. *The Journal of Information Systems*, **1**(4), 178–180.

Gilder, G. (1993). The death of telephony. *Economist*, **328**(7828), 75–78.

Hansson, A. (1995). Evolution of intelligent network concepts. *Computer Communications*, **18**(11), 793–809.

Heilmeier, G.H. (1993). Strategic technology for the next ten years and beyond. *IEEE Communications Magazine*, **31**(3), 30–34.

Huws, U. (1991). Telework: projections. *Futures*, **23**(1), 135–157.

Jobs, S. (1996). The next insanely great thing. *Wired*, **4**(2), 102–107, 158–163.

Keyes, J. (1991). AI on a chip. *AI Expert*, **6**(4), 33–38.

Knorr, E. (1990). Software's nest ware: putting the user first. *PC World*, **8**(1), 134–143.

Levy, S. (1994). E-money: that's what I want. *Wired*, **2**(12), 174–177, 213–215.

Lillis, N. and Herman, J. (eds.) (1991). Supplement on 'The internetwork decade.' *Data Communications*, **16**(1), S2–S29.

Lowry, M.R. (1992). Software engineering in the twenty-first century. *AI Magazine*, **13**(9), 71–87.

Malhotra, Y. (1994). Controlling copyright infringements of intellectual property: Part 2. *Journal of Systems Management*, **45**(7), 12–17.

Montague, A and Snyder, S. (1972). *Man and the Computer*. Auerbach.

Motiwala, J (1991). Artificial intelligence in management: future challenges. *Transactions on Knowledge and Data Engineering*, **3**(2), 125–159.

Mujjender, D.D. (1989). Fifth generation computer systems. *Telematics India*, November, pp. 25–31.

Nash, J. (1993). State of the market, art, union and technology. *AI Expert*, **8**(1), 45–51.

Niederman, F., Brancheau, sJ.C. and Wetherbe, J. (1991). Information systems management issues for the 1990s. *MIS Quarterly*, **15**(4), 475–502.

Nora, S. and Minc, A. (1980). *The Computerization of Society*. MIT Press.

Reed, S.R. (1990). Technologies in the 1990s. *Personal Computing*, **14**(1), 66–90.

Rheingold, H. (1991). *Virtual Reality: The Revolution of Computer Generated Artificial World – and How it Promises and Threatens to Transform Business and Society*. Summit Books.

Rockhart, J.F. and Short, J.E. (1989). IT in the 1990s: managing organizational independence. *Management Review*, **30**(2), 7–12.

Rowe, A.J. and Watkins, P.R. (1992). Beyond expert systems reasoning, judgement, and wisdom. *Expert Systems with Applications*, **4**(1), 1–10.

Postman, N. (1993). *Technopology: Technology and the Surrender of Culture*. Knopf.

Schroth, R. and Mui, C. (1995). Ten major trends in strategic networking. *Telecommunications*, **29**(10), 33–42.

Schnaars, S. (1989). *Megamistakes. Forecasting and the Myth of Rapid Change*. Collier Macmillan.

Seitz, K. (1991). Creating a winning culture. *Siemens Review*, **58**(2), 37–39.

Spectrum, **32**(1), (1995). 26–51. Special Issue on 'Technology, 1995'.

Spence, M.D. (1990). A look into the 21st century: people, business and computer. *Information Age*, **12**(2), 91–99.

von Simpson, E. (1993). Customers will be the innovators. *Fortune*, **128**(7), 105–107.

Toffler, A. (1980). *The Third Wave*. William Murrow.

Tatsuno, S. (1986). *Technopolis Strategy: Japan, High Technology, the Control of the Twenty-First Century*. Prentice-Hall.

Wilkes, M.V. (1996). Computers then and now – Part 2. *Proceedings of the 1996 Computer Conference*, pp. 115–119.

Yoon, Y. and Peterson, L.L. (1992). Artificial neural networks: an emerging new technique. *Database*, **23**(1), 55–58.

GLOSSARY OF ACRONYMS AND TERMS IN TELECOMMUNICATIONS AND NETWORKING

Computing is notorious for its jargon and acronyms. Telecommunications and networking should take more than their share of the blame. They have introduced many new terms and acronyms to the computing vocabulary. Some of these are innocent looking words used in daily life but mean something different in networking like backbone, bonding, bridge, bus, cloud, flag, host, Java layer, open, master—slave, packet, peer, platform, robust, token, transparent and virtual.

There are also terms that are downright confusing because they have already been used in the computing vocabulary. Examples are SDLC which does not mean Systems Development Life Cycle but Synchronous Data Link Control; and ATM does not mean Automatic Teller Machine but Asynchronous Transfer Mode. Some terms are duplicates even in telecommunications like MHS for Message Handling Service and for a product by Novell, a telecommunications product vendor. Some words have different meanings in computer science, like encapsulation in OO (object-oriented) methodology and in networking. Similar words that are confusing are T-1 and T1. To avoid confusion, most of these terms are defined in the text. All are defined in this glossary.

There are a number of books and dictionaries on computing which include telecommunications and networking terms. There are also books on acronyms alone, like the 1986 oversized book *Computer & Telecommunication Acronyms* by Julie E. Towell and Helen E. Sheppard.

ACRONYMS IN TELEPROCESSING AND NETWORKING

ANSI American National Standards Institute.
API Applications Programming Interface.
APPN Advanced Peer-to-Peer Networking.
ASCII American Standard Code for Information Exchange.
ATM Asynchronous Transfer Mode.
B channel Barrier Channel.
B-ISDN see BISDN.
BBS Bulletin Board System.
BISDN Broad-based ISDN.
BRI Basic Rate Interface.
BSI British Standards Institute.
CAT Computerized Axial Tomography.
CBDS Connectionless Broadband Data Service (SDMS in the US).
CCIR International Radio Consultative Committee.
CCITT International Telegraph and Telephone Consultative Committee, now ITU.
CD-ROM Compact disk for read-only-memory.
CDDT Copper Distributed Data Exchange.
CDMA Code Division Multiple Access.
CDPD Cellular Digital Packet Data.
CEN European Committee on Standards.
CENELEC European Committee for Electromechanical Standardization.
CEPT Conference European des Administrations des Postes et des Télécommunications.
CIX Commercial Internet Exchange.
CNMA Communications Network Manufacturing Association.
CSMA Carrier-Sense Multiple Access.
CT-2 Cordless Telephone, 2nd generation.
DIN Name of the National Standards Organization in Germany.
E-mail electronic mail.
ECMA European Computer Manufacturers Association.

EDH Electronic Data Handling.
EDI Electronic Data Interchange.
EDIFACT EDI for Administration, Commerce and Transportation.
EIUF European Computer Manufacturers Association.
ESPIRIT European Strategic Programme for Research and Development in Information Technology.
ETSI European Telecommunications Standards Institute.
EUF European ISDN Users Forum.
EuroCAIRN European Cooperation for Academic and Industrial Research Networking.
FAQ Frequently Asked Questions.
FDDI Fibre Distributed Data Interface.
FLOPS Floating-Point Operations per Second.
FTP File Transfer Protocol.
GUI Graphical User Interface.
HTML HyperText Markup Language.
HTTP HyperText Transfer Protocol.
Hz Hertz, cycles per second.
ICMP Internet Control Message Protocol.
IEC International Electrical Technical Committee.
IEEE Institute of Electrical and Electronics Engineers.
IP Internet Protocol.
IPX Protocol for transmitting and moving information over a network.
ISDN Integrated Services Digital Network.
ISO International Organization for Standardization (in Switzerland).
ISP Internet service Provider.
ITU International Telecommunications Union.
ITU-T ITU-Telecommunications (standards).
IXC Inter eXchange Carriers.

JTC1 Joint Technical Committee 1.
Kbps Kilobits per second.
LAN Local Area Network.
LEC Local Exchange Carriers.
LINX London INternet eXchange.
LEO Low Earth Orbit satellite.
MAN Metropolitan Area Network.
Mbps Megabits per second.
MHS Mail Handling Systems.
MIME Multipurpose Internet Mail Extensions.
MIPS Millions of instructions per second.
MPEG Motion Picture Experts Group.
MRI Magnetic Resonance Imaging.
NII National Information Infrastructure (in US).
NNTP Network News Transfer Protocol.
NOS Network Operating system.
OSI model Open Systems Interconnection model developed by the international organization for standards (ISO).
PAL Phase Alternating Line.
PBX Private Branch Exchange.
PC Personal Computer, also referred to as a 'home computer'.
PCN Personal Communication Network.
PCS Personal Communications Service.
PIPEX Public IP Exchange.
PoP Point of Presence.
PnP Plug-and-Play.
PPP Point-to-Point protocol.
PRI Primary Rate Interface.
PRI Prime Rate Interface.
PTM Packet Transfer Mode (by IBM).
PTT Poste de Téléphony (and) Télégraph.
PVC Permanent Virtual Connection.
RACE R&D in Advanced Computer Technology (in Europe).
RBOC Regional Bell Operating Companies.
RPOA Regional Private Operating Agency.
SDH A framing format chosen by B-ISDN. It is also the European name for SONET in America.

SDH Synchronous Digital Hierarchy.
SDLC Synchronous Data Link Control.
SGML Standard Generalized Markup Language.
SIO Scientific and Industrial Organizations.
SLIP Serial Line Internet Protocol.
SMDS Switched Multimegabit Data Service (same as CBDS).
SMR Specialized Mobile Radio.
SMTP Simple Mail Transfer Protocol.
SNA Systems Network Architecture.
SNMP Simple Network Management Protocol.
SONET Synchronous Optical Network.
STM Synchronous Transfer Mode.
SVC Switched Virtual Connection.
T1 Transmission link for distances up to 50 miles at 9.6 to 1.544 mbps
T1 Committee on Standards in the US.
T3 Transmission link up to 500 miles at speeds up to 44.7366.
TCP/IP Transmission Control Protocol/Internet Protocol.
TERENA Trans-European Research and Education Networking Association.
TTC Telecommunications Technology Committee (in Japan).
URL Uniform Resource Locator.
V.xx Designation for standards in the field of integrated circuit equipment. The xx are numerals. Each for a different standard.
VERONICA Very Easy Rodent-Oriented Net-wide Index to Computer Archives.
WAIS Wide Area Information Service.
WAN Wide Area Network.
Web short for WWW.
WWW World Wide Web.
X.11 The dominant windowing system on Internet.
X.25 A packet switching protocol defined by CCITT.
X.400 A common protocol for standard mail messaging.

GLOSSARY

access provider   a company that sells access to the Internet.

access time   interval of time between the instant at which a call for data is initiated and at which it is delivered.

agent   programming code designed to handle background tasks and perform actions when a specific event occurs; see daemon.

analogue   a transfer method that uses continuously variable physical quantities for transmitting data and voice signals over conventional lines.

asynchronous   not derived from the same clock, therefore not having a fixed timing relationship.

asynchronous transfer mode   a packet oriented transfer protocol that is asynchronous.

Archie   A system on the Internet that allows searching of files on public servers by autonomous FTP.

B-channel   Bearer channel. A circuit switched digital channel that sends and receives voice and data signals at speeds of 64 kbps.

backbone   the top level in a hierarchical network. Stub and transit networks which connect to the same backbone are guaranteed to be interconnected.

backbone network   a central network to which other networks connect.

backplane   a pathway in which electrical signals travel between devices conceptually similar to a bus.

bandwidth   the information carrying capacity of a telecommunications media; also the upper and lower limits of a frequency range available for transmission (e.g. the bandwidth would be 4000 Hz for a range of 400 to 4400 Hz).

bandwidth-on-demand   contract for bandwith as it is needed not a contract for fixed capacity (see SMDS in US and CBDS in Europe).

baud   a measure the speed with which a moderm transmits data; and named after Emile Baud, a telecommunications pioneer.

baud rate   the speed rate of a data channel expressed at bits/second.

beta   as in a beta test is the preliminary testing stage.

Bitnet   is a subset of the Internet.

broadband   a method of transmitting large amounts of data.

brouter   a bridge and a router.

browser   a program that allows you to download and display documents from the World Wide Web.

bulletin board   a system that allows users to post messages and receive replies electronically like on the Internet of an information service provider.

bus   configuration in which all nodes are connected to one main connection line.

byte   8 bits.

cable TV   a broadcasting system that uses giant antennae.

cable   a flexible metal of glass wire or group of wires.

carrier   a public transmission system in the US and Canada corresponding to the PT&T in Europe.

cell   term used in switches D 8 octets.

channel   a path between sender and receiver that carries a stream of data; also a pathway between two computers or between computer and control unit or devices.

client   it is a user system accessing services as in a 'client—server-system'.

cloud   boundary of a packet switching service. Often used as a symbol for a network.

cluster addresses   represents a domain as a single address rather than a set of individual addresses.

codec   short of **co**der/**dec**oder in multimedia and is the equivalent of a modem for non-multimedia transmission.

compatibility the degree to which there is an understanding of the same commands, format, and languages of each other.

configuration shape, arrangement or parts.

connection-less service a type of service in which no particular path is established for the transfer of information.

connection-oriented service a type of service in which, for any given call or session, information traverses only one path from sender to receiver.

contention arises when two or more devices attempt to use a single resource at any one time.

connection type specifies whether an application has a long-term relationship (permanent connection), a bounded short-term relationship (switched connection), or a boundary-less connection (connection-less).

cyber see cyberspace and cyberpunk and then deduce the meaning of cyberculture, cyberworld, cybersex, etc.

cyberpunk coined by Gardnier Dozois to evoke the combination of anarchy and high tech.

cyberspace a term coined by William Gibson, a science eviction writer to represent the counterculture outlaws who survived on the edge of the information highway.

D channel a digital channel that carries control signals and customer data in a packet switched mode.

daemon a program that runs in the background on a Unix workstation waiting to handle requests. It is usually an unattended process initiated at startup. See agent.

dark filtering part of bandwidth held in reserve.

Dante is an organization working towards a seamless integration of all Europe's networks. Dante's services complement those provided by the national research networks in Europe.

dedicated line same as leased line.

Demon a UK-based Internet access provider.

digital the representation of data in on/off signals of 0's and 1's. Digital transmission lines offer faster speeds, greater speed and more flexibility in transmission than analog.

domain a part of naming hierarchy on the Internet; an organizational area.

dot looks like a period but is used in a telecommunications address as a separator of the address subfields.

download transfer of data, offer from larger to smaller computers.

end-to-end delay the time span between the generation a data unit at its origin and its presentation a the destination.

e-mail electronically transmitted messages.

encapsulation inserting a frame header and data from a higher level protocol into a data frame of a lower level protocol.

engine the portion of a program that determines how the program manages and manipulates data; also called a processor.

Ethernet a network cabling and signalling scheme using a bus architecture.

filtering elimination of unwanted network traffic.

finger a service that provides data about users logged on the local system or on a remote system and used for security purposes.

firewall software that controls network traffic through a node.

frame relay a networking technology that exploits the high quality fibre optics to deliver data up to 10 times faster than today's packet switching.

gateway a communication program (or device) which passes data between networks having similar functions but dissimilar implementations.

gigabyte a billion bytes.

Gopher a tool on the Internet that searches fields of a hierarchical nature from a menu.

hacker describes a skilled programmer who has a mischievous bent and likely to intrude into computer files without authority to do so.

home-page an explanation of the database on the Internet and may include a description of the content and an explanation of how to access and use the database on the server.

host a system that has at least one Internet address associated with it; a large computer that serves other computers or peripherals.

hub a central switching device for communication in a store-and-forward mechanism. It can regenerate signals as well as monitor signals.

hypermedia video and text files transmitted by way of the Web.

hypertext a link between one document and another, related documents elsewhere in a collection.

icons graphical symbols that represents a function or a subroutine.

infobahn   the European version of the 'information highway'.

information highway   a term used in the US for communications across telecommunications networks to transfer information; also called an information.

information infrastructure   a publicly owned facility like the roads or electricity utilities to be shared and used by others infrastructure.

interface   connection and interaction between hardware, software and user.

intranet   a private network using the protocols and infrastructure of the Internet.

isochronous   time dependent, e.g. real-time video or telemetry data.

Java   is a programming language designed for network computing. Any program written in Java can be executed in any computer or digital device ranging from a machine tool to a computer.

jitter   the variation a packet may get through a service.

JPEG   a way of compressing still images and video which is widely used on the Internet.

killer app   short for killer application which is a very successful computer application of computers.

kluge   a clumsy but serviceable solution.

Local Area Network (LAN)   a network offering connection services within a very limited area typically the same building or campus.

latency   the time interval between the instant at which an instruction control unit initiates a call for data and the instant at which the actual transfer of the data starts. See access time.

layer   a group of services that is complete from a conceptual point of view, that is one of a set of hierarchically agreed arranged groups, and extends across all systems that conform to the architecture.

leased line   a private line for dedicated access to a network; also referred to as private line or dedicated line.

link   a line channel or circuit over which data is transmitted.

looping   a condition in packet switching networks when packets are travelling around in a circle.

LINX   is an organization set up to provide interconnectivity for UK Internet service providers and also to further the cause of the UK within Europe.

mail gateway   a machine that connects two or more electronic mail systems and transfers messages between them.

master   primary and controlling unit.

master–slave   a communication in which one (the master) initiates and controls the session. The slave responds to the master's commands.

megabits   just over 1 million bits of binary digits (0 or 1).

message   in information theory it is the ordered series of characters intended to convey information; in telecommunications it is a unit of data transmitted.

Metropolitan Area Network (MAN)   a network that connects computers within a metropolitan city area.

modem   a device that allows a computer to transmit information over a telephone line. A modem performs the function of a **mo**dulator and **dem**odulator.

Mosaic   Hypermedia browser on the Internet that allows searching using hypertext and a GUI, Graphical User Interface.

multilink   a software technique of adding channels to networks.

multimedia   information that may be in one or more forms including data, text, audio, graphics, animated graphics or full motion video.

mung   to destroy data, usually accidentally.

network   An arrangement of nodes and connecting branches.

newsgroup   these are bulletin boards on the Internet.

octet   8 bits.

open protocols   protocols that do not purposefully favour any single manufacturer.

open systems   ability to connect any two systems that conform to a reference model and its associated standards.

packet   a unit of information travelling as a whole from one device to another on a network.

packet switches   small computers linked to form a network.

peer   a functional unit that is on the same protocol layer as another.

peer-to-peer   a network architecture where a user's PC doubles as a server rather than accessing centralized file or print servers.

peer-to-peer communication   communication in which both sides have equal responsibility for initiating a session compared to a master-and-slave relationship.

peer-to-peer networking is where files could be exchanged and terminal sessions established on a non-hierarchical basis.

pel short for **pixel**.

photonic switch device that switches optically rather than convert signals to an electronic path as in conventional semiconductor technology.

pipeline allows for simultaneous or parallel processing within a computer.

pixel a smallest element on a video display screen. It could be a dot and is sometimes called a pel. Pixel is short for picture element, **pix**(picture) **el**ement.

platform the principles on which an operating system is based.

plug-and-play usually referred to for hardware that can be attached (plugged-in) almost anywhere and starts operating (playing) without the need of special interfaces.

portal a meeting point between local and long-distance services.

posting a message on the bulletin board.

private line same as leased line.

propagation delay a delay in the transmission of information from source to destination.

processor see engine.

protocol sets of rules and agreements (on format and procedures).

real-time processing with updated database. All real-time is on-line, but not all on-line is real-time

resource allocation a mechanism to allocate resources according to a promised resource reservation.

roamer a subscriber to a telecom in locations remote from the home-service area.

robust refers to a solid program that works properly under all normal but not abnormal conditions.

router a system responsible for making decisions about which of several paths traffic will follow. In OSI terminology, a router is a network layer intermediate system; a system responsible for selecting the path for the flow of data from among many alternative paths.

scalability capability of being changes in size and configuration.

seamless smooth without awkward transitions.

seamlessly blending smoothly.

server a central computer which makes services and data available.

service provider an on-line service that lets users connect to the Internet and which in turn is a gateway to the Internet.

sounding a hardware technique of adding channels to networks (same as bandwidth-on-demand).

sniffer synonymous with network analyser and used for diagnostics.

standard describes how things should be.

surfing exploring the Internet.

switching means of relaying information from one path to another.

switched 56 digital service are 56 kbps provided by local telephone companies and long distance carriers.

switching system consists of hardware and software, a switching system's primary purpose is to form dynamic connections between channels.

synchronous derived from the same clock, therefore having a fixed timing relationship.

T-1 carrier this system uses a time-division multiplexing to carry voice channels.

teleprocessing is a computer-supported technique for providing a number of remote users access to a computer system.

Telnet a tool for interaction communication with remote computers.

terrestrial earthbound.

token a series of bits which when grabbed by a user allows the use by sending packets across the network.

token ring a signalling device where a special message, passed from node to node, gives a node permission to enter a message or frame into the ring.

transceiver a physical device that connects a host to a LAN.

transparent virtually "invisible". Of no significance.

Universal Resource Locator a URL is the technical name of the World Wide Web address.

upload opposite of download.

Veronica an index to computerized archives.

video-on-demand an interactive system that allows you to point your remote control at the screen and select a desired program.

virtual pertaining to a functional unit that appears to be real but whose functions are accomplished any other means.

voice line   a communications link usually limited to transmitting data at the bandwidth of the human voice.

voice mail   a service that stores voice messages for users and enables them to retrieve and hear their messages in various ways.

Web   short for World Wide Web.

Web site   a collection of files on the web built around a common subject or theme.

Wide Area Network (WAN)   a network that serves a geographic area larger than a city or metropolitan area.

INDEX